RA1242.P

01341

PAHs : a
 ecotoxicological
 c2003.

PAHs: An Ecotoxicological Perspective

Ecological and Environmental Toxicology Series

Series Editors

Jason M. Weeks
National Centre for
Environmental Toxicology,
WRc-NSF, UK

Sheila O'Hare
Scientific Editor
Hertfordshire,
UK

Barnett A. Rattner
Patuxent Wildlife
Research Center, USGS
Laurel, MD, USA

The fields of environmental toxicology, ecological toxicology and ecotoxicology are rapidly expanding areas of research. This explosion of interest within the international scientific community demands comprehensive and up-to-date information that is easily accessible to both professionals and an increasing number of students with an interest in these subject areas.

Books in the series will cover a diverse range of relevant topics ranging from taxonomically based handbooks of ecotoxicology to current aspects of international regulatory affairs. Publications will serve the needs of undergraduate and postgraduate students, academics and professionals with an interest in these developing subject areas.

The Series Editors will be pleased to consider suggestions and proposals from prospective authors or editors in respect of books for future inclusion in the series.

Published titles in the series

Environmental Risk Harmonization:
Federal and State Approaches to Environmental Hazards in the USA
Edited by Michael A. Kamrin (ISBN 0 471 97265 7)

Handbook of Soil Invertebrate Toxicity Tests
Edited by Hans Løkke and Cornelius A. M. van Gestel (ISBN 0 471 97103 0)

Pollution Risk Assessment and Management:
A Structured Approach
Edited by Peter E. T. Douben (ISBN 0 471 97297 5)

Statistics in Ecotoxicology
Edited by Tim Sparks (ISBN 0 471 96851 X CL, ISBN 0 471 97299 1 PR)

Demography in Ecotoxicology
Edited by Jan E. Kammenga and Ryszard Laskowski (ISBN 0 471 49002 4)

Forecasting the Environmental Fate and Effects of Chemicals
Edited by Philip S. Rainbow, Steve P. Hopkin and Mark Crane (ISBN 0 471 49179 9)

Ecotoxicology of Wild Mammals
Edited by Richard F. Shore and Barnett A. Rattner (ISBN 0 471 97429 3)

Environmental Analysis of Contaminated Sites
Edited by Geoffrey I. Sunahara, Agnès Y. Renoux, Claude Thellen, Connie L. Gaudet and Adrien Pilon (ISBN 0 471 98669 0)

Behavioural Ecotoxicology
Edited by Giacomo Dell'Omo (ISBN 0 471 96852 8)

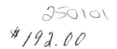
PAHs: An Ecotoxicological Perspective

Edited by

Peter E. T. Douben
Unilever Colworth R&D, Safety and Environmental Assurance Centre,
Sharnbrook, Bedford, UK

WILEY

Other Wiley Editorial Offices

John Wiley & Sons Inc., 111 River Street, Hoboken, NJ 07030, USA

Jossey-Bass, 989 Market Street, San Francisco, CA 94103-1741, USA

Wiley-VCH Verlag GmbH, Boschstr. 12, D-69469 Weinheim, Germany

John Wiley & Sons Australia Ltd, 33 Park Road, Milton, Queensland 4064, Australia

John Wiley & Sons (Asia) Pte Ltd, 2 Clementi Loop #02-01, Jin Xing Distripark, Singapore 129809

John Wiley & Sons Canada Ltd, 22 Worcester Road, Etobicoke, Ontario, Canada M9W 1L1

Wiley also publishes its books in a variety of electronic formats. Some content that appears
in print may not be available in electronic books.

Library of Congress Cataloging-in-Publication Data

PAHs : an ecotoxicological perspective / edited by Peter E.T. Douben.
    p. cm. — (Ecological and environmental toxicology series)
  Includes bibliographical references and index.
  ISBN 0-471-56024-3 (alk. paper)
   1. Polycyclic aromatic hydrocarbons – Toxicology. 2. Polycyclic aromatic
hydrocarbons – Environmental aspects. I. Douben, Peter E. T. II. Ecological &
environmental toxicology series.

RA1242.P73.P346 2003
615.9'02 — dc21

                   2002192409

British Library Cataloguing in Publication Data

A catalogue record for this book is available from the British Library

ISBN 0-471-56024-3

Typeset in 10/12pt Garamond by Laserwords Private Limited, Chennai, India
Printed and bound in Great Britain by Biddles Ltd, Guildford and King's Lynn
This book is printed on acid-free paper responsibly manufactured from sustainable forestry
in which at least two trees are planted for each one used for paper production.

Contents

List of Contributors

Rudolf K. Achazi
Biology Department, Ecotoxicology and Biochemistry, Freie Universität Berlin, Ehrenbergstrasse 26–28, D-14195 Berlin, Germany

Michael J. Ahrens
National Institute of Water and Atmospheric Research, Gate 10, Silverdale Road, Box 11-115, Hamilton, New Zealand

Farida Akcha
IFREMER, Rue de l'Ile d'Yeu, BP21105, 44311 Nantes cedex 03, France

Peter H. Albers
USGS Patuxent Wildlife Research Center, 12011 Beech Forest Road, Laurel, MD 20708-4041, USA

Rolf Altenburger
Department of Chemical Ecotoxicology, UFZ Centre for Environmental Research, Leipzig-Halle, Permoserstrasse 15, 04318 Leipzig, Germany

Gerald T. Ankley
Mid-continent Ecology Division, National Health and Environmental Effects Research Laboratory, US Environmental Protection Agency, 6201 Congdon Boulevard, Duluth, MN 55804, USA

Janet Arey
Air Pollution Research Center, University of California, Riverside, CA 92521, USA

Roger Atkinson
Department of Environmental Sciences, University of California, Riverside, CA 92521, USA

Anton M. Breure
National Institute for Public Health and the Environment, Laboratory for Ecotoxicology, PO Box 1, 3720 BA Bilthoven, The Netherlands

Thierry Burgeot
IFREMER, Rue de l'Ile d'Yeu, BP21105, 44311 Nantes Cedex 03, France

Robert M. Burgess
United States Environmental Protection Agency, ORD/NHEERL Atlantic Ecology
Division, 27 Tarzwell Drive, Narragansett, RI 02882, USA

Lawrence P. Burkhard
Mid-continent Ecology Division, National Health and Environmental Effects
Research Laboratory, US Environmental Protection Agency, 6201 Congdon
Boulevard, Duluth, MN 55804, USA

Tracy K. Collier
Northwest Fisheries Science Center, 2725 Montlake Boulevard East, Seattle, WA
98112, USA

Philip M. Cook
Mid-continent Ecology Division, National Health and Environmental Effects
Research Laboratory, US Environmental Protection Agency, 6201 Congdon
Boulevard, Duluth, MN 55804, USA

Pieter J. den Besten
Institute for Inland Water Management and Waste Water Treatment (RIZA),
Ministry of Transport, Public Works and Water Management, PO Box 17, 8200
AA Lelystad, The Netherlands

Stephen A. Diamond
Mid-continent Ecology Division, National Health and Environmental Effects
Research Laboratory, US Environmental Protection Agency, 6201 Congdon
Boulevard, Duluth, MN 55804, USA

Peter E. T. Douben
Unilever Colworth R&D, Safety and Environmental Assurance Centre, Sharnbrook, Bedford MK44 1LQ, UK

Russell J. Erickson
Mid-continent Ecology Division, National Health and Environmental Effects
Research Laboratory, US Environmental Protection Agency, 6201 Congdon
Boulevard, Duluth, MN 55804, USA

Philippe Garrigues
Laboratory of Physico- and Toxicochemistry of Natural Systems (LPTC), UMR
5472 CNRS, University of Bordeaux I, 351 Cours de la Libération, 33405 Talence
Cedex, France

Bruce M. Greenberg
Department of Biology, University of Waterloo, Waterloo, ON N2L 3G1, Canada

Christopher W. Hickey
National Institute of Water and Atmospheric Research, Gate 10, Silverdale Road,
Box 11-115, Hamilton, New Zealand

Christopher G. Ingersoll
US Geological Survey, Biological Resources Division, Columbia Environmental Research Center, 4200 New Haven Road, Columbia, MO 65201, USA

James S. Latimer
US Environmental Protection Agency, Office of Research and Development, National Health and Environmental Effects Research Laboratory, Atlantic Ecology Division, Narragansett, RI, USA

Gordon Lethbridge
Shell Global Solutions, Cheshire Innovation Park, Chester CH1 3SH, UK

Thomas R. Loughlin
National Marine Mammal Laboratory, Alaska Fisheries Science Center, National Marine Fisheries Service, 7600 Sandpoint Way NE, Seattle, WA 98115-6349, USA

Heath M. Malcolm
Centre for Ecology and Hydrology, Monks Wood, Abbots Ripton, Huntingdon, Cambridgeshire PE28 2LS, UK

Eric Martin
Concawe, Brussels, Belgium

Anne Mathieu
Oceans Ltd, 31 Temperance Street, St. John's, Newfoundland A1C 3J3, Canada

Joy A. McGrath
HydroQual Inc., One Lethbridge Plaza, Mahwah, NJ 07430, USA

James P. Meador
Environmental Conservation Division, Northwest Fisheries Science Center, National Oceanic and Atmospheric Administration, 2725 Montlake Boulevard East, Seattle, WA 98112, USA

David R. Mount
Mid-continent Ecology Division, National Health and Environmental Effects Research Laboratory, US Environmental Protection Agency, 6201 Congdon Boulevard, Duluth, MN 55804, USA

Jean-François Narbonne
Laboratory of Physico- and Toxicochemistry of Natural Systems (LPTC), UMR 5472 CNRS, University of Bordeaux I, 351 Cours de la Libération, 33405 Talence Cedex, France

Vikram Paul
Shell Global Solutions, Cheshire Innovation Park, Chester CH1 3SH, UK

Jerry F. Payne
Science Oceans and Environment Branch, Department of Fisheries and Oceans, 1 White Hills Road, St. John's, Newfoundland A1C 5X1, Canada

Helmut Segner
Centre for Fish and Wildlife Health, Department of Animal Pathology, University of Berne, Laenggass-Strasse 122, CH-3012 Berne, Switzerland

Richard F. Shore
Centre for Ecology and Hydrology, Monks Wood, Abbots Ripton, Huntingdon, Cambridgeshire PE28 2LS, UK

Dorien ten Hulscher
Institute for Inland Water Management and Waste Water Treatment (RIZA), Ministry of Transport, Public Works and Water Management, PO Box 17, 8200 AA Lelystad, The Netherlands

Ron van der Oost
OMEGAM Environmental Research Institute, Environmental Toxicology, PO Box 94685, 1090 GR Amsterdam, The Netherlands

Cornelis A.M. Van Gestel
Department of Animal Ecology, Institute of Ecological Science, Vrije Universiteit, De Boelelaan 1085, 1081 HV Amsterdam, The Netherlands

Bert van Hattum
Institute for Environmental Studies, Vrije Universiteit Amsterdam, De Boelelaan 1115, 1081 HV Amsterdam, The Netherlands

Frank Volkering
Tauw bv, Research & Development Department, PO Box 122, 7400 AK Deventer, The Netherlands

Graham Whale
Shell Global Solutions, Cheshire Innovation Park, Chester CH1 3SH, UK

Jinshu Zheng
Department of Biology and Chemistry, City University of Hong Kong, Kowloon, Hong Kong, China

Series Foreword

Polycyclic aromatic hydrocarbons (PAHs) are compounds with two or more (up to seven) condensed aromatic and other cyclic rings. PAHs occur naturally as a by-product of incomplete combustion, for example, from road traffic, coal fires, heating, the summer barbecue, and are frequent and troublesome environmental pollutants. Some members of this chemical class such as benzo(a)pyrene are also prominent and strong carcinogens. PAHs are ubiquitous in their distribution, impacting on all media, and are considered a serious pollution problem.

This book brings together in a comprehensive fashion the many aspects of the ecotoxicology of PAHs. We congratulate the editor and contributing authors of this much needed book in managing to combine and present this work in an informative, detailed, yet legible fashion.

We are pleased to have this latest contribution as part of the *Ecological and Environmental Toxicology Series* and wish to thank the editor for his considerable effort in its production.

This book crosses many different boundaries and disciplines and will be of interest to students, chemists, biologists, toxicologists, risk assessors and managers, regulators and those scientists working within industry.

Jason M. Weeks
Sheila O'Hare
Barnett A. Rattner

PART I

Introduction and Rationale

1

Introduction

PETER E. T. DOUBEN*
*Unilever Colworth R&D, Safety and Environmental Assurance Centre, Sharnbrook,
Bedford, UK*

1.1 OUR INTEREST IN PAHs

Polycyclic aromatic hydrocarbons (PAHs) are a unique class of persistent
organic pollutants (POPs) constituted by hundreds of individual substances.
These compounds contain two or more fused aromatic rings made up of carbon
and hydrogen atoms.

They can be formed naturally by low-temperature, high-pressure reactions
of natural organic matter and in this way constitute a significant fraction
of petroleum hydrocarbons. Incomplete combustion of wood and petroleum
products are responsible for a large proportion of their formation. PAHs may
also be released from petrogenic and natural sources (Eisler 1987). The largest
emissions of PAHs result from industrial processes and other human activities
(WHO 1998). With the successful abatement of point sources in the past
decades, non-point sources such as atmospheric deposition and surface runoff
nowadays constitute major inputs into the environment.

The physical and chemical characteristics of key PAHs are given in the
next chapter in Table 2.1. At ambient temperatures they are solid. The general
characteristics are high melting and boiling points, low vapor pressure, and very
low water solubility, which tends to decrease with increasing molecular mass.

PAHs occur mostly in mixtures, and it is therefore impossible to cover
all combinations. Attempts have been made to identify a specific source by
determining the concentration profile of characteristic PAHs, which is only
partially successful. Reactions that are of interest governing their fate and
possible loss during atmospheric transport are photodecomposition and reactions with nitrogen oxides, nitric acid, sulfur oxides, sulfuric acid, ozone and
hydroxyl radicals.

---

* The views expressed in this chapter are those of the author and do not represent those of Unilever
or its companies.

PAHs: An Ecotoxicological Perspective. Edited by Peter E.T. Douben.
© 2003 John Wiley & Sons Ltd

PAHs entering the environment via the atmosphere are adsorbed onto particulate matter. The hydrosphere and geosphere are affected by wet and dry deposition. Other routes are release of creosote-preserved wood PAH into the hydrosphere, and the deposition of contaminated refuse contributing to emissions into the geosphere. Several distribution and transformation processes determine the fate of PAHs. Partitioning between water and air, between water and sediment, and between water and biota are important processes. As the affinity of PAH for organic phases is greater than that for water, their partition coefficients between organic solvents (e.g. octanol) and water are high. In line with this their affinity for organic fractions in sediments, soil, and biota is also high, with consequences for bioaccumulation.

PAHs are degraded by photodegradation, biodegradation by microorganisms and metabolism in higher biota. While the latter is less important for the overall fate, it is crucial in terms of effects through the formation of carcinogenic metabolites. PAHs have been of environmental concern because of the mutagenic and carcinogenic properties of the metabolites of several compounds, and are included in most environmental monitoring programs described in various monographs (Neff 1979; Meador *et al.* 1995; Neilson and Hutzinger 1998).

Skin cancers were documented in young chimney sweeps in London as early as 1775 and in German coal tar workers in the late 1800s (Eisler 1987). Various soots, tars, and oils were subsequently shown to be carcinogenic to humans and laboratory animals, and such complex mixtures were later found to contain rich sources of polycyclic aromatic hydrocarbons (PAHs), including the potent carcinogen, benzo[a]pyrene.

1.2   THIS BOOK

The interest in PAHs remains firmly present from a regulatory and academic perspective. While much focus has been on human health-related issues, ecotoxicological aspects gradually come to the fore. Since the publication of the IPCS monograph (WHO 1998), much work has been done. It is therefore appropriate to concentrate on these environmental aspects.

The set-up of this book is that after this chapter (i.e. **Part I, Introduction**), **Part II** covers the **General Characteristics of PAHs**. Chapter 2 covers the sources relevant to the marine environment, and their fate and transport. It is oriented to the large-scale (global) distribution processes in several varying media. Complementing this topic, Chapter 3 deals with the geochemistry of PAHs in aquatic systems, focusing specifically on the microscale geochemistry. Photochemical reactions of PAHs are dealt with in Chapter 4. The atmospheric chemistry of PAHs is fundamental to our understanding of their fate in the environment. The light-induced formation of free radicals (such as hydroxyl- and nitrate-radicals) and subsequent reactions with other molecules, such as PAHs, cause them to have significant effect on ambient mutagenic activity. It is vital to note that PAHs differ in their susceptibility to photochemical degradation.

Moving to more biological aspects, Chapter 5 deals with PAH metabolism and related chemical reactions in chemical reactions inside an organism that may alter the toxicity and impact of PAHs. Essentially it covers the pathways involving the formation of DNA adducts and their biological significance, both in general terms and in the context of environmental protection. The information stems mostly from studies for human health purposes but is completed with other information so that it can be interpreted for that environmental context. How this knowledge is being used to indicate a biological response to exposure of invertebrates to PAHs is covered later, in Chapter 10, and its usefulness for the development of biomarkers is addressed in Chapter 16.

Concluding Part II, Chapter 6 covers degradation and general aspects of bioavailability. The conditions required for degradation and mineralization are first discussed as part of the general principles, followed by the degradation of PAHs in soil. This is complemented by the impact of bioavailability on the biodegradation processes for the clean-up of PAH-polluted soil.

Part III on **Bio-availability, Exposure and Effects** in environmental compartments continues the theme introduced by the final chapter of the previous part: Chapter 7 covers fundamental aspects of partitioning and bioavailability in relation to exposure and effects. This is very much an introduction to key aspects relevant to the other chapters in this part, dealing with compartment-specific exposure and effects: factors influencing partitioning, bioavailability and bioaccumulation. The obvious link with substance-specific characteristics is made to facilitate predictions of those facets.

Chapter 8 focuses on the uptake and effects of PAHs in aquatic invertebrates. Data on PAH levels in foodchains will be presented that were obtained from field studies in the delta of the rivers Rhine and Meuse. Special attention will be given to different approaches that can be used to assess the bioavailability of PAHs. Finally, the contribution of PAHs in effects on aquatic invertebrates is discussed. Completing this work is Chapter 9 on invertebrates in the marine environment. Topics covered include the role of organic carbon and lipid in the control of PAH partitioning, the different modes of exposure, the rates of uptake and elimination, and seasonal variation. Chapter 10 concludes the invertebrate theme with those in the terrestrial compartment. Where possible, behavior and effects are linked. This suite of chapters is obviously built on the fundamental aspects of partitioning and bioavailability covered in Chapter 7.

Ecotoxicological studies focusing on marine and freshwater fish are covered in Chapter 11. It covers the different types of effects, such as biochemical effects at the enzyme level, histopathological, immunological, reproduction, developmental and behavioral effects. Continuing the theme of other organisms, Chapter 12 deals with effects of PAHs on terrestrial and freshwater birds, mammals and amphibians. The issues are exposure, metabolism, and toxicity of PAHs. Studies on laboratory species are used to evaluate potential effects of exposure in wild species. Chapter 13 is a companion to the previous one for the marine environment, obviously where the food source makes that pertinent for exposure. It includes an evaluation of our knowledge of the biological consequences of such exposure.

Chapter 14 deals with uptake and effects of PAHs in vegetation. PAH impairment of plant growth and/or development will limit primary productivity, constraining total biological activity in an ecosystem. Assays and examination of PAH mechanisms of plant toxicity are also covered. Then follows the use of plants as part of a remediation strategy. Chapter 15 points out that the toxicity of some PAHs can be greatly enhanced through mechanisms that involve molecular activation or excitation. UV radiation can greatly increase the toxicity of PAHs in a broad phylogenetic spectrum of aquatic species, including bacteria. Mechanisms and possible prediction of toxicity through modeling are covered, as well as the role of mixture toxicity. The importance of this photoactivation in relation to risk assessment of contaminated sites concludes the chapter.

Chapter 16 builds on previous chapters, in particular Chapter 5, and provides an overview of biomarker studies for the assessment of exposure of aquatic organisms to PAHs and their molecular, biochemical and cellular actions, as well as their organismic and ecological effects. The focus is on aquatic systems.

Part IV is aimed at some studies focusing on **Integration of Information on PAHs**. It is not the aim of this part to cover every environmental compartment but it merely highlights the issues involved with integrating our knowledge for a specific purpose. First, Chapter 17 looks at the various approaches to developing sediment quality guidelines for PAHs. Chapter 18 completes the book, rounding it off with a practical example of managing risk from PAHs, highlighting the problems and solutions in a case study. It summarizes some of the methods used by a multinational oil company to assess and manage environmental risks posed by pollutants such as PAHs. Remediation forms a critical element of that approach. Critical to appreciating the contents of the individual chapters is the fact that in the environment PAHs are not present as single compounds: invariably complex mixtures exist, which makes an assessment more difficult. Furthermore, in attempts to reduce risk, as point sources are more and more under control, diffuse sources become important.

REFERENCES

Eisler R (1987) *Polycyclic Aromatic Hydrocarbon Hazards to Fish, Wildlife, and Invertebrates: A Synoptic Review*, vol.11. Fish and Wildlife Service, US Department of the Interior.

Meador JP, Stein JE, Reichert WL and Varanasi U (1995) Bioaccumulation of polycyclic aromatic hydrocarbons by marine organisms. *Reviews of Environmental Contamination and Toxicology*, **143**, 79–165.

Neff JM (1979) *Polycyclic Aromatic Hydrocarbons in the Aquatic Environment*. Applied Science Publishers, London.

Neilson A and Hutzinger O (eds) (1998) *PAHs and Related Compounds. The Handbook of Environmental Chemistry*, vol. 3, Part J. Springer Verlag, Berlin.

WHO (1998) *Environmental Health Criteria 202: Selected Non-heterocyclic Polycyclic Aromatic Hydrocarbons*. World Health Organization, Geneva, p. 883.

PART II

General Characteristics of PAHs

2

The Sources, Transport, and Fate of PAHs in the Marine Environment*

JAMES S. LATIMER[1] AND JINSHU ZHENG[2]

[1]*US Environmental Protection Agency, Office of Research and Development, National Health and Environmental Effects Research Laboratory, Atlantic Ecology Division, Narragansett, RI, USA*
[2]*Department of Biology and Chemistry, City University of Hong Kong, Kowloon, Hong Kong, China*

2.1 INTRODUCTION

PAH inputs to the coastal marine environment are primarily from two sources: (a) the movement of water containing dissolved and particulate constituents derived from watersheds; and (b) atmospheric deposition both in precipitation and dry deposition from airsheds of the coastal ocean. PAHs have been observed to be most concentrated in estuaries and coastal environments near urban centers, where inputs from the watersheds and airsheds are most localized. The major sources of PAHs to the coastal marine environment include urban runoff, wastewater effluents, industrial outfalls, atmospheric deposition, and spills and leaks during the transport and production of fossil fuels.

Some of these compounds are environmentally important because they are, or can become, carcinogenic or mutagenic. The US Environmental Protection Agency (EPA) has identified 16 PAHs as particularly important due to their toxicity to mammals and aquatic organisms (EPA 1987). The 16 priority pollutant PAHs, the primary focus of this chapter, are: acenaphthylene, acenaphthene, anthracene, benzo[a]anthracene, benzo[a]pyrene, benzo[b]fluoranthene,

---

* Although the synthesis described in this chapter has been funded wholly by the US Environmental Protection Agency, it has not been subjected to Agency-level review. Therefore, it does not necessarily reflect the views of the Agency. This publication is ORD/NHEERL/Atlantic Ecology Division Contribution No. AED-02-30.

TABLE 2.1 Physical, chemical[1], toxicological, and sediment quality values for selected PAHs

Abbr.	No. of rings	MW	MP (°C)	BP (°C)	S (g/m³ or mg/L)	V_p (Pa, solid)	H (Pa m³/mol)	log (K_{ow})	Avg. log (BCF)	Avg. K_{oc} (log K_{oc})	FCV^2 (µg/L)	$C_{oc,PAHi,FCVi}{}^3$ (µg/g$_{oc}$)	ERL^4 (ng/g dw[5])	ERM^4 (ng/g dw)
NAP	2	128	80.5	218	31	10.4	43.01	3.37	2.33	3.11	193.5	385	160	2100
ACL	3	150	92	270	16.1	0.9	8.4	4.00	2.79	3.64	306.9	452	44	640
ACT	3	154	96.2	277	3.8	0.3	12.17	3.92	2.63	4.02	55.85	491	16	500
TMN[6]	2	176	64	185	2.1	0.681	57.14	5.00			9.488	584		
FLU	3	166	116	295	1.9	0.09	7.87	4.18	3.10	4.35	39.30	538	19	540
PHE	3	178	101	339	1.1	0.02	3.24	4.57	3.27	4.31	19.13	596	240	1500
ANT	3	178	216	340	0.045	0.001	3.96	4.54	3.14	4.39	20.73	594	85.3	1100
FLR	4	202	111	375	0.26	0.00123	1.037	5.22	2.87	5.04	7.109	707	600	5100
PYR	4	202	156	360	0.132	0.0006	0.92	5.18	3.41	4.86	10.11	697	665	2600
BAA	4	228	160	435	0.011	2.80E-05	0.581	5.91	4.13	5.33	2.227	841	261	1600
CHR	4	228	255	448		5.70E-07	0.065	5.86	3.19	5.14	2.042	844	384	2800
BBF	5	252	168	481	0.0015			5.80	2.58	5.72	0.6774	979		
BKF	5	252	217	481	0.0008	5.20E-08	0.016	6.00	2.67	5.73	0.6415	981		
BEP	5	252	178		0.004	7.40E-07	0.02			5.60	0.9008	967		
BAP	5	252	175	495	0.0038	7.00E-07	0.046	6.04	3.74	6.24	0.9573	965	430	1600
PER	5	252	277	495	0.0004	1.40E-08	0.003	6.25	2.91		0.9008	967		
INP	6	278									0.2750	1115		
DBA	5	278	267	524	0.0006	3.70E-10		6.75	3.91	5.96	0.2825	1123	63.4	260
BPR	6	268	277		0.00026		0.075	6.50	5.00	6.23	0.4391	1095		
Σ PAHs													4022	44792

PAH abbreviations: naphthalene (NAP), acenaphthylene (ACL), acenaphthene (ACN), 2,3,5-trimethylnaphthalene (TMN), fluorene (FLU), phenanthrene (PHE), anthracene (ANT), 1-methylphenanthrene (1MP), fluoranthene (FLR), pyrene (PYR), benzo[a]anthracene (BAA), chrysene (CHR), benzo[b]fluoranthene (BBF), benzo[k]fluoranthene (BKF), benzo[e]pyrene (BEP), benzo[a]pyrene (BAP), perylene (PER), indeno[1,2,3-cd]pyrene (INP), dibenzo[ah]anthracene (DBA), benzo[ghi]perylene (BPR).

[1] Physical and chemical data: S, water solubility; V_p, vapor pressure; H, Henry's law constant; K_{ow}, octanol–water partition coefficient; K_{oc}, organic carbon partition coefficient (Mackay et al. 1992).

[2] FCV, final chronic value for specific PAH (EPA 2000).

[3] $C_{oc,PAHi,FCVi}$: effect concentration of a PAH in sediment on an organic carbon basis calculated from the product of its FCV and K_{oc} (EPA 2000).

[4] Sediment quality data: ERL, effects range low; ERM, effects range medium (Long et al. 1995).

[5] dw, dry weight.

[6] Actually presented, 1,4,5-trimethylnaphthalene.

benzo[k]fluoranthene, benzo[ghi]perylene, chrysene, dibenzo[ah]anthracene, fluoranthene, fluorene, indeno[1,2,3-cd]pyrene, naphthalene, phenanthrene, and pyrene. These compounds have 2–6 fused rings and molecular weights (MWs) of 128–278 g/mol. Solubility (S) and vapor pressure (Vp) characteristics of PAHs are the major physical/chemical factors that control their distribution between the soluble and particle components of the atmosphere, hydrosphere, and biosphere. Solubility values range from highly insoluble (e.g. benzo[ghi]perylene, 0.003 mg/L) to slightly soluble (e.g. naphthalene, 31 mg/L) and vapor pressures range from highly volatile (naphthalene) to relatively non-volatile (dibenzo[ah]anthracene) (Table 2.1). These constituents range from moderately to highly lipophilic, having logarithmic octanol–water partition coefficients (log K_{ow}) of 3.37–6.75.

This chapter will provide an overview of the important sources, transport, and fate processes of PAHs in the marine environment. The following topics will be included: anthropogenic sources; spatial source trends; major transport routes and processes; global mass balances; spatial distributions in seawater; and marine sediments.

2.2 DISTRIBUTION OF PAHs IN SOURCES TO THE MARINE ENVIRONMENT

2.2.1 POINT SOURCES

Point sources are those inputs that directly discharge into the marine environment from identifiable pipes or outfalls. Municipal wastewater facilities and industrial outfalls are the major point sources for most estuaries. The concentrations of total PAHs in wastewater from North American and European municipalities range from < 1 µg/L to over 625 µg/L (Table 2.2). A high value of 216.5 µg/L is observed for the Los Angeles County Sanitation District which, unlike most other wastewater facilities, receives a large fraction of effluent from petroleum refineries (Eganhouse and Gossett 1991). This effluent is dominated by two- and three-ringed PAHs (naphthalenes, phenanthrenes, and substituted naphthalenes, phenanthrenes) which is consistent with a strong signal from petroleum, either derived from refinery effluents or from urban runoff, and has also been observed in Rhode Island, USA, wastewater effluents (Hoffman *et al.* 1984). High concentrations (i.e. 625 µg/L) associated with oil shipping and refinery operations have also been reported for effluents in the Eastern Mediterranean Sea (Yilmaz *et al.* 1998). Concentration ranges for wastewater treatment plants that serve primarily domestic wastes are generally less than 5 µg/L (Table 2.2).

Mass emission rates for wastewater effluents ranged from < 2 kg/year (Paris) to over 100 Mtonnes/year of PAHs discharging into receiving waters (Table 2.2). For Narragansett Bay, RI, wastewater effluent represents 23% of the total inputs from point and non-point sources (Latimer 1997). On a per capita basis, these

TABLE 2.2 Concentrations and mass fluxes of PAHs in various point and non-point sources

Source/Location	Concentration (μg/L)	Flux (Mtonne/year)	References
Wastewater treatment facilities WWTF			
Los Angeles	216.5	110.5 (83 mg/cap/day)	Eganhouse and Gossett (1991)
Montreal	0.42	0.26 (0.5 mg/cap/day)	Pham and Proulx (1997)
Narragansett Bay	0.017–4.91	0.23 (0.6 mg/cap/day)[1]	Latimer (1997) Quinn et al. (1988)
N. Mediterranean cities	0.01–625	nr	Yilmaz et al. (1998)
Paris environs (influent)	1.99 ± 0.86	0.002–1.71[2]	Blanchard et al. (2001)
Urban runoff			
USA	0.24–560	nr	Walker et al. (1999)
South Carolina	0.040–16.3	nr	Ngabe et al. (2000)
Rhode Island	0.293–49.1[3]	0.681 (3 mg/cap/day)	Hoffman et al. (1984)
Paper factory (Mediterranean Sea)	40–1404	nr	Yilmaz et al. (1998)
Fertilizer factory (Mediterranean Sea)	nd–4451	nr	Yilmaz et al. (1998)
Iron and steel factory (Mediterranean Sea)	nd–1390	nr	Yilmaz et al. (1998)
WWTF sewage sludge	(ng/g dry weight)		
Manresa (Spain)[4]	4730 ± 1117	nr	Perez et al. (2001)
Igualada (Spain)[5]	1240 ± 312	nr	Perez et al. (2001)
Ripoll (Spain)[4]	4270	nr	Perez et al. (in press)
Porto (Spain)[6]	3710	nr	Perez et al. (in press)
Montornés (Spain)[7]	4120	nr	Perez et al. (in press)
Alrera (Spain)[4]	1750	nr	Perez et al. (in press)
Paris (environs)	5300 ± 2640	nr	Blanchard et al. (2001)

nr, not reported; nd, none detected.
[1] Using 1×10^6 population (estimated).
[2] Wet and dry conditions.
[3] Flow weighted means.
[4] Domestic waste.
[5] Industrial waste.
[6] 65% industrial waste, 35% urban runoff.
[7] 60% industrial waste.

emission rates translate to < 1–83 mg/cap/day for major North American cities, with the highest value, for Los Angeles, likely due to effluents from oil refineries (Eganhouse and Gossett 1991; Latimer 1997; Pham and Proulx 1997; Quinn et al. 1988).

Sewage sludges generated from the treatment of wastewaters contain levels of PAHs in the order of 1.6 μg/g dry weight (dw), range 1.2–5.3 μg/g, depending upon the fraction of industrial waste in the treatment stream (Table 2.2).

Little current data are available on the concentrations of PAHs in industrial effluents, save what we know about petroleum inputs to municipal sewer systems (see above). Concentrations range from undetectable to 4.4 mg/L for paper, fertilizer, iron and steel effluents (Table 2.2).

2.2.2   NON-POINT SOURCES

In the USA, the Clean Water Act of 1972 (enacted by Public Law 92–500, October 18 1972, 86 Stat. 816; 33 U.S.C. 1251 *et seq.*) specified, in Sections 401 and 402, that all point source discharges into ambient water bodies must be regulated by a permitting process. The aim of this legislation was to reduce pollutants and ultimately restore surface water quality by the control of point source discharges. Non-point sources, those inputs to ambient waterbodies that are not from discrete pipes, have only recently been scrutinized for their contribution to coastal pollution (enacted as an amendment of the Clean Water Act in 1987, PL 100-4). Such inputs as urban runoff, surface water runoff from various land use types, as well as atmospheric inputs, have recently been quantified (Hoffman *et al.* 1984; Ngabe *et al.* 2000; Walker *et al.* 1999).

2.2.2.1   Urban runoff

Urban runoff, the water that washes off urban surfaces during rainstorms, has not been studied extensively for PAHs. Concentrations of $< 0.05–560$ µg/L have been reported (Table 2.2). Some of the most extensive work was done in Narragansett Bay, where the total PAH contribution from urban runoff was estimated at 681 kg/year or 2.7 mg/cap/day. This comprised nearly 40% of the total estimated PAH inputs to Narragansett Bay at the time (i.e. early 1980s, including only direct wastewater and atmospheric inputs) (Hoffman *et al.* 1984).

2.2.2.2   Atmosphere

Sources of PAHs to the atmosphere include the combustion of fossil fuel oils, gasoline, wood, and refuse. In addition, volatilization of certain PAH components from petroleum products into the atmosphere may also be a source. These contribute to the levels discharged into the coastal marine environment via dry and wet deposition from coastal watersheds, as well as directly onto estuarine surfaces.

Concentrations of total PAHs in air vary markedly across the globe due presumably to the myriad local sources in urban environments. Concentrations in North American areas are in the range $3.7–450$ ng/m^3, with the highest values observed in urban environments (Table 2.3). European concentrations are similar with values ranging from 0.2 ng/m^3 (Germany) to 137 (London) ng/m^3. The Asian concentrations tend to be skewed toward higher levels, with values of $25–500$ ng/m^3 in China and India; however, the differences are not statistically significant.

The PAH distributions in air show a wide range in concentrations (Figure 2.1a) and are generally dominated by phenanthrene, fluoranthene, and pyrene

TABLE 2.3 Concentrations of PAHs in air (pg/m^3)

PAHs[1]	Narragansett Bay, RI (1993,1994)	Galveston Bay, TX[2]	Chesapeake Bay (three sites)	Chicago, IL	London, UK	Munich, Germany	Algiers City, Algeria	Guangzhou City, China	Tianjin, China	Beijing, China	Mumbai, India
ACL	532–963	166 ± 236	nr	nr	nr	nr	nr	nr	nr	nr	nr
ACN	744–895	527 ± 685	nr	76900	63400[3]	nr	nr	nr	nr	nr	nr
TMN	971–990	nr	nr	nr	nr	nr	nr	nr	nr	nr	nr
FLU	2923–2987	1593 ± 1559	696–1110	74800	nr	nr	nr	nr	nr	nr	nr
PHE	5116–6427	11936 ± 12161	1830–2640	200300	46200	nr	nr	nr	nr	nr	nr
ANT	769–2737	543 ± 889	29.5–47.2	14100	3400	nr	nr	nr	nr	nr	nr
1MP	616–972	nr	nr	nr	nr	nr	nr	nr	nr	nr	nr
FLR	1363–1676	5740 ± 7602	332–480	44100	10300	40–2120	400–3800	nr	nr	nr	nr
PYR	949–975	3164 ± 4331	403–761	24600	7500	40–2200	100–1800	nr	nr	nr	nr
BAA	85–108	48 ± 63	23.1–37.5	2100	3,000	nr	100–1900	nr	nr	nr	nr
CHR	181–357	281 ± 375	81.3–109	3600	950	10–2010	220–1800	nr	nr	nr	nr
BBF	293–294	109 ± 156	65.8–115	2300	nr	60–4510	680–5300	nr	nr	nr	nr
BKF	183–307	28 ± 26	52.9–78.8	1900	nr	nr	nr	nr	nr	nr	nr
BEP	173–207	57 ± 90	50.4–79.6	nr	nr	nr	160–3100	nr	nr	nr	nr
BAP	101–147	47 ± 90	29.9–44.6	1600	650	20–1660	70–2100	nr	41100	66500	nr
PER	30–47	22 ± 23	nr	nr	nr	nr	nr	nr	nr	nr	nr
INP	80–96	53 ± 87	58.3–81.1	1200	nr	nr	30–4900	nr	nr	nr	nr
DBA	18–28	13 ± 18	12.7–22.4	nr	nr	nr	< 10–300	nr	nr	nr	nr
BPR	175–194	58 ± 65	58.7–85.9	1100	1480	20–2750	220–4800	nr	nr	nr	nr
ΣPAHs	15776–19928	51778 ± 41700	3724–5692	448600	137000	150–15130	5500–43400	27400–228600	465000	500000	25000–40000
References	Latimer (1997)	Park et al. (2001)	Baker et al. (1997)	Odabasi et al. (1999)	Wild and Jones (1995)	Schnelle-Kreis et al. (2001)	Yassaa et al. (2001)	Cheng et al. (1998)	Zhu et al. (1998)	Zhu et al. (1998)	Kulkarni and Venkataraman (2000)

nr, not reported.
[1] See Table 2.1 for abbreviations.
[2] Sum of vapor and particulate phases.
[3] Includes fluorene.

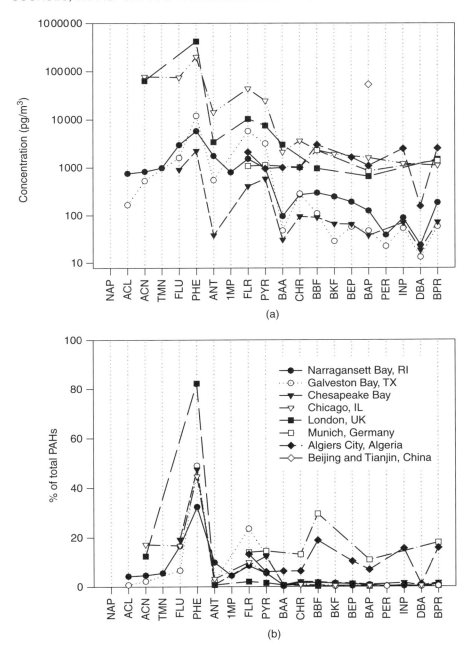

Figure 2.1 PAH concentrations (a) and composition (b) in air samples from around the world (see Table 2.1 for abbreviations)

(Figure 2.1b). This type of PAH distribution, dominance by three- and four-ringed compounds, suggests a mixture of combustion and petroleum sources. In addition, it is clear that the Chinese signal for benzo[a]pyrene, an important carcinogen, is much higher than reported elsewhere in the world (Table 2.3, Figure 2.1a; Zhu *et al.* 1998). It is not clear why benzo[a]pyrene is so elevated, but it may be due to the dominance of coal as a fuel source.

2.2.2.3 Precipitation

PAHs have also been detected in precipitation. Concentrations of total PAHs range from 84 to 1397 ng/L (Table 2.4). Although it is difficult to compare the PAH concentrations (Figure 2.2a) and distributions (Figure 2.2b) across different regions because not all the same constituents are reported in all the areas, the concentrations are dominated by naphthalene, phenanthrene, fluoranthene, and pyrene. In common with air samples, the reported distributions are symptomatic of mixed combustion and petroleum sources.

In an extensive study, atmospheric deposition (wet and dry) comprised nearly 7% of the nearly 1 Mtonne of PAHs that enters Narragansett Bay, USA, on an annual basis (Table 2.5). This is equivalent to 250 $\mu g/m^2/year$, which is twice as large as the areal flux entering the Chesapeake Bay (Table 2.5). The differences in areal fluxes are likely due to the proximity of Narragansett Bay to large urban centers (i.e. New York and Boston), as well to the prevailing west-to-east winds which move air from upper mid-west USA, an area with large power generation effluents, to the north-east.

2.2.2.4 Summary of sources

Total inputs of PAHs to Narragansett and Chesapeake Bays, two of the most studied estuaries in the USA, have been estimated to be approximately 1 and 10 Mtonne/year, respectively (Table 2.5). In both estuaries, the sources are dominated by the inputs from rivers that discharge into the estuary. In Narragansett Bay, riverine inputs represent approximately 70% of the total (river + municipal/industrial + direct atmosphere) (Latimer 1997). In Chesapeake Bay, riverine input of PAHs comprise 87% of the total, and atmospheric deposition comprises the remainder (Baker *et al.* 1997). It must be appreciated that sources to rivers and municipal wastewater treatment facilities themselves are derived from atmospheric deposition and direct petroleum inputs onto the watersheds and urban surfaces (Latimer *et al.* 1990). For Narragansett Bay, these inputs comprise approximately 60% of the total mass of PAHs being deposited to Narragansett Bay sediments (i.e. 1.4–2.2 mton/year) (Latimer and Quinn 1996), which indicates that the sources and sinks are not as accurately characterized as they need to be.

TABLE 2.4 Concentrations of PAHs in precipitation (ng/L)

PAH	Narragansett Bay, RI[1]	Portland, OR[2]	Portland, OR[3]	Great Lakes USA[5]	Galveston Bay, TX[4]	Northern Greece[5]
Naphthalene	nr	nr	nr	nr	19.01	32–781
Acenaphthylene	1.1–4.3	nr	nr	nr	0.893	nr
Acenaphthene	0.6–1.7	nr	nr	nr	1.03	nr
2,3,5-trimethylnaphthalene	0.4–1.5	nr	nr	nr	nr	nr
Fluorene	2.3–6.5	nr	nr	nr	2.10	nr
Phenanthrene	11–31	90	4.1	3.2–11	10.77	29–598
Anthracene	0.4–6.6	5.1	nr	nr	0.456	1.7–96
1-Methylphenanthrene	0.8–4.5	nr	nr	nr	nr	nr
Fluoranthene	16–28	48	4.4	nr	8.40	14–54
Pyrene	15–30	39	4.1	2.7–8.2	5.78	2.0–76
Benzo[a]anthracene	1.0–2.6	3.3	1.5	nr	1.13	0.9–12
Chrysene	2.6–6.1	7.9	3.6	nr	3.62	0.8–36
Benzo[b]fluoranthene	2.5–5.7	1.6	9.2	nr	3.96	1.0–9.5
benzo[k]fluoranthene	3.0–5.2			2.9–5.1	1.08	0.3–5.4
benzo[e]pyrene	1.7–11	0.37	3.0	nr	2.12	15–63
benzo[a]pyrene	1.6–10	nr	2.8	2.9–5.1	1.41	0.9–11
Perylene	3.5–11	nr	nr	nr	0.342	nr
Indeno[1,2,3-cd]pyrene	1.1–2.9	nr	nr	nr	2.187	2.5–11
Dibenzo[a,h]anthracene	0.3–1.1	nr	nr	nr	0.390	1.3–2.3
Benzo[ghi]perylene	1.1–4.3	nr	6.0	nr	2.60	2.1–14
ΣPAHs	84–162	nr	nr	nr	106	143–1397
References	Latimer (1997)	Ligocki et al. (1985a)	Ligocki et al. (1985b)	Hoff et al. (1996)	Park et al. (2001)	Manoli et al. (2000)

nr, not reported.
[1] Dissolved + particulate (range of means for 3 years).
[2] Dissolved phase.
[3] Particulate phase.
[4] Volume weighted mean.
[5] Bulk precipitation.

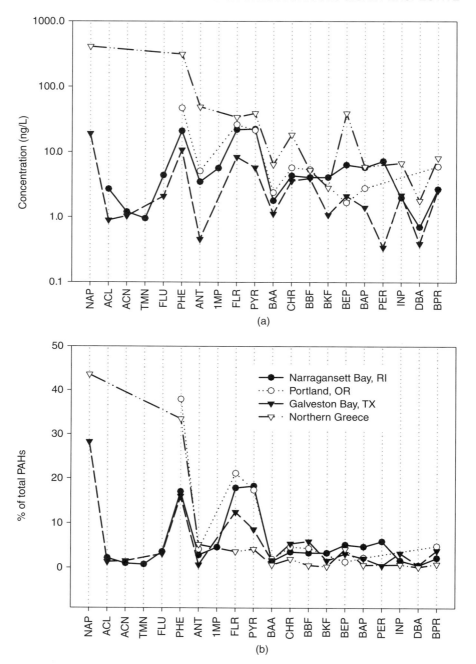

Figure 2.2   PAH concentrations (a) and composition (b) in precipitation samples from around the world (see Table 2.1 for abbreviations)

TABLE 2.5 Annual input and areal flux of PAHs to the Chesapeake and Narragansett Bay estuaries

Source/location	Narragansett Bay[1]		Chesapeake Bay[2]	
	Input (kg/year)	Areal Flux[3] (μg/m^2/year)	Input (kg/year)	Areal Flux[3] (μg/m^2/year)
Riverine inputs				
Wet conditions	220	834		
Dry conditions	460	1744		
Total riverine	680	2578	8682	755
Point source inputs				
Wet conditions	140	531		
Dry conditions	89	337		
Total point sources	229	868		
Atmospheric inputs				
Wet deposition	38	144		
Dry deposition	28	106		
Total atmosphere	66	250	1303	113
Grand totals	975	3697	9985	868

[1]Adapted from Latimer (1997).
[2]Adapted from Baker *et al*. (1997).
[3]Inputs normalized to surface area of waterbodies.

2.3 DISTRIBUTION OF PAHs IN THE MARINE ENVIRONMENT

Thus far, we have discussed the major sources of PAHs to the coastal marine environment. Next, we will review the distributions of PAHs in the water and sediments of the coastal marine environment itself.

2.3.1 SEAWATER

The concentrations of total PAHs in marine waters are quite variable ranging from undetectable to 11 μg/L (Table 2.6). The concentration ranges are so large that even within a relatively small geographic region it is difficult to distinguish between areas. In general, however, there is a gradient in which offshore concentrations are lowest, followed by inshore, and lastly, the sea surface microlayer (SSM). For samples collected at the same time and in the same place, the SSM values are over a factor of 10 as large as the bulk seawater (Table 2.6; Cincinelli *et al*. 2001). It is noteworthy that under no circumstances do total PAH concentrations ever exceed the US EPA's marine lowest observed effects level (LOEL) of 300 μg/L (EPA 1987).

PAHs in water partition between dissolved and particulate fractions, depending upon the solubility of the individual PAHs and the availability of binding

TABLE 2.6 Concentrations of PAHs in marine water (ng/L)

PAH[1]	Narragansett Bay[2]		England and Wales[3]	Eastern Mediterranean Sea[4]	Tyrrhenian Sea (Central Mediterranean Sea)	Tyrrhenian Sea (Central Mediterranean Sea)[5]	Hong Kong	
	Mean	SD					Mean	SD
NAP	0.50	0.81	<6–6850	nr	nr	nr	199.3	164.8
ACL	0.06	0.30	nr	nr	nr	nr	23.3	26.6
ACN	0.22	0.40	<1–1740	nr	nr	nr	128.6	103.9
TMN	nr	nr	nr	nr	nr	nr	nr	nr
FLU	0.29	0.63	<1–1400	nr	nr	nr	48.9	33.6
PHE	3.30	3.76	<3–1170	nr	nr	nr	66.5	54.6
ANT	0.46	1.15	<1–157	nr	nr	nr	28.0	24.0
1MP	nr	nr	nr	nr	nr	nr	nr	nr
FLR	6.14	7.92	<1–940	nr	nr	nr	30.2	28.4
PYR	5.95	7.23	<1–1090	nr	nr	nr	54.1	65.5
BAA	2.04	3.61	<1–609	nr	nr	nr	nr	nr
CHR	5.37	8.64	<1–726	nr	nr	nr	24.9	30.4
BBF	6.99	9.71	<1–621	nr	nr	nr	18.9	17.1
BKF			<1–250	nr	nr	nr	17.0	15.9
BEP	nr	nr	<1–207	nr	nr	nr	nr	nr
BAP	2.70	4.30	<1–909	nr	nr	nr	70.4	69.5
PER	nr	nr	nr	nr	nr	nr	nr	nr
INP	1.96	3.12	nr	nr	nr	nr	17.2	16.8
DBA	0.48	1.17	<1–126	nr	nr	nr	29.8	36.1
BPR	2.60	5.37	<1–627	nr	nr	nr	15.3	16.2
ΣPAHs	39.05	58.13	nd–10724	10–4140	3400 ± 2780	47510 ± 56100	769.4	596.9
References	Quinn et al. (1988)		Law et al. (1997)	Yilmaz et al. (1998)	Cincinelli et al. (2001) (9 stations)	Cincinelli et al. (2001) (9 stations)	Zheng (unpublished data)	

nr, not reported; nd, none detected, SD, standard deviation.
[1] See Table 2.1 for abbreviations.
[2] Particulate fraction only.
[3] Estuarine and offshore, unfiltered water.
[4] Surface.
[5] Sea surface microlayer.

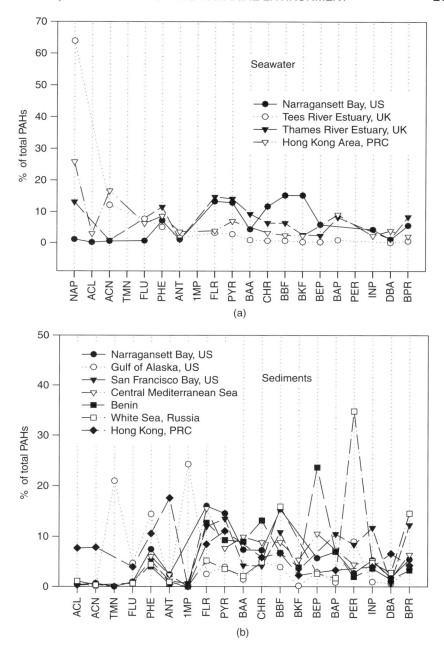

Figure 2.3 PAH composition in seawater (a) and sediment (b) samples from around the world (see Table 2.1 for abbreviations)

substrates such as suspended particulates. Figure 2.3a shows the distribution of PAHs in seawater samples from the USA, UK, and China. Naphthalene is either the most, or nearly the most, dominant PAH in the samples from the Rivers Tees and Thames estuaries (UK; Law *et al.* 1997), as well as in samples from the Hong Kong vicinity. This suggests that PAHs from Hong Kong and the Tees River estuary are dominated by petroleum sources, since the naphthalene/phenanthrene ratio is much greater than 1 (Steinhauer and Boehm 1992). Data from Narragansett Bay show much lower levels of naphthalene, relative to high molecular weight PAHs, than for other regions. This is because only the particulate fraction of water was analyzed, and naphthalene is not expected to be present in large quantities due to its higher water solubility. Conversely, the larger relative magnitude of higher molecular weight PAHs, evident in the Narragansett Bay water, is expected because these species are highly associated with particulate matter.

2.3.2   MARINE SEDIMENTS

The distribution of PAHs in the environment is largely controlled by their solubility and hydrophobicity, making sediments their primary repository. As such, PAHs have been measured in sediments from a great number of regions around the globe. Table 2.7 provides data from over 20 different studies, representing hundreds of sediment samples, from four continents. Current total PAHs in North America are in the range 2.17-> 170, 000 ng/g dw. European and African data are in the range 0.32-48,060 ng/g dw and Asian values 9.1-11,740 ng/g dw. In general, concentrations are highest in harbors and embayments near urban centers and decrease markedly with distance from these centers (Baumard *et al.* 1998; Kim *et al.* 1999; Pruell and Quinn 1985; Soclo *et al.* 2000; Zheng and Richardson 1999). This is borne out in data from the Hong Kong area, where the highest values (To Kwa Wan, 3072 ng/g dw and Tai Po Market, 4420 ng/g dw) are found in a harbor and near a coal and oil production facility, respectively. In addition, the highest concentrations of PAHs in sediments around the USA are located in urban embayments, whereas the lowest levels are located in more distant, well-flushed areas.

The distribution of individual PAH compounds in sediment samples is generally dominated by the 4-, 5-, and 6-ring species (Table 2.8, Figure 2.3b). This is in contrast to the constituents in the dissolved or vapor phases of air, precipitation and seawater, which tend to be dominated by 2- and 3-ring species. The specific ratios of parent PAHs and parent to alkyl homolog distributions of PAHs are useful to determine the dominance of petrogenic vs. pyrogenic sources. Parent compounds, which do not contain alkyl constituents, are indicative of pyrogenic sources such as soot and other combustion-derived materials whereas PAH distributions dominated by alkyl constituents are representative of petroleum sources (Sporstol *et al.* 1983). Table 2.9 gives a listing of the most

TABLE 2.7 Concentrations of total PAHs in marine sediment of North America, Europe, Africa and Asia (ng/g dry weight)

Location	Concentration	Reference
North America		
Entire US coast	13.4–40453	EPA EMAP Program[1]
Entire US coast	4.87–30674	NOAA Status and Trends Program[2]
Pales Verdes Shelf, CA, USA	1252–7037	Eganhouse and Gossett (1991)
Alaska Stations	2.17–733	Valette-Silver et al. (1999)
Gulf of Alaska (pre-Exxon Valdez)	1096	Bence et al. (1996)
West Beaufort Sea (Polar Star sediments)	159–1092	Valette-Silver et al. (1999)
Fraser Estuary (BC, Canada)	180–620 (combustion) 220–660 petroleum)	Yunker et al. (1999)
Burrard Inlet (BC, Canada)	430–91800 (combustion) 70–39500 (petroleum)	Yunker et al. (1999)
Strait of Georgia (BC, Canada)	300–8470 (combustion) 560–4300 (petroleum)	Yunker et al. (1999)
San Francisco Bay embayments (1800s–present)	40–6300 (pyrogenic)	Pereira et al. (1999)
South Carolina estuaries	33–9630	Kucklick et al. (1997)
Narragansett Bay, RI	100–29300	Quinn et al. (1992)
New Bedford Harbor, MA	14000–170000	Pruell et al. (1990)
Europe		
Eastern Mediterranean Sea	20–18700	Yilmaz et al. (1998)
Baltic Sea	3.16–30100	Baumard et al. (1999)
Irish estuaries	83–22960	Guinan et al. (2001)
Gironde Estuary (France)	3.5–853	Soclo et al. (2000)
Arcachon Bay (France)	293	Soclo et al. (2000)
Cretan Sea (Eastern Mediterranean)	14.6–158.5 (73% combustion derived)	Gogou et al. (2000)
Lazaret Bay (Central Mediterranean)	86.5–48060	Benlahcen et al. (1997)
Near-coastal Spain and France (Mediterranean Sea)	0.32–8400	Baumard et al. (1998)
Africa		
Cotonou Coast (Benin)	80–1411	Soclo et al. (2000)
Asia		
White Sea (Russia, Arctic Ocean)	13–208	Savinov et al. (2000)
Kyeonggi Bay (Korea)	9.1–1400	Kim et al. (1999)
Hong Kong (surface)	7.25–4420	Zheng and Richardson (1999)
South China Sea	24.7–275.4	Yang (2000)
Yangtze River Estuary (core)	122–11740	Liu et al. (2000)
Bohai Bay	31–2513	Ma et al. (2001)
Yellow Sea	20–5734	Cheng et al. (1998)

[1]http://www.epa.gov/emap/nca/html/data/(downloaded 12/01).
[2]http://ccmaserver.nos.noaa.gov/nsandtdata/NSandTdatasets/benthicsurveillance/welcome.html (downloaded 12/01).

TABLE 2.8 Concentrations of individual PAHs in marine sediments (ng/g dry weight)

PAH[1]	Narragansett Bay (USA) (26 stations) Mean	Narragansett Bay (USA) (26 stations) Min–max	Gulf of Alaska (USA) (18 stations)	San Francisco Bay (USA) (1 station)	Arcachon Gironde Bays (France) (7 stations)	Lazaret Bay (Central Mediterranean Sea) (9 stations)	Cretan Sea (Mediterranean Sea) (10 stations)	Cotonou Coast (Benin) (6 stations)	White Sea (Russia, Arctic Ocean) (11 stations)	Kyeonggi Bay (Korea) (66 stations)	Hong Kong (China) (20 stations) Mean	Hong Kong (China) (20 stations) SD
ACL	29	nd–140	1.90	2.4	nr	nr	nr	nr	0.1–2.0	nd–8.8	38.1	99.4
ACN	55	nd–270	0.27	3.9	nr	nr	nr	nr	0.1–0.5	nd–23	38.8	90.7
TMN	nr	nr	81.1	nr	nr	nr	nr	nr	nr	nd–14		
FLU	82	nd–300	18.0	7.8	nr	nr	nr	nr	0.1–1.1	nd–66	19.4	46.7
PHE	652	nd–2800	55.7	47	0.5–74	4–3190	0.13–4.18	5–66	0.4–8.2	0.71–220	52.4	117.6
ANT	207	nd–820	2.24	10	0.1–18	0.5–1230	nd–0.32	0.1–9	0.1–1.7	nd–43	87.4	185.7
1MP	nr	nr	93.9	5.7	nr	nr	0.36–5.07	nr	nr	nd–69		
FLR	1401	20–5940	9.60	108	0.5–100	1–8230	1.13–11.1	4–224	0.6–9.5	2.5–300	42.1	144.1
PYR	1271	20–4488	15.1	122	1–102	8–4060	0.50–6.35	4–163	0.4–6.4	2.3–240	55.1	132.9
BAA	643	10–2570	5.56	38	0.3–68	12–5250	0.51–4.82	1–160	0.2–3.9	nd–71		
CHR	635	10–2490	19.3	37	0.2–45	8–4670	0.97–7.30	6–230	0.5–9.0	nd–92	29.0	89.3
BBF	1345	30–5660	15.0	98	0.2–79	12–4690	2.93–22.12[2]	2–119	3.0–28	nd–27	33.2	91.9
BKF			0.53	30	< 0.1–24	6–2830		1–68		nd–53	11.2	20.3
BEP	497	20–2830	10.9	nr	1–103	13–5590	0.90–8.90	112–313	0.1–4.9	nd–13		
BAP	610	20–2350	3.26	95	0.1–52	8–3900	0.70–6.65	1–124	0.2–3.1	nd–5.5	16.3	34.3
PER	236	10–1180	34.9	76	< 0.1–52	nd–2230	0.43–18.35	1–34	0.8–67	nd–27		
INP	462	20–1710	3.58	106	1.5–46	6–2790	1.39–9.85	1–64	0.7–8.5	nd–6.3	19.5	80.5
DBA	151	nd–490	2.51	6.9	0.1–12	1–830	0.37–2.46	1–24	0.4–5.0	nd–8.8	32.7	106.4
BPR	474	20–1580	13.4	111	1–73	3–3380	0.79–8.91	2–56	1.4–27	nd–6.9	21.0	59.9
References	Quinn et al. (1992)		Bence et al. (1996)	Pereira et al. (1999)	Soclo et al. (2000)	Benlahcen et al. (1997)	Gogou et al. (2000)	Soclo et al. (2000)	Savinov et al. (2000)	Kim et al. (1999)	Zheng and Richardson (1999)	

nr, not reported; nd, none detected; SD, standard derivation.

[1] See Table 2.1 for abbreviations.

[2] Includes benzo[j]fluoranthene.

TABLE 2.9 PAH chemical markers used to distinguish combustion from petroleum sources

Indicator	References	Notes
Parent compound distribution (PCD) MW: 178[1], 202, 228, 252, 276, 278	Lake et al. (1979), Steinhauer and Boehm (1992)	Used to characterize different distributions of unsubstituted parent PAHs. PAHs formed under high T conditions (combustion derived = pyrogenic)
Alkyl-homolog distribution (AHD) NAP + C1[2], C2, C3, C4, FLU + C1, C2, C3, PHE + C1, C2, C3, C4, FLR + C1, C2, C3, CHR + C1, C2, C3, C4	Lake et al. (1979), Steinhauer and Boehm (1992), Sporstol et al. (1983)	Used to characterize different distributions. PAHs formed under low T conditions (petroleum derived = petrogenic)
NAP + alkyl NAP (NAP + C1 + C2 + C3 + C4 alkyl homologs)	Steinhauer and Boehm (1992)	Associated with unweathered petroleum, rarely found in clean sediments
FLU + alkyl-FLU (FLU + C1 + C2 + C3 alkyl homologs)	Steinhauer and Boehm (1992)	The more highly alkylated, the more indicative of petroleum
PHE + alkyl-PHE (PHE + C1 + C2 + C3 + C4 alkyl homologs)	Steinhauer and Boehm (1992)	The more highly alkylated, the more indicative of petroleum
NAP/PHE	Steinhauer and Boehm (1992)	$\gg$ 1 for petroleum; 0.2–1.5 for clean sediment
PHE/ANT	Colombo et al. (1989), Steinhauer and Boehm (1992)	$\geq$ 50 fuel oil; 3–26 sediment
DbT[3] + alkyl-DbT (DbT + C1 + C2 + C3 alkyl homologs)	Steinhauer and Boehm (1992)	These sulfur heterocycles are distinct in petroleum
TPHE/tDbT[4]	Steinhauer and Boehm (1992)	With increasing oil input, this ratio approaches the value for oil (e.g. Pudroe crude = 1.1). In unoiled sediments this value can be in the range 10-> 100
FLR/PYR	Colombo et al. (1989)	0.6–1.4 for Kuwait and Louisiana crudes and No. 2 fuel oil
Σ4,5-ringed PAHs	Steinhauer and Boehm (1992)	Indicative of pyrogenic influences (i.e. FLR, PYR, BAA, CHR, BBF, BKF, BAP, BEP)
Σ4,5-ringed PAHs/ΣPAHs (i.e. tNAP, etc.)	Steinhauer and Boehm (1992)	High value indicates petrogenic influences
FFPI (Fossil Fuel Pollution Index) [(tNAP + tPHE× (0.5 (tPHE + C1-PHE + tDbT))/ΣPAHs] × 100	Steinhauer and Boehm (1992)	High value indicates petrogenic influences

[1] See Table 2.1 for molecular weight (MW) and abbreviation for compounds.
[2] Designates the alkyl group position on parent PAH.
[3] DbT, Dibenzothiophene.
[4] Total of phenanthrene + alkyl homologs/total of dibenzothiophene + alkyl homologs.

prominent indices that have been used to distinguish between petrogenic and pyrogenic sources of PAHs. Using these ratios it is evident, for example, that PAHs from the Gulf of Alaska are dominated by petroleum sources (Figure 2.3b).

Using a method that correlates concentrations of contaminants to biological effects, it has been estimated that benthic organisms living in sediments with PAH values below 4022 ng/g dw exhibit no risk of adverse effects, and exhibit minimal risks with values of 4022–44,792 (Long *et al.* 1995). Thus, using these guidelines, many of the sediments in these estuaries and embayments are not expected to be excessively harmful to benthic organisms (Table 2.7, Figure 2.4). However, highly urbanized embayments, such as the Burrard Inlet, Canada (Yunker *et al.* 1999) and New Bedford Harbor, USA (Pruell *et al.* 1990), are expected to contain concentrations that may affect the biota. It must be noted that, while these guidelines are based on the weight of evidence from large data sets, they do not take into consideration the synergistic effects of multiple contaminants that are present in estuarine sediments, and as such may underestimate risks to the environment from contaminated sediments. Moreover, a statistical method relies on correlation rather than causation. In addition, this approach does not explicitly take into consideration the fraction of the total PAHs that is bioavailable. Bioavailability is considered using the equilibrium partitioning approach (EPA 2000), which is based on the magnitude of PAHs in the organic carbon phase that would cause dissolved PAHs in porewater to be detrimental to exposed organisms ($C_{oc,PAHi,FCVi}$; see Table 2.1).

Temporal trends in PAH inputs to the marine environment are not very evident from current monitoring programs because of their use of surficial sediments. However, by using sediment cores it is possible to trace the inputs to the marine environment over many decades and longer periods (Latimer *et al.*, in preparation). PAHs determined in sediment cores from throughout the USA generally show subsurface maxima (Figure 2.5), although current levels are still elevated above background or reference values (Latimer *et al.*, in preparation; Valette-Silver 1993). In Western countries this is generally attributed to a reduction in emissions due to legislation aimed at reducing atmospheric inputs of other constituents (e.g. SO_x, NO_x and other ozone precursors). This has led to a shift to cleaner burning fuels which is reflected in lower levels of PAHs in sediments (Latimer *et al.* 1999; Pereira *et al.* 1999). However, this may not be true in other parts of the globe, where emerging economic trends have only recently affected environmental priorities, e.g. it is not as evident in the vicinity of Hong Kong where the recent concentrations tend to be increasing (Figure 2.5).

CONCLUSION

Polycyclic aromatic hydrocarbons are ubiquitous contaminants in the marine environment. Contained in petroleum hydrocarbons and formed from the combustion of fossil and bio-fuels, PAHs are present in the greatest concentrations

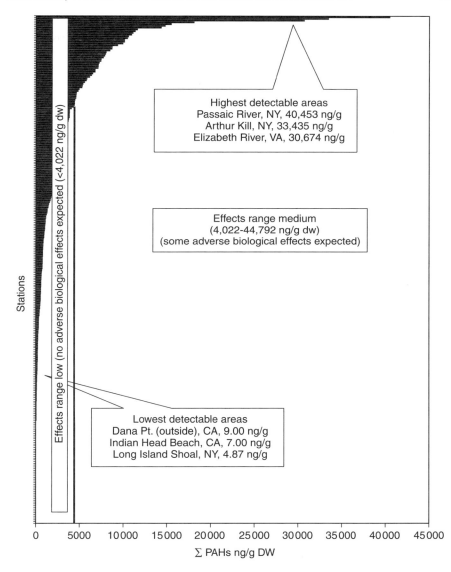

Figure 2.4 Total PAH concentrations in sediments from US estuaries from NOAA and EPA national monitoring programs (288 stations sampled)

near urban centers and in areas where there are large petroleum transport operations or near natural or anthropogenic oil sources. Inputs of PAHs to the atmosphere have been estimated for the USA, UK, Sweden and Norway (Table 2.10). The earliest estimate for US inputs did not include mobile (vehicular emissions including diesel engines) and industrial sources (Eisler 1987). All

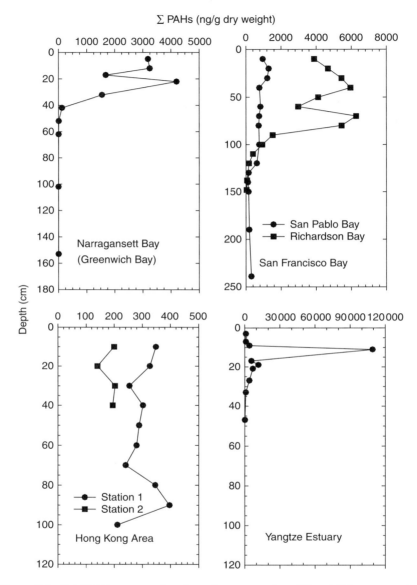

Figure 2.5 Total PAH concentrations in sediment cores from US and Chinese estuaries

the subsequent US and European budgets, while absolute magnitudes differed markedly, ranked mobile sources, heating and power generation and industrial inputs as the most important sources of PAHs to the atmosphere (Baek *et al.* 1991; Wild and Jones 1995). Eisler (1987) estimated that inputs from petroleum spillage and atmospheric deposition are the most important source of PAHs to

TABLE 2.10　Sources of PAHs to the atmosphere and aquatic environment of North America and Europe (Mtonne/year)

Ecosystem and Source	USA	USA	USA	UK	Sweden and Norway
Atmosphere					
Forest and prairie fires	19513	600–1478		—	1–5
Agricultural burning	13009	400–1190		6.3	1–2
Refuse and other open burning	4769	1357–1428		—	—
Enclosed incineration	3902	50–56[1]		0.06	1–2
Heating and power	2168	1468–4357		610	64–139
Industrial processes[2]	—	640–3497		19	203–312
Mobile sources	—	2170–2266		80	20–47
Total	43361	6685–14272		715	290–507
Aquatic environments					
Natural seeps	—		2500		
Petroleum spillage	170000		790		
Atmospheric deposition	50000		1600		
Wastewater	4400		—		
Surface runoff	2940		—		
Biosynthesis	2700		—		
References	Eisler (1987)	Baek et al. (1991)	NAS (in press)	Wild and Jones (1995)	Baek et al. (1991)

[1] Commercial only.
[2] Mostly coke production.

the aquatic environment. However, recent estimates suggest that natural seeps (into the deep ocean waters) outrank anthropogenic petroleum spillage in total inputs (NAS in press).

Environmental distributions of these constituents follow their physical and chemical characteristics. For example, vapor and dissolved phase PAH distributions in air, precipitation, and water are dominated by 2- and 3- ringed species, whereas aerosol and particulate phases and sediments are generally dominated by 4-, 5- and 6-ringed species, which are typical of pyrogenic sources. Geographic differences in the distribution of PAHs in air, water, biota and sediments across the globe from East to West are generally less significant than differences within each region and tend to be concentrated in a manner that decreases with distance from local and regional sources. Using sediment cores, it is generally observed that current PAH levels in the marine environment are less than historic highs, although they are still well above preindustrial reference conditions. Despite technical and regulatory controls, PAHs are still being deposited to the environment in significant quantities.

ACKNOWLEDGMENTS

Thanks are extended to Drs Richard Pruell, Robert Burgess and Mark Cantwell and three anonymous reviewers for their contributions to improve the content of this chapter, and special thanks to Kiki Latimer for editorial comments.

REFERENCES

Baek SO, Field RA, Goldstone ME, Kirk PW and Kester JN (1991) A review of atmospheric polycyclic aromatic hydrocarbons: sources, fate and behavior. *Water, Air and Soil Pollution*, **60**, 279–300.

Baker JE, Poster DL, Clark CA, Church TM, Scudlark JR, Ondov JM, Dickut RM and Cutter G (1997) Loadings of atmospheric trace elements and organic contaminants to the Chesapeake Bay. In Baker JE (ed.), *Atmospheric Deposition of Contaminants to the Great Lakes and Coastal Waters*. SETAC Press, Pensacola, FL, pp. 171–194.

Baumard P, Buzinski H, Michon Q, Garrigues P, Burgeat T and Bellocq J (1998) Origin and bioavailability of PAHs in the Mediterranean Sea from mussel and sediment records. *Estuarine, Coastal and Shelf Science*, **47**, 77–90.

Baumard P, Budzinski H, Garrigues P, Dizer H and Hansen PD (1999) Polycyclic aromatic hydrocarbons in recent sediments and mussels (*Mytilus edulis*) from the Western Baltic Sea: occurrence, bioavailability and seasonal variations. *Marine Environmental Research*, **47**(1), 17–47.

Bence AE, Kvenvolden KA and Kennicutt MC II (1996) Organic geochemistry applied to environmental assessments of Prince William Sound, Alaska, after the Exxon Valdez oil spill–a review. *Organic Geochemistry*, **24**(1), 7–42.

Benlahcen KT, Chaoui A, Budzinski H, Bellocq J and Garrigues P (1997) Distribution and sources of polycyclic aromatic hydrocarbons in some Mediterranean coastal sediments. *Marine Pollution Bulletin*, **34**(5), 298–305.

Blanchard M, Teil MJ, Ollivon D, Garban B, Chesterikoff C and Chevreuil Marc (2001) Origin and distribution of polyaromatic hydrocarbons and polychlorobiphenyls in urban effluents to wastewater treatment plants of the Paris area (France). *Water Research*, **35**(15), 3679–3687.

Cheng Y, Chen L, Sheng G, Min Y, Jiamo F and Bo S (1998) Distribution, seasonal change and source identifications of PAHs in aerosols from Guangzhou, China. *China Environmental Science*, **18**(2), 136–139.

Cincinelli A, Stortini AM, Perugini M, Checchini L and Lepri L (2001) Organic pollutants in sea-surface microlayer and aerosol in the coastal environment of Leghorn (Tyrrhenian Sea). *Marine Chemistry*, **76**(1–2), 77–98.

Colombo JC, Pelletier E, Brochu C, Khalil M and Catoggio JA (1989) Determination of hydrocarbon sources using *n*-alkane and polyaromatic hydrocarbon distribution indices. Case study: Rio de La Plata Estuary, Argentina. *Environmental Science and Technology*, **23**, 888–894.

Eganhouse RP and Gossett RW (1991) Historical deposition and biogeochemical fate of polycyclic aromatic hydrocarbons in sediments near a major submarine wastewater outfall in Southern California. In Baker RA (ed.), *Organic Substances and Sediments in Water*. Lewis, Chelsea, MI, pp. 191–220.

Eisler R (1987) *Polycyclic Aromatic Hydrocarbon Hazards to Fish, Wildlife, and Invertebrates: A Synoptic Review*, vol. 11. Fish and Wildlife Service, US Department of the Interior, Washington, DC.

EPA (1987) *Quality Criteria for Water 1986*. EPA 440/5-86-001. US Environmental Protection Agency, Washington, DC.

EPA (2000) Equilibrium Partitioning Sediment Guidelines (ESGs) for the Protection of Benthic Organisms: PAH Mixtures (draft). US Environmental Protection Agency, Office of Water, Office of Science and Technology, Office of Research and Development, Washington, DC.

Gogou A, Bouloubassi I and Stephanou EG (2000) Marine organic geochemistry of the Eastern Mediterranean: 1. aliphatic and polyaromatic hydrocarbons in Cretan Sea surficial sediments. *Marine Chemistry*, **68**(4), 265–282.

Gschwend PM, Chen PH and Hites RA (1983) On the formation of perylene in recent sediments: kinetic models. *Geochimica et Cosmochimica Acta*, **47**, 2115–2119.

Guinan J, Charlesworth M, Service M and Oliver T (2001) Sources and geochemical constraints of polycyclic aromatic hydrocarbons (PAHs) in sediments and mussels of two Northern Irish sea-loughs. *Marine Pollution Bulletin*, **42**(11), 1073–1081.

Hoff RM, Strachan WMJ, Sweet CW, Chan CH, Shackleton M, Bidleman TF, Brice KA, Burniston DA, Cussion S and Gatz DF (1996) Atmospheric deposition of toxic chemicals to the great lakes: a review of data through 1994. *Atmospheric Environment*, **30**(20), 3505–3527.

Hoffman EJ, Mills GL, Latimer JS and Quinn JG (1984) Urban runoff as a source of polycyclic aromatic hydrocarbons to coastal waters. *Environmental Science and Technology*, **18**, 580–586.

Kim GB, Maruya KA, Lee RF, Lee JH, Koh CH and Tanabe S (1999) Distribution and sources of polycyclic aromatic hydrocarbons in sediments from Kyeonggi Bay, Korea. *Marine Pollution Bulletin*, **38**(1), 7–15.

Kucklick JR, Sivertsen SK, Sanders M and Scott GI (1997) Factors influencing polycyclic aromatic hydrocarbon distributions in South Carolina estuarine sediments. *Journal of Experimental Marine Biology and Ecology*, **213**(1), 13–29.

Kulkarni P and Venkataraman C (2000) Atmospheric polycyclic aromatic hydrocarbons in Mumbai, India. *Atmospheric Environment*, **34**(17), 2785–2790.

Lake JL, Norwood C, Dimock C and Bowen R (1979) Origins of polycyclic aromatic hydrocarbons in estuarine sediments. *Geochimica et Cosmochimica Acta*, **43**, 1847–1854.

Latimer JS (1997) The significance of atmospheric deposition as a source of PCBs and PAHs to Narragansett Bay. In Baker JE (ed.), *Atmospheric Deposition of Contaminants to the Great Lakes and Coastal Waters*. SETAC Press, Pensacola, FL, pp. 227–243.

Latimer JS and Quinn JG (1996) Historical trends and current inputs of hydrophobic organic contaminants in an urban estuary: the sedimentary record. *Environmental Science and Technology*, **30**(2), 623–633.

Latimer JS, Hoffman EJ, Hoffman G, Fasching JL and Quinn JG (1990) Sources of petroleum hydrocarbons in urban runoff. *Water, Air and Soil Pollution*, **52**, 1–21.

Latimer JS, Jayaraman S and McKinney RA (1999) Historical reconstruction of pollution stress and recovery in an urban estuary: organic contaminants. Estuarine Research Federation Conference. New Orleans, LA.

Latimer JS, Boothman WS, Jayaraman S and Pesch C (in preparation). Environmental stress and recovery: the geochemical record of human disturbance in New Bedford Harbor and Apponagansett Bay.

Law RJ, Dawes VJ, Woodhead RJ and Matthiessen P (1997) Polycyclic aromatic hydro-carbons (PAH) in seawater around England and Wales. *Marine Pollution Bulletin*, **34**(5), 306-322.

Ligocki MP, Leuenberger C and Pankow JF (1985a) Trace organic compounds in rain — II. Gas scavenging of neutral organic compounds. *Atmospheric Environment*, **19**(10), 1609-1617.

Ligocki MP, Leuenberger C and Pankow JF (1985b) Trace organic compounds in rain — III. Particle scavenging of neutral organic compounds. *Atmospheric Environment*, **19**(10), 1619-1626.

Liu M, Baugh PJ, Hutchinson SM, Yu L and Xu S (2000). Historical record and sources of polycyclic aromatic hydrocarbons in core sediments from the Yangtze Estuary, China. *Environmental Pollution*, **110**(2), 357-365.

Long ER, MacDonald DD, Smith SL and Calder FD (1995) Incidence of adverse biological effects with ranges of chemical concentrations in marine and estuarine sediments. *Environmental Management*, **19**(1), 81-97.

Ma M, Feng Z, Guan C, Ma Y, Xu H and Li H (2001) DDT, PAH and PCB in sediments from the intertidal zone of the Bohai Sea and the Yellow Sea. *Marine Pollution Bulletin*, **42**(2), 132-136.

Mackay D, Shiu WY and Ma KC (1992) *Illustrated Handbook of Physical-Chemical Properties and Environmental Fate for Organic Chemicals*, vol. II, *Polycyclic Aromatic Hydrocarbons, Polychlorinated Dioxins and Dibenzofurans.* Lewis, Boca Raton, FL.

Manoli E, Samara C, Konstantinou I and Albanis T (2000) Polycyclic aromatic hydrocar-bons in the bulk precipitation and surface waters of Northern Greece. *Chemosphere*, **41**(12), 1845-1855.

Ngabe B, Bidleman TF and Scott GI (2000) Polycyclic aromatic hydrocarbons in storm runoff from urban and coastal South Carolina. *Science of the Total Environment*, **255**(1-3), 1-9.

NAS (in press). Oil in the Sea III: Inputs, Fates, and Effects. Committee on Oil in the Sea: Inputs, Fates, and Effects, National Research Council, Washington, DC.

Odabasi M, Vardar N, Sofuoglu A, Tasdemir Y and Holsen TM (1999) Polycyclic aromatic hydrocarbons (PAHs) in Chicago air. *Science of the Total Environment*, **227**(1), 57-67.

Park JS, Wade TL and Sweet S (2001) Atmospheric distribution of polycyclic aromatic hydrocarbons and deposition to Galveston Bay, TX, USA. *Atmospheric Environment*, **35**(19), 3241-3249.

Pereira WE, Hostettler FD, Luoma SN, van Geen A, Fuller CC and Anima RJ (1999) Sedi-mentary record of anthropogenic and biogenic polycyclic aromatic hydrocarbons in San Francisco Bay, CA. *Marine Chemistry*, **64**(1-2), 99-113.

Perez S, Marinel la Farre MJG and Barcelo D (2001) Occurrence of polycyclic aromatic hydrocarbons in sewage sludge and their contribution to its toxicity in the ToxAlert(R) 100 bioassay. *Chemosphere*, **45**(6-7), 705-712.

Perez S, Guillamon M and Barcelo D (in press) Quantitative analysis of polycyclic aro-matic hydrocarbons in sewage sludge from wastewater treatment plants. *Journal of Chromatography A*.

Pham TT and Proulx S (1997) PCBs and PAHs in the Montreal urban community (Quebec, Canada) wastewater treatment plant and in the effluent plume in the St Lawrence river. *Water Research*, **31**(8), 1887-1896.

Pruell RJ and Quinn JG (1985) Geochemistry of organic contaminants in Narragansett Bay sediments. *Estuarine, Coastal and Shelf Science*, **21**, 295-312.

Pruell RJ, Norwood CB, Bowen RD, Boothman WS, Rogerson PF, Hackett M and Butter-worth BC (1990) Geochemical study of sediment contamination in New Bedford Harbor, Massachusetts. *Marine Environmental Research*, **29**, 77-101.

Quinn JG, Latimer JS, Ellis JT, LeBlanc LA and Zheng J (1988) *Analysis of Archived Water Samples for Organic Pollutants*. NBP-88-04, Narragansett Bay Project, Providence, RI.

Quinn JG, Latimer JS, LeBlanc LA and Ellis JT (1992) *Assessment of Organic Contaminants in Narragansett Bay Sediments and Hard Shell Clams*. NBP 92 – 111, Graduate School of Oceanography, University of Rhode Island, Narragansett, RI.

Savinov VM, Savinova TN, Carroll JL, Matishov GG, Dahle S and Naes K (2000) Polycyclic aromatic hydrocarbons (PAHs) in sediments of the White Sea, Russia. *Marine Pollution Bulletin*, **40**(10), 807 – 818.

Schnelle-Kreis J, Gebefugi I, Welzl G, Jaensch T and Kettrup A (2001) Occurrence of particle-associated polycyclic aromatic compounds in ambient air of the city of Munich. *Atmospheric Environment*, **35**(1), 71 – 81.

Soclo HH, Garrigues P and Ewald M (2000) Origin of polycyclic aromatic hydrocarbons (PAHs) in coastal marine sediments: case studies in Cotonou (Benin) and Aquitaine (France) areas. *Marine Pollution Bulletin*, **40**(5), 387 – 396.

Sporstol S, Gjos N, Lichtenthaler RG, Gustavsen KO, Urdal K, Oreld F and Skei J (1983) Source identification of aromatic hydrocarbons in sediments using gc/ms. *Environmental Science and Technology*, **17**, 282 – 286.

Steinhauer MS and Boehm PD (1992) The composition and distribution of saturated and aromatic hydrocarbons in nearshore sediments, and coastal peat of the Alaskan Beaufort Sea: implications for detecting anthropogenic hydrocarbon input. *Marine Environmental Research*, **33**, 223 – 253.

Valette-Silver NJ (1993) The use of sediment cores to reconstruct historical trends in contamination of estuarine and coastal sediments. *Estuaries*, **16**(3B), 577 – 588.

Valette-Silver N, Hameedi Jawed M, Efurd DW and Robertson A (1999) Status of the contamination in sediments and biota from the Western Beaufort Sea (Alaska). *Marine Pollution Bulletin*, **38**(8), 702 – 722.

Walker WJ, McNutt RP and Maslanka CAK (1999) The potential contribution of urban runoff to surface sediments of the Passaic River: sources and chemical characteristics. *Chemosphere*, **38**(2), 363 – 377.

Wild SR and Jones KC (1995) Polynuclear aromatic hydrocarbons in the United Kingdom environment: a preliminary source inventory and budget. *Environmental Pollution*, **88**(1), 91 – 108.

Yang GP (2000) Polycyclic aromatic hydrocarbons in the sediments of the South China Sea. *Environmental Pollution*, **108**(2), 163 – 171.

Yassaa N, Meklati BY, Cecinato A and Marino F (2001) Particulate *n*-alkanes, *n*-alkanoic acids and polycyclic aromatic hydrocarbons in the atmosphere of Algiers City Area. *Atmospheric Environment*, **35**(10), 1843 – 1851.

Yilmaz K, Yilmaz A, Yemenicioglu S, Sur M, Salihoglu I, Karabulut Z, Telli Karrakoc F, Hatipoglu E, Gaines AF, Philips D and Hewer A (1998) Polynuclear aromatic hydrocarbons (PAHs) in the Eastern Mediterranean Sea. *Marine Pollution Bulletin*, **36**(11), 922 – 925.

Yunker MB, Macdonald RW, Goyette D, Paton DW, Fowler BR, Sullivan D and Boyd J (1999) Natural and anthropogenic inputs of hydrocarbons to the Strait of Georgia. *Science of the Total Environment*, **225**(3), 181 – 209.

Zheng GJ and Richardson BJ (1999) Petroleum hydrocarbons and polycyclic aromatic hydrocarbons (PAHs) in Hong Kong marine sediments. *Chemosphere*, **38**(11), 2625 – 2632.

Zhu T, Sun R, Zhang L and Jiang L (1998) Study on identifying the distribution and pollution sources of PAHs in airborne particulates in Dagang, Tianjin China. *China Environmental Science*, **18**(4), 289 – 292.

3

Geochemistry of PAHs in Aquatic Environments: Source, Persistence and Distribution*

ROBERT M. BURGESS[1], MICHAEL J. AHRENS[2] AND CHRISTOPHER W. HICKEY[2]

[1]*US EPA, ORD/NHEERL Atlantic Ecology Division, Narragansett, RI, USA*
[2]*NIWA, Hamilton, New Zealand*

3.1   INTRODUCTION

On the basis of their distributions, sources, persistence, partitioning and bioavailability, polycyclic aromatic hydrocarbons (PAHs) are a unique class of persistent organic pollutants (POPs) contaminating the aquatic environment. They are of particular interest to geochemists and environmental toxicologists for several reasons. First, PAHs are not released into the environment from a single source of origin, as are most industrial, petro- or agricultural chemicals. Rather, PAHs are formed in at least three ways: pyrogenically, petrogenically and diagenetically. Second, when released, PAHs are not in a relatively pure chemical form like most other human-produced organic pollutants entering the environment, e.g. pyrogenic PAHs are frequently associated with soot carbon formed during the combustion process of fossil fuels. Furthermore, petrogenic PAHs in petroleum escape into the environment in complex mixtures of thousands of aromatic and aliphatic compounds. These differences greatly affect their persistence and bioavailability (see Chapter 6 for persistence and Chapter 7 for partitioning and bioavailability). Third, PAHs, unlike many other POPs, are often metabolized by aquatic organisms, thus reducing their persistence in the environment. Fourth,

* Mention of trade names or commercial products does not constitute endorsement or recommendation for use. Although the research described in this article has been funded wholly by the US Environmental Protection Agency, it has not been subjected to Agency-level review. Therefore, it does not necessarily reflect the views of the Agency. This is Contribution #AED-02-024 of the Office of Research and Development National Health and Environmental Effects Research Laboratory's Atlantic Ecology Division.

PAHs: An Ecotoxicological Perspective. Edited by Peter E.T. Douben.
© 2003 John Wiley & Sons Ltd

there are thousands of possible PAH structures present in the environment as compared to the relatively limited numbers of other POPs. Finally, unlike many other harmful organic chemicals that have been banned or limited in discharges, PAHs continue to be released into the environment because of their widespread formation during the burning of fossil fuels and escape during petroleum recovery, transport and use. Ongoing utilization of fossil fuels on a global basis guarantees the continued release of PAHs into the environment. PAHs do share some qualities with other POPs. For example, PAHs are hydrophobic and lipophilic, interact strongly with sedimentary organic carbon, are often sparingly soluble in water, commonly have relatively low volatility, and readily bioaccumulate and are toxic to some aquatic organisms.

Several excellent reviews written over the last 30 years have addressed the geochemistry and bioavailability of PAHs in aquatic environments (e.g. Andelman and Suess 1970; Suess 1976; Neff 1979; McElroy *et al*. 1989; Meador *et al*. 1995). Therefore, we provide a current synthesis of the factors affecting PAH bioavailability in aquatic environments rather than a comprehensive review of PAH geochemistry. The most critical aspect of this synthesis is an emphasis on the importance of the freely dissolved form of PAH. This form of PAH appears to be readily bioavailable relative to the forms associated with sediments, colloids, soot carbon and other matrices. To understand and estimate the manifestations of PAH, such as bioavailability, toxicity and bioaccumulation, one must know the quantity of freely dissolved PAH that an organism may encounter. Throughout this overview of the processes affecting PAH sources, persistence, distributions, partitioning and bioavailability, we will re-emphasize this salient point.

3.2 SOURCES AND PERSISTENCE

As noted above, PAHs found in aquatic environments originate from three possible sources: pyrogenic, petrogenic and diagenetic (see Chapter 2 for more detail). *Pyrogenic* PAHs result from the incomplete but high-temperature, short-duration combustion of organic matter (Neff 1979; Meyer and Ishiwatari 1993). These pyrogenic PAHs are believed to form from the breakdown or 'cracking' of organic matter to lower molecular weight radicals during pyrolysis, followed by rapid reassembly into non-alkylated PAH structures (Neff 1979). Importantly, this pyrolysis also has another product, namely soot carbon formed as an agglomeration of pericondensed PAHs (Thomas *et al*. 1968; Neff 1979). The newly formed PAHs and soot carbon establish a very strong interaction that continues from formation through deposition in the aquatic environment and has significance for PAH partitioning and bioavailability. *Petrogenic* PAHs are created by diagenetic processes at relatively low temperatures over geologic time scales, leading to the formation of petroleum and other fossil fuels containing PAHs (Meyer and Ishiwatari 1993; Boehm *et al*. 2001). PAHs formed at relatively low temperatures ($\sim$150°C) over long periods of time will be primarily alkylated molecules. The alkylated structure of petrogenic PAHs reflects the ancient plant material from which the compounds formed (Neff 1979). Diagenetic PAHs refer to PAHs from biogenic precursors, like plant terpenes, leading to the formation of compounds such as retene and derivatives of phenanthrene and chrysene (Hites *et al.* 1980; Meyer and Ishiwatari 1993; Silliman *et al.* 1998). Perylene is another common diagenetic PAH. Although its exact formation process remains unclear, an anaerobic process appears to be involved (Gschwend *et al.* 1983; Venkatesan 1988; Silliman *et al.* 1998). While diagenetic PAHs are frequently found at background levels in recent sediments (i.e. deposited over the last 150 years), they often dominate the assemblage of PAHs present in older sediments deposited before human industrial activity (Gschwend *et al.* 1983). A potential fourth source of PAHs is biogenic, i.e. purely from bacteria, fungi, plants or animals in sedimentary environments without any contributions from diagenetic processes. However, attempts to produce biogenic PAHs have arguably failed, indicating this source is not significant (Hase and Hites 1976; Neff 1979).

Meinschein (1959) first observed that the composition of PAH molecules in sediments differed from that commonly found in petroleum. This discrepancy continued to be observed, even as measurement of PAHs in sediments became more routine, supporting the conclusion that most PAHs in aquatic environments originate from pyrogenic sources (Blumer 1976; Suess 1976; Hites *et al.* 1977; NRC 1985; Wu *et al.* 2001). However, petrogenic PAHs do also occur alone or in combination with pyrogenic PAHs (Lake *et al.* 1979; Wakeham *et al.* 1980; NRC 1985; Gschwend and Hites 1981; Readman *et al.* 1992). In general, petrogenic PAHs appear to be associated with a local or point sources, such as refineries and other petroleum industries, and adjacent to roads and navigational routes. This contrasts with the distribution of pyrogenic PAHs, which occur on broader geographic scales. Finally, diagenetic PAHs occur at background levels, although anthropogenic sources (e.g. perylene) can contribute to these types of PAHs. Thus, a strong relationship exists between PAH source and distribution, with pyrogenic PAHs often dominating in the aquatic environment in terms of concentration and geographical distribution.

Another important difference between pyrogenic and petrogenic PAHs is that petrogenic PAHs do not occur in the sedimentary record to the same extent as pyrogenic PAHs. Clearly, part of the explanation for this observation is that fewer petrogenic PAHs are released into the environment. However, pyrogenic PAHs associated with soot carbon also appear to be far more persistent and largely protected from various forms of environmental degradation. Studies of PAHs adsorbed to soot carbon show little photochemical oxidation compared to PAHs on other matrices (Butler and Crossley 1981; Korfmacher *et al.* 1981; Yokley *et al.* 1986). Furthermore, pyrogenic PAHs may also be resistant to microbial degradation (Farrington *et al.* 1983). Thus, the fate of pyrogenic PAHs appears to be largely unaffected by degradation or removal processes other than sedimentation (Figure 3.1). Conversely, petrogenic PAHs

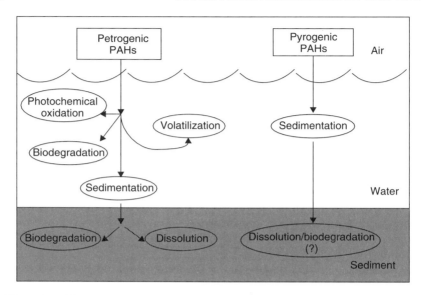

Figure 3.1   Major processes affecting the fate of petrogenic and pyrogenic PAHs upon introduction to aquatic environments (based on Lee 1980)

appear to be more prone to biogeochemical alteration, being susceptible to volatilization, dissolution, photochemical oxidation (see Chapter 4), microbial breakdown and sedimentation (Blumer and Sass 1972; Gearing *et al.* 1979; Lee 1980; Jordan and Payne 1980) (Figure 3.1). In general, sedimentation accounts for most of the initial removal of petrogenic PAHs from the water column and is initiated by adsorption to various types of particles. Once in the sediment, particle-associated PAHs are mixed throughout the surficial sediment by physical and biological processes; e.g. tidal suspension and bioturbation (Gordon *et al.* 1978; Lee and Swartz 1980; Aller 1982; Reynoldson 1987). Some of the PAHs are microbially degraded during this process, with the most rapid transformations/modifications occurring in the aerobic zone (Gardner *et al.* 1979; Gearing *et al.* 1980; Hinga *et al.* 1980; Lee *et al.* 1981; Cerniglia 1992). Some of the PAHs dissolve back into the overlying or interstitial waters (Pruell and Quinn 1985a). A general view of petrogenic PAH fate divides the compounds into two groups, based on molecular weight (Herbes and Schwall 1978; Readman *et al.* 1982; Heitkamp and Cerniglia 1987). The low molecular weight PAHs are readily degraded microbially, while higher molecular weight PAHs tend to be removed from solution by sedimentation. As discussed above, the basic differences in PAH source and persistence contribute significantly to how PAHs are distributed in aquatic environments which, in turn, affects directly their bioavailability.

3.3 DISTRIBUTIONS

Many physicochemical processes affect PAH distributions in aquatic environments. However, the majority of PAHs will rapidly associate with suspended or resuspended organic particles and settle to the sediment, once introduced to a watery setting (Neff 1979; Schwarzenbach *et al.* 1993; Maliszewska-Kordybach 1999). Regardless of their origin, the low solubility and high hydrophobicity of PAHs, together with the polar structure of water, ensures they will ultimately be associated with sediments in depositional environments. This efficient removal process is demonstrated by the generally very low concentrations of PAHs found dissolved in overlying waters (Neff 1979; Maliszewska-Kordybach 1999).

Measurement of PAHs in environmental samples (e.g. soils, sediments) arguably started in the 1940s with Kern's discovery of chrysene in a German soil (Kern 1947; Blumer 1961). He related the chrysene to components of pyrogenic materials including coal and tar (Kern 1947; Blumer 1961). Several contemporary investigations of PAH distributions provide comprehensive data on their concentration and composition throughout the world (Grimmer and Bohnke 1975; Youngblood and Blumer 1975; Hites *et al.* 1980; Wakeham *et al.* 1980; Maher and Aislabie 1992; Holland *et al.* 1993; Daskalakis and O'Connor 1995). Due to their many sources, PAHs are nearly ubiquitous environmental contaminants. Background concentrations are typically in the tens to hundreds of parts per billion, depending upon the conditions, while the environmental concentration range covers parts per trillion in sediments from the Great Barrier Reef, Australia (Smith *et al.* 1985; Maher and Aislabie 1992; Haynes and Johnson 2000) to parts per thousand in contaminated urban sediments from North America (Neff 1979; Huggett *et al.* 1987).

Surface sediment PAH concentrations provide some useful information on the magnitude of contamination. However, downcore profiles of PAHs from freshwater and marine sediments provide a means for elucidating trends in deposition. A selection of total PAH concentrations in sediment cores from several locations around the world is presented in Figure 3.2. At least for PAHs measured in sediment cores from industrially developed regions of the world, a 'common' pattern in deposition is apparent: deep in the sediment cores, little PAH is observed, although a background level of PAHs is always present. PAH concentrations are maximal for intermediate depths, and concentrations decrease again in surface layers, albeit not to background levels. By dating sediment cores, it is possible to determine the time frame associated with these patterns while also quantifying PAH flux into the sediments (Charles and Hites 1987; Valette-Silver 1993). In appearance, the flux of PAHs into sediments resembles the concentration distribution observed in Figure 3.2. However, the temporal trends can now be explained or at least speculated upon (Figure 3.3). The very low fluxes of PAHs in the deeper portions of the sediment cores are generally presumed to have been caused by preindustrialization sources (before 1850) of PAHs, such as the diagenetic breakdown of organic matter and other

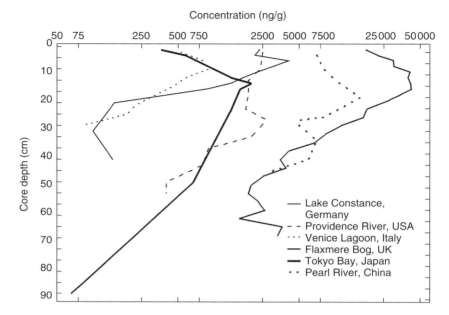

Figure 3.2 Concentration (ng/g) of total PAHs vs. depth in a selection of sediment cores from sites around the world (Muller *et al*. 1977; Pruell and Quinn 1985b; Pavoni *et al*. 1987; Sanders *et al*. 1995; Yamashita *et al*. 2000; Bixian *et al*. 2001)

natural phenomena (e.g. burning of biomass and petroleum seeps; Gschwend and Hites 1981; Zepp and Macko 1997). By the time of the industrial revolution in Europe, North America, Australasia and parts of Asia in the late nineteenth century, PAHs formed via the burning of coal and other fossil fuels began to appear in the sediment record. The trend of increased PAH fluxes continued until approximately the last part of the twentieth century, when PAH fluxes were observed, in general, to decrease again. This decrease is often associated with the reduction in use of coal as a fuel for home heating and industrial processes in the 1950s in exchange for relatively cleaner burning petroleum and natural gas (Charles and Hites 1987; Juttner *et al*. 1997). Furthermore, by the 1970s, many countries were enforcing air emission regulations that sought to reduce PAH and other pollutants released from power plants and automobiles (Charles and Hites 1987; Sanders *et al*. 1995). Whatever the individual cause(s), there is a clear and common trend of reduced PAH fluxes to sediments during this period in most dated sediment cores. Several regions of the world lack deposition and flux data and require investigations to determine the extent and universality of these trends. Interestingly, countries that continue to undergo heavy industrialization and/or lack stringent regulation of air emissions demonstrate contemporary increased PAH fluxes into sediments (e.g. Pearl River and Estuary, China; Bixian *et al*. 2001). In conclusion, the distributions of PAHs vary considerably as a result

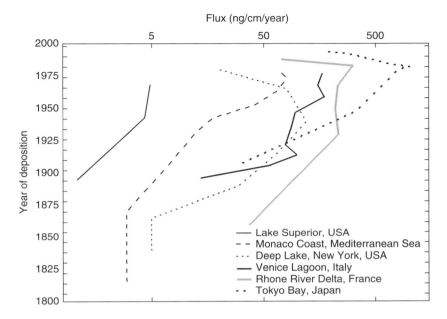

Figure 3.3   Flux (ng/cm²/year) of total PAHs vs. year of deposition (from dated cores) in a selection of sediment cores from sites around the world (Gschwend and Hites 1981; Burns and Villeneuve 1983; Furlong *et al*. 1987; Pavoni *et al*. 1987; Tolosa *et al*. 1996; Yamashita *et al*. 2000)

of different sources and persistences. As discussed in Chapter 7, these factors, along with PAH partitioning behavior, ultimately determine the bioavailability to organisms.

REFERENCES

Aller RC (1982) The effects of macrobenthos on chemical properties of marine sediment and overlying water. In McCall PL and Tevesz MJS (eds), *Animal - Sediment Relations*. Plenum, New York, pp. 53 - 102.

Andelman JB and Suess MJ (1970) Polynuclear aromatic hydrocarbons in the water environment. *Bulletin of the World Health Organization*, **43**, 479 - 508.

Bixian M, Jiamo F, Gan Z, Zheng L, Yushun M, Guoying S and Xingmin W (2001) Polycyclic aromatic hydrocarbons in sediments from the Pearl River and Estuary, China: spatial and temporal distribution and sources. *Applied Geochemistry*, **16**, 1429 - 1445.

Blumer M (1961) Benzpyrenes in soil. *Science*, **134**, 474 - 475.

Blumer M (1976) Polycyclic aromatic compounds in nature. *Scientific American*, **234**, 35 - 45.

Blumer M and Sass J (1972) Oil pollution: persistence and degradation of spilled fuel oil. *Science*, **176**, 1120 - 1122.

Boehm PD, Page DS, Burns WA, Bence AE, Mankiewicz PJ and Brown JS (2001) Resolving the origin of the petrogenic hydrocarbon background in Prince William Sound, Alaska. *Environmental Science and Technology*, **35**, 471 - 479.

Burns KA and Villeneuve J-P (1983) Biogeochemical processes affecting the distribution and vertical transport of hydrocarbon residues in the coastal Mediterranean. *Geochimica et Cosmochimica Acta*, **47**, 995–1006.

Butler JD and Crossley P (1981) Reactivity of polycyclic aromatic hydrocarbons adsorbed on soot particles. *Atmospheric Environment*, **15**, 91–94.

Cerniglia CE (1992) Biodegradation of polycyclic aromatic hydrocarbons. *Biodegradation*, **3**, 351–368.

Charles MJ and Hites RA (1987) Sediment as archives of environmental pollution trends. In Hites RA and Eisenreich SJ (eds), *Sources and Fates of Aquatic Pollutants*. American Chemical Society, Advances in Chemistry Series No. 216. Washington, DC, pp. 365–389.

Daskalakis KD and O'Connor TP (1995) Distribution of chemical concentrations in US coastal and estuarine sediment. *Marine Environmental Research*, **40**, 381–398.

Farrington JW, Goldberg ED, Risebrough RW, Martin JH and Bowen VT (1983) US 'Mussel Watch' 1976–1978: an overview of the trace metal, DDE, PCB, hydrocarbon, and artificial radionuclide data. *Environmental Science and Technology*, **17**, 490–496.

Furlong ET, Cessar LR and Hites RA (1987) Accumulation of polycyclic aromatic hydrocarbons in acid-sensitive lakes. *Geochimica et Cosmochimica Acta*, **51**, 2965–2975.

Gardner WS, Lee RF, Tenore KR and Smith LW (1979) Degradation of selected polycyclic aromatic hydrocarbons in coastal sediments: importance of microbes and polychaete worms. *Water, Air, and Soil Pollution*, **11**, 339–347.

Gearing JN, Gearing PJ, Wade T, Quinn JG, McCarty HB, Farrington J and Lee RF (1979) The rates of transport and fates of petroleum hydrocarbons in a controlled marine ecosystem, and a note on analytical variability. In *Proceedings of the 1979 Oil Spill Conference (Prevention, Behavior, Control and Clean-up)*, American Petroleum Institute, Washington, DC, pp. 555–564.

Gearing PJ, Gearing JN, Pruell RJ, Wade TL and Quinn JG (1980) Partitioning of No. 2 fuel oil in controlled estuarine ecosystems, sediments and suspended particulate matter. *Environmental Science and Technology*, **14**, 1129–1136.

Gordon DC, Dale J and Keiser PD (1978) Importance of sediment working by the deposit-feeding polychaete *Arenicola marina* on the weathering rate of sediment-bound oil. *Journal of the Fisheries Research Board of Canada*, **35**, 591–603.

Grimmer G and Bohnke H (1975) Profile analysis of polycyclic aromatic hydrocarbons and metal content in sediment layers of a lake. *Cancer Letters*, **1**, 75–84.

Gschwend PM and Hites RA (1981) Fluxes of polycyclic aromatic hydrocarbons to marine and lacustrine sediments in the north-eastern United States. *Geochimica et Cosmochimica Acta*, **45**, 2359–2367.

Gschwend PM, Chen PH and Hites RA (1983) On the formation of perylene in recent sediments: kinetic models. *Geochimica et Cosmochimica Acta*, **47**, 2115–2119.

Hase A and Hites RA (1976) On the origin of polycyclic aromatic hydrocarbons in recent sediments: biosynthesis by anaerobic bacteria. *Geochimica et Cosmochimica Acta*, **40**, 1141–1143.

Haynes D and Johnson JE (2000) Organochlorine, heavy metal and polyaromatic hydrocarbon pollutant concentrations in the Great Barrier Reef (Australia) environment: a review. *Marine Pollution Bulletin*, **41**, 267–278.

Heitkamp MA and Cerniglia CE (1987) Effects of chemical structure and exposure on the microbial degradation of polycyclic aromatic hydrocarbons in freshwater and estuarine ecosystems. *Environmental Toxicology and Chemistry*, **6**, 535–546.

Herbes SE and Schwall LR (1978) Microbial transformations of polycyclic aromatic hydrocarbons in pristine and petroleum-contaminated sediments. *Applied and Environmental Microbiology*, **35**, 306–316.

Hinga KR, Pilson MEQ, Lee RF, Farrington JW, Tjessem K and Davis AC (1980) Bio-geochemistry of benzanthracene in an enclosed marine ecosystem. *Environmental Science and Technology*, **14**, 1136–1143.

Hites RA, LaFlamme RE and Farrington JW (1977) Sedimentary polycyclic aromatic hydro-carbons: the historical record. *Science*, **198**, 829–831.

Hites RA, LaFlamme RE and Windsor J.G (1980) Polycyclic aromatic hydrocarbons in the marine/aquatic sediments: their ubiquity. In Petrakis L and Weiss FT (eds), *Petroleum in the Marine Environment*. American Chemical Society, Washington, DC, pp. 289–311.

Holland PT, Hickey CW, Roper DS and Trower TM (1993) Variability of organic contam-inants in intertidal sandflat sediments from Manukau Harbour, New Zealand. *Archives of Environmental Contamination and Toxicology*, **25**, 456–463.

Huggett RJ, Bender ME and Unger MA (1987) Polynuclear aromatic hydrocarbons in the Elizabeth River, Virginia. In Dickson KL, Maki AW and Brungs WA (eds), *Fate and Effects of Sediment-bound Chemicals in Aquatic Systems*. Pergamon, New York, pp. 327–341.

Jordan RE and Payne JR (1980) *Fate and Weathering of Petroleum Spills in the Marine Environment*. Ann Arbor Science, Ann Arbor, MI, p. 174.

Juttner I, Lintelmann J, Michalke B, Winkler R, Steinberg CEW and Kettrup A (1997) The acidification of the Herrenwieser See, Black Forest, Germany, before and during industrialisation. *Water Research*, **31**, 1194–1206.

Kern W (1947) On the occurrence of chrysene in soil. *Helvetica Chimica Acta*, **30**, 1595–1599 (in German).

Korfmacher WA, Mamantov G, Wehry EL, Natusch DFS and Mauney T (1981) Non-photochemical decomposition of fluorene vapor-adsorbed on coal fly ash. *Environmental Science and Technology*, **15**, 1370–1375.

Lake JL, Norwood C, Dimock C and Bowen R (1979) Origins of polycyclic aromatic hydrocarbons in estuarine sediments. *Geochimica et Cosmochimica Acta*, **43**, 1847–1854.

Lee H and Swartz RC (1980) Biological processes affecting the distribution of pollu-tants in marine sediments, Part II. Biodeposition and bioturbation. In Baker RA (ed.), *Contaminants and Sediments*, vol. 2, *Analysis, Chemistry and Biology*. Ann Arbor Science, Ann Arbor, MI, pp. 555–606.

Lee RF (1980) Processes affecting the fate of oil in the sea. In Geyer RA (ed.), *Marine Environmental Pollution. 1. Hydrocarbons*. Elsevier Scientific, Amsterdam, pp. 337–351.

Lee RF, Dornseif B, Gonsoulin F, Tenore K and Hanson R (1981) Fate and effect of a heavy fuel oil spill on a Georgia salt marsh. *Marine Environmental Research*, **5**, 125–143.

Maher, WA and Aislabie J (1992) Polycyclic aromatic hydrocarbons in nearshore marine sediments of Australia. *Science of the Total Environment*, **112**, 143–164.

Maliszewska-Kordybach B (1999) Persistent organic contaminants in the environment: PAHs as a case study. In Baveye P *et al.* (eds), *Bioavailability of Organic Xenobiotics in the Environment*. Kluwer Academic, Amsterdam, pp. 3–34.

McElroy AE, Farrington JW and Teal JM (1989) Bioavailability of polycyclic aromatic hydrocarbons in the aquatic environment. In Varanasi U (ed.), *Metabolism of Poly-cyclic Aromatic Hydrocarbons in the Aquatic Environment*. CRC Press, Boca Raton, FL, pp. 1–39.

Meador JP, Stein JE, Reichert WL and Varanasi U (1995) Bioaccumulation of polycyclic aromatic hydrocarbons by marine organisms. *Reviews of Environmental Contami-nation and Toxicology*, **143**, 79–165.

Meinschein WG (1959) Origin of petroleum. *Bulletin of the American Association of Petroleum Geologists*, **43**, 925–943.

Meyer PA and Ishiwatari R (1993) Lacustrine organic geochemistry — an overview of indicators of organic matter sources and diagenesis in lake sediments. *Organic Geochemistry*, **20**, 867–900.

Muller G, Grimmer G and Bohnke H (1977) Sedimentary record of heavy metals and polycyclic aromatic hydrocarbons in Lake Constance. *Naturwissenschaften*, **64**, 427–431.

NRC (National Research Council) (1985) *Oil in the Sea: Inputs, Fates, and Effects*. National Academy Press, Washington, DC, p. 601.

Neff JM (1979) *Polycyclic Aromatic Hydrocarbons in the Aquatic Environment*. Applied Science Publishers Ltd., London.

Pavoni B, Sfriso A and Marcomini A (1987) Concentration and flux profiles of PCBs, DDTs and PAHs in a dated sediment core from the lagoon in Venice. *Marine Chemistry*, **21**, 25–35.

Pruell RJ and Quinn JG (1985a) Polycyclic aromatic hydrocarbons in surface sediments held in experimental mesocosms. *Toxicological and Environmental Chemistry*, **10**, 183–200.

Pruell RJ and Quinn JG (1985b) Geochemistry of organic contaminants in Narragansett Bay sediments. *Estuarine, Coastal, and Shelf Science*, **21**, 295–312.

Readman JW, Mantoura RFC, Rhead MM and Brown L (1982) Aquatic distribution and heterotrophic degradation of polycyclic aromatic hydrocarbons (PAH) in the Tamar Estuary. *Estuarine, Coastal and Shelf Science*, **14**, 369–389.

Readman JW, Fowler SW, Villeneuve J-P, Cattini C, Oregioni B and Mee LD (1992) Oil and combustion-product contamination of the Gulf marine environment following the war. *Nature*, **358**, 662–665.

Reynoldson TB (1987) Interactions between sediment contaminants and benthic organisms. *Hydrobiologia*, **149**, 53–66.

Sanders G, Jones KC and Hamilton-Taylor J (1995) PCB and PAH fluxes to a dated UK peat core. *Environmental Pollution*, **89**, 17–25.

Schwarzenbach RP, Gschwend PM and Imboden DM (1993) *Environmental Organic Chemistry*. Wiley, New York, 681 pp.

Silliman JE, Meyers PA and Eadie BJ (1998) Perylene: an indicator of alteration processes or precursor materials? *Organic Geochemistry*, **29**, 1737–1744.

Smith JD, Hauser JY and Bagg J (1985) Polycyclic aromatic hydrocarbons in sediments of the Great Barrier Reef region, Australia. *Marine Pollution Bulletin*, **16**, 110–114.

Suess MJ (1976) The environmental load and cycle of polycyclic aromatic hydrocarbons. *Science of the Total Environment*, **6**, 239–250.

Thomas JF, Mukai M and Tebbens BD (1968) Fate of airborne benzo[e]pyrene. *Environmental Science and Technology*, **2**, 33–39.

Tolosa I, Bayona JM and Albaiges J (1996) Aliphatic and polycyclic aromatic hydrocarbons and sulfur/oxygen derivatives in northwestern Mediterranean sediments: spatial and temporal variability, fluxes, and budgets. *Environmental Science and Technology*, **30**, 2495–2503.

Valette-Silver N (1993) The use of sediment cores to reconstruct historical trends in contamination of estuarine and coastal sediments. *Estuaries*, **16** (3B), 577–588.

Venkatesan MI (1988) Occurrence and possible sources of perylene in marine sediments — a review. *Marine Chemistry*, **25**, 1–27.

Wakeham SG, Schaffner C and Giger W. (1980) Polycyclic aromatic hydrocarbons in Recent lake sediments — I. compounds having anthropogenic origins. *Geochimica et Cosmochimica Acta*, **44**, 403–413.

Wu Y, Zhang J, Mi T-Z and Li B (2001) Occurrence of n-alkanes and polycyclic aromatic hydrocarbons in the core sediments of the Yellow Sea. *Marine Chemistry*, **76**, 1–15.

Yamashita N, Kannan K, Imagawa T, Villeneuve DL, Hashimoto S, Miyazaki A and Giesy JP (2000) Vertical profile of polychlorinated dibenzo-*p*-dioxins, dibenzofurans, naphthalenes, biphenyls, polycyclic aromatic hydrocarbons, and alkylphenols in a sediment core from Tokyo Bay, Japan. *Environmental Science and Technology*, **34**, 3560–3567.

Yokley RA, Garrison AA, Wehry EL and Mamantov G (1986) Photochemical transformation of pyrene and benzo[a]pyrene vapor deposited on eight coal stack ashes. *Environmental Science and Technology*, **20**, 86–90.

Youngblood WW and Blumer M (1975) Polycyclic aromatic hydrocarbons in the environment: homologous series in soils and recent marine sediments. *Geochimica et Cosmochimica Acta*, **39**, 1303–1314.

Zepp RG and Macko SA (1997) Polycyclic aromatic hydrocarbons in sedimentary records of biomass burning. In Clark JS, Cachier H, Goldammer JG and Stocks B (eds), *Sediment Records of Biomass Burning and Global Change*. NATO ASI Series No. 151. Springer-Verlag, Berlin, Heidelberg, pp. 145–166.

Photochemical Reactions of PAHs in the Atmosphere

JANET AREY AND ROGER ATKINSON
Air Pollution Research Center, University of California, and Department of Environmental Sciences, University of California, Riverside, CA, USA

4.1 INTRODUCTION

The fate of polycyclic aromatic hydrocarbons (PAHs) in the environment, and the routes of wildlife or human exposures to them, is influenced by the environmental media, whether air, water or soil, in which the PAHs reside. In this chapter the atmospheric photochemical reactions of PAHs are examined. PAHs, predominantly the products of incomplete combustion, are ubiquitous in the atmosphere as the result of natural events, such as forest fires, and widespread anthropogenic sources, such as biomass burning (e.g. slash-and-burn agriculture), vehicle traffic, home heating and industrial emissions (Baek *et al.* 1991; Howsam and Jones 1998; see Chapters 2 and 3 for details). PAHs are of concern because certain of them are classified as probable human carcinogens (IARC 1983) and show tumorigenic activity in mammals (Cavalieri and Rogan 1998) and fish (de Maagd and Vethaak 1998). The fate of the PAHs in the atmosphere is strongly influenced by whether the PAH is present in the gaseous form or is particle-associated. Small particles are known to remain in the atmosphere for (on average) 1–2 weeks, thereby allowing long-range transport of particle-associated PAHs. Furthermore, association with sub-micron-sized particles may enhance inhalation exposure to adsorbed PAHs as they become trapped, along with the particles, deep in lung alveoli.

While the atmospheric photochemical PAH reactions described in this chapter will limit long-range transport of gaseous PAHs, a plethora of products are produced, some of which may be more toxic than the parent PAHs. Among the products described here are the highly mutagenic nitro-PAHs and nitro-PAH lactones (Atkinson and Arey 1994; Arey 1998). The nitro-PAH 2-nitrofluoranthene (2-NF), formed as the result of gas-phase atmospheric reactions of fluoranthene (Atkinson and Arey 1994), has been found worldwide (Arey *et al.* 1987; Ciccioli

PAHs: An Ecotoxicological Perspective. Edited by Peter E.T. Douben.
© 2003 John Wiley & Sons Ltd

et al. 1995; Arey 1998 and references therein; Dimashki *et al.* 2000; Feilberg *et al.* 2001). Although 2-NF is formed by a gas-phase reaction, 2-NF rapidly condenses onto particles and exposures to 2-NF then occur through particle inhalation. Thus, atmospheric photochemical reactions transform gas-phase PAHs into polar PAH derivatives, which are less volatile and more water-soluble and potentially more toxic and bioavailable.

4.2 PHASE DISTRIBUTION AND THE FATE OF PAHs IN THE ATMOSPHERE

The chemistry of the atmosphere is driven by the sun and, as shown in Figure 4.1, photolysis in the troposphere (the lower part of the atmosphere from the ground to the base of the stratosphere) results in the formation of hydroxyl (OH) radicals, ozone (O_3) and nitrate (NO_3) radicals. In turn these species may react with organic compounds, including PAHs, and notably the OH radical with any gaseous hydrogen-containing organic compound. These radical and O_3 reactions comprise the tropospheric loss processes for gas-phase PAHs shown in Figure 4.1 and determine what is known as the PAH's 'lifetime', τ, the time required for the PAH to decay to $1/e$ (37%) of its initial concentration, and thereby determine the radius of impact of the chemical. As noted on Figure 4.1, the most important loss process for a particular PAH molecule will differ not only with its phase (gaseous or particle-associated), but during daytime and night-time.

Photochemical formation of reactive radicals and ozone

Hydroxyl radical formation	Urban ozone formation	Nitrate radical formation
$O_3 + h\nu \rightarrow O_2 + O(^1D)$	$RH + OH \rightarrow R^{\bullet} + H_2O$ $R^{\bullet} + O_2 \rightarrow RO_2^{\bullet}$	$NO + O_3 \rightarrow NO_2 + O_2$
$(\lambda \sim 290\text{--}335\ nm)$	$RO_2^{\bullet} + NO \rightarrow RO^{\bullet} + NO_2$	$NO_2 + O_3 \rightarrow NO_3 + O_2$
$O(^1D) + H_2O \rightarrow 2\ OH$	$NO_2 + h\nu \rightarrow NO + O(^3P)$	
	$O(^3P) + O_2 + M \rightarrow O_3 + M$	
	(where M = air)	

Tropospheric loss processes

Gas-phase PAHs	Particle-associated PAHs
• Reaction with OH radicals (daytime)	• Photolysis ($\lambda > 290\ nm$)
• Reaction with O_3	• Reaction with O_3
• Reaction with NO_3 radicals (night-time)	• Wet and dry deposition of particles

Figure 4.1 Photochemical generation of the reactive species in the troposphere (see Graedel and Crutzen 1993 for more details) and loss processes for gas-phase and particle-associated polycyclic aromatic hydrocarbons. RH, a hydrocarbon

In the atmosphere the PAHs are partitioned between the gas and particle phases, with the gas/particle partitioning depending on a number of factors, including the liquid-phase (or subcooled liquid-phase) vapor pressure of the PAH at the ambient atmospheric temperature, the surface area of the particles per unit volume of air, and the nature of the particles (Wania and Mackay 1996; Pankow 1987; Bidleman 1988). To a first approximation, chemical compounds with liquid-phase vapor pressures of $P_L < 10^{-5}$ Pa ($< 10^{-7}$ Torr) at the ambient atmospheric temperature are present in the particle phase, and those with values of $P_L > 10^{-2}$ Pa ($> 10^{-4}$ Torr) at the ambient atmospheric temperature are essentially totally in the gas phase (Bidleman 1988). Chemicals with intermediate values of P_L are present in both the gas and particle phases and are often termed 'semi-volatile organic compounds'.

The subcooled liquid vapor pressures of the 2–4-ring PAHs are greater than or equal to 10^{-4} Pa (10^{-6} Torr) at 298 K, classifying them as volatile or semi-volatile, and ambient air measurements have shown that the 2–4-ring PAHs (Arey et al. 1987; Coutant et al. 1988; Ligocki and Pankow 1989), as well as the 2-ring nitro-PAHs (Wilson et al. 1995), are largely gas-phase species. PAHs with five and more rings and 4-ring nitro-PAHs are particle-associated under typical ambient conditions.

4.2.1 LOSS PROCESSES FOR PARTICLE-ASSOCIATED PAHs

The major loss processes for particle-associated PAHs are noted in Figure 4.1. The upper limit to the atmospheric lifetime of a particle-associated PAH is the lifetime of the particle on or in which the PAH is adsorbed or absorbed. During precipitation events, particles are efficiently removed from the atmosphere by wet deposition, but the lifetime of particles due to dry deposition varies with the particle size, with small respirable size particles having a typical lifetime for dry deposition of around 10 days (Graedel and Weschler 1981), sufficient for long-range transport.

Photolysis is probably the most important loss process for particle-associated PAHs (Kamens et al. 1988) and nitro-PAHs (Fan et al. 1996), although lifetimes measured for adsorbed PAHs vary over at least an order of magnitude, with the shorter lifetimes comparable to the lifetimes measured due to gas-phase OH radical-initiated reaction of PAHs (Korfmacher et al. 1980; Daisey et al. 1982; Yokley et al. 1986; Behymer and Hites 1988; Kamens et al. 1988; Dunstan et al. 1989). Photolysis of PAHs depends upon the light intensity, temperature and relative humidity, and the nature of the particle surface, including the presence and composition of an organic layer on the aerosol (McDow et al. 1994; Odum et al. 1994; Jang and McDow 1995). Carbonaceous material seems to protect PAHs from photolysis while the presence of certain organic species, such as the methoxyphenols abundant in wood combustion particles, seems to enhance the rates of photolysis.

Studies of benzo[a]pyrene adsorbed on various substrates or in ambient particles and exposed to O_3 have produced disparate results (Peters and Seifert 1980; Grosjean *et al*. 1983; Pitts *et al*. 1986; Coutant *et al*. 1988). The results can generally be rationalized by assuming that benzo[a]pyrene will react with O_3, but that in ambient particles not all of the benzo[a]pyrene is at, or sufficiently near, the surface to be available for reaction.

4.2.2   LOSS PROCESSES FOR GASEOUS PAHs

Unlike particle-associated PAHs, direct photolysis of gaseous PAHs by wavelengths of light which reach the lower troposphere ($\lambda > 290$ nm) has not been reported. The gas-phase atmospheric reactions of several 2–4-ring PAHs have been simulated in the laboratory (see Arey 1998 and references therein) and these kinetic studies allow the atmospheric lifetimes of PAHs due to reaction with OH radicals, NO_3 radicals and ozone to be predicted (Calvert *et al*. 2002). Calculated lifetimes are inversely proportional to the concentration of OH, NO_3, or O_3 assumed, e.g. the lifetime of a PAH due to reaction with the OH radical is:

$$\tau_{OH} = (k_{OH}[OH])^{-1} \qquad (4.1)$$

where k_{OH} is the reaction rate constant measured in laboratory kinetic studies and [OH] is the ambient atmospheric concentration of OH radicals. As noted previously, virtually all organic compounds react with the OH radical, while alkenes (containing C=C bonds) also generally react at a significant rate with O_3 and with NO_3 radicals (Atkinson 1994). Among the semi-volatile PAHs, acenaphthylene, and to a lesser extent phenanthrene and perhaps pyrene, undergo 'alkene-like' reactions. Atmospheric lifetimes calculated for reactions of selected PAHs with OH radicals, NO_3 radicals and O_3 are given in Table 4.1 for hypothetical clean air and polluted air conditions, using assumptions briefly discussed below.

4.2.2.1   Hydroxyl radical

Photolysis of O_3 in the troposphere by light of wavelengths 290–335 nm leads to the electronically excited oxygen atom $O(^1D)$, which either reacts with water vapor to form the OH radical, as shown in Figure 4.1, or is deactivated by reaction with N_2 or O_2 to form the ground state oxygen atom, $O(^3P)$, which then rapidly recombines with O_2 to re-form O_3. The OH radical concentration will therefore vary with time of day, latitude and the Sun's seasonal position, but annual global average tropospheric concentrations of the OH radical have been derived from the emissions, atmospheric concentrations, and OH radical reaction rate constant for methyl chloroform (CH_3CCl_3), resulting in 24 h average OH radical concentration estimates of $(9.4–10.3) \times 10^5$ radical/cm^3

TABLE 4.1 Calculated atmospheric lifetimes of selected polycyclic aromatic hydrocarbons (PAHs) due to gas-phase reactions with hydroxyl (OH) radicals, nitrate (NO_3) radicals, and ozone (O_3) for hypothetical summertime conditions in clean air and in a polluted atmosphere. Values have been calculated using rate constants taken from Calvert *et al.* (2002) unless noted otherwise

| PAH | Lifetime due to gas-phase reaction with | | | | | |
| | OH^a | | $NO_3{}^b$ | | $O_3{}^c$ | |
	Clean air	Polluted air	Clean air	Polluted air	Clean air	Polluted air
Naphthalene	5.7 h^d	2.3 h	14 years	2.6 days	>80 days	e
1-Methylnaphthalene	3.4 h^d	1.3 h	7 years	1.3 days	>125 days	e
2-Methylnaphthalene	2.8 h^d	1.1 h	5 years	11 h	>40 days	e
1-Ethylnaphthalene	4.1 h^d	1.7 h	5 years	1.0 years	f	e
2-Ethylnaphthalene	3.4 h^d	1.4 h	7 years	1.2 years	f	e
Dimethylnaphthalenes	1.8–2.4 h^d	40–60 min	0.2 −4 years^g	0.5 −9 h^g	>40 days	e
Acenaphthylene	1.1 h^h	30 min	6 min	0.6 min	2.5 h^h	20 min
Acenaphthene	1.8 h^h	40 min	1.4 h	8 min	>30 days	e
Biphenyl	1.6 days	7.8 h	>800 years	>160 days	>80 days	e
Fluorene	9.9 h	4.0 h	1.2 days	1.4 h	>80 days	e
Phenanthrene	7.7 h	3.1 h	4.3 h	30 min	41 days	e

[a] Assuming 12 h average OH radical concentrations of 2×10^6 radical/cm^3, based on a global annually averaged 24 h concentration of 1×10^6 radical/cm^3 (Prinn *et al.* 1995, 2001; Hein *et al.* 1997) for clean air and assuming 5×10^6 radical/cm^3 (see e.g. George *et al.* 1999) for polluted air.
[b] Assuming for clean air a 12 h average NO_3 radical concentration of 5×10^8 molecule/cm^3 and 1 ppbv of NO_2 and assuming for polluted air 5×10^9 molecule/cm^3 for a 12 h average NO_3 radical concentration and 200 ppbv of NO_2 (Atkinson *et al.* 1986; Atkinson 1991).
[c] Assuming a 24 h tropospheric average O_3 concentration of 7×10^{11} molecule/cm^3 (30 ppbv) (Logan 1985) for clean air and assuming 1 h at 200 ppbv O_3 for polluted air (a value which prompts a first-stage smog alert in California, USA).
[d] Rate constants from Phousongphouang and Arey (2002).
[e] Even in polluted areas the O_3 will not be continually elevated and therefore the PAH lifetime will still be measured in weeks.
[f] No reaction expected.
[g] Rate constants from Phousongphouang and Arey (2003).
[h] Rate constants from Reisen and Arey (2002).

(Prinn *et al.* 1995, 2001; Hein *et al.* 1997). Therefore, a 12 h daytime average tropospheric OH radical concentration of 2×10^6 radical/cm^3 is used for the PAH lifetime calculations in 'clean air' given in Table 4.1.

One of the few ground-level direct measurements of the OH radical in an urban area was made by George *et al.* (1999) in the Los Angeles basin. Their measurements occurred during a September air pollution episode and, because values for the OH radical peaked at $\sim 5 \times 10^6$ radical/cm^3, this value is used to calculate the lifetimes of PAHs in a polluted atmosphere. As may be seen from Table 4.1, reaction with the OH radical is the dominant loss process for the gas-phase PAHs.

4.2.2.2   Ozone

Ozone is present in the troposphere due to downward transport from the stratosphere and *in situ* chemical formation from the reaction of hydrocarbons and NO_x under the influence of sunlight (see Urban ozone formation in Figure 4.1). Background tropospheric O_3 levels of ~ 30 ppbv have been measured (Logan 1985) and the equivalent concentration of 7×10^{11} molecule/cm^3 is used to calculate the 'clean air' atmospheric lifetimes in Table 4.1. Note that only acenaphthylene, which contains a five-membered cyclopenta-fused ring, reacts rapidly with O_3, suggesting significant double-bond character for the bond in the cyclopenta-fused ring. The slow reaction of O_3 with phenanthrene also indicates significant double-bond character for its 9,10-bond.

In extremely polluted urban areas such as the Los Angeles basin in the 1960s, O_3 values over 500 ppbv were recorded. A value of 200 ppbv, which, if maintained for 1 h would currently trigger a 'first stage smog alert' in California [for perspective, the World Health Organization guideline is 60 ppbv (120 $\mu g/m^3$) for a maximum of 8 h], is used to calculate the lifetime of acenaphthylene for a 'polluted' atmosphere. Clearly, depending upon the O_3 concentration, reaction with O_3 can be the dominant loss process for acenaphthylene and presumably for similarly structured PAHs such as acephenanthrylene, but generally OH radical reaction dominates as a daytime PAH loss process.

4.2.2.3   Nitrate radical

NO_3 radical concentrations are highly variable because their formation requires the presence of both O_3 and NO_2 (see Figure 4.1). Furthermore, NO_3 radical reaction is only a night-time loss process because the NO_3 radical rapidly photolyzes. Analogous to the O_3 reactions, PAHs with double-bond character such as acenaphthylene and phenanthrene react most rapidly with the NO_3 radical. For those PAHs in which the NO_3 radical is adding to a ring that is fully aromatic, the radical formed apparently reacts with NO_2 (rather than with O_2) and the rates of these reactions, therefore, also depend on the NO_2 concentration. Concentrations of oxides of nitrogen vary widely between urban and remote areas and we have utilized 1 ppbv of NO_2 as typical of continental concentrations for the clean air calculations and 200 ppbv, a level reached in some urban areas, to calculate lifetimes in a polluted atmosphere. There is a relatively small database of NO_3 radical concentrations and the 5×10^9 molecule/cm^3 used to represent a polluted area is 50% of the maximum measured concentration reported to date (Atkinson *et al.* 1986). Although reaction with NO_3 is generally not an important loss process for PAHs, as discussed below the high yield of nitro-PAHs from certain PAHs can make night-time NO_3 chemistry an important source of ambient nitro-PAHs.

4.3 AMBIENT PAH MEASUREMENTS

Given in Table 4.2 are ambient PAH concentrations measured at three sites in California, chosen to be representative of different emission sources and sampled at night under meteorological conditions likely to result in high ambient concentrations. As seen from the table, independent of the emission source the more volatile PAHs were more abundant than the particle-associated PAHs such as benzo[a]pyrene and benzo[e]pyrene. The profiles of the PAHs were generally similar, with the exception of the high retene concentration at the ski resort site impacted by wood smoke combustion.

Because of their higher concentrations and volatility, Tenax-adsorbent samplers at relatively low flow rates were used to collect the 2- and 3-ring PAHs (naphthalene through phenanthrene in Table 4.2). Particles were sampled with a modified high-volume sampler that had a polyurethane foam plug (PUF) located downstream of the filter to collect gas-phase PAHs and PAHs that volatilized off the filter. Although the PAHs collected on the PUFs may operationally be considered gas-phase, it should be realized that both positive (blow-off from particles on the filter) and negative (gas-phase species adsorbed onto the filter) artifacts preclude strict phase distinction. However, the influence of temperature on the gas/particle distribution of the PAHs may be seen in the distribution of the 4-ring PAHs, fluoranthene and pyrene, between the PUF and the particles

TABLE 4.2 Ambient PAH concentrations (ng/m^3) at three sites in California impacted by different combustion sources. Three sampling media were utilized for different volatility PAHs. Data from Atkinson *et al.* (1988)

PAH	Sampling media	Ambient concentrations (ng/m^3) at sites impacted by		
		Vehicle traffic[a]	Wood smoke[b]	Industrial emissions[c]
Naphthalene	Tenax	4800	1400	4600
1-Methylnaphthalene	Tenax	390	210	370
2-Methylnaphthalene	Tenax	730	340	780
Fluorene	Tenax	37	75	52
Phenanthrene	Tenax	51	230	79
Fluoranthene	PUF (Filter)	6.2 (0.3)	16 (30)	18 (0.5)
Pyrene	PUF (Filter)	5.5 (0.3)	12 (30)	16 (0.6)
Retene[d]	Filter	0.1	86	1.4
Benzo[e]pyrene	Filter	1.5	8	4.2
Benzo[a]pyrene	Filter	0.6	12	6.2

[a]Glendora, CA, USA, night-time sample, August 13–14 1986.
[b]Mammoth Lakes, CA, USA, Tenax sample, night-time February 20–21, 1987. PUF plug and filter samples were composites of February 16–17, 17–18, 20–21, 27–28 and February 28–March 1 1987.
[c]Concord, CA, USA, Tenax sample, night-time December 6–7 1986. PUF plug and filter samples were composites of December 6–7, 7–8, 8–9, 1986.
[d]Retene (1-methyl-7-isopropylphenanthrene) has been reported as a marker of wood combustion (Ramdahl, 1983).

(filter). Only at the wood smoke-impacted site, which was typically below 0°C, was the majority of these semi-volatile PAHs collected on the filter sample rather than on the PUFs.

Recent work has shown that unburned fuel may contribute significantly to the high concentrations of volatile PAHs measured at urban sites, in particular alkyl-PAHs originating from diesel fuels (Truex *et al.* 1998). In the Los Angeles air basin, we have observed a 'traffic signature' for the volatile PAHs present with naphthalene > 2-methylnaphthalene > 1-methylnaphthalene > C_2-naphthalenes (and 2-ethylnaphthalene > 1-ethylnaphthalene, and with all possible dimethylnaphthalenes except 1,8-dimethylnaphthalene present). Evidence that OH radical reaction is the dominant loss process for gas-phase PAHs, as predicted from laboratory kinetic studies, can be seen through the diurnal differences in the volatile PAH concentrations. PAHs emitted at night will be essentially unreacted, while the decreases in concentrations due to OH radical reaction during daytime should be proportional to the rate constants for reactions of the PAHs with the OH radical. Thus, as seen from Figure 4.2, the night-time/daytime concentration ratios of naphthalene and alkylnaphthalenes plotted against their rate constants for reaction with the OH radical show a

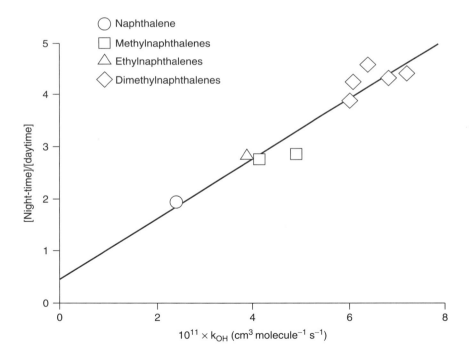

Figure 4.2 Plot of the ambient night-time/daytime concentrations for naphthalene and alkylnaphthalenes against their OH radical reaction rate constants. Data are from Riverside, CA, taken in August, 1997

linear relationship. Because the abundant volatile PAHs will react rapidly in the atmosphere, the health and ecological consequences of their reaction products need to be evaluated.

4.4 PRODUCTS OF ATMOSPHERIC REACTIONS OF PAHs

4.4.1 LABORATORY PRODUCT STUDIES

4.4.1.1 Hydroxyl radical

The OH radical reactions with PAHs proceed by two reaction pathways: OH radical addition to the aromatic ring to form an initially energy-rich hydroxycyclohexadienyl-type radical (OH–PAH adduct; see naphthalene reaction scheme, Figure 4.3) and OH radical interaction with substituent groups, either through H-atom abstraction from C–H bonds (such as the C–H bonds in the methyl groups on alkylnaphthalenes) or OH radical addition to $>C=C<$

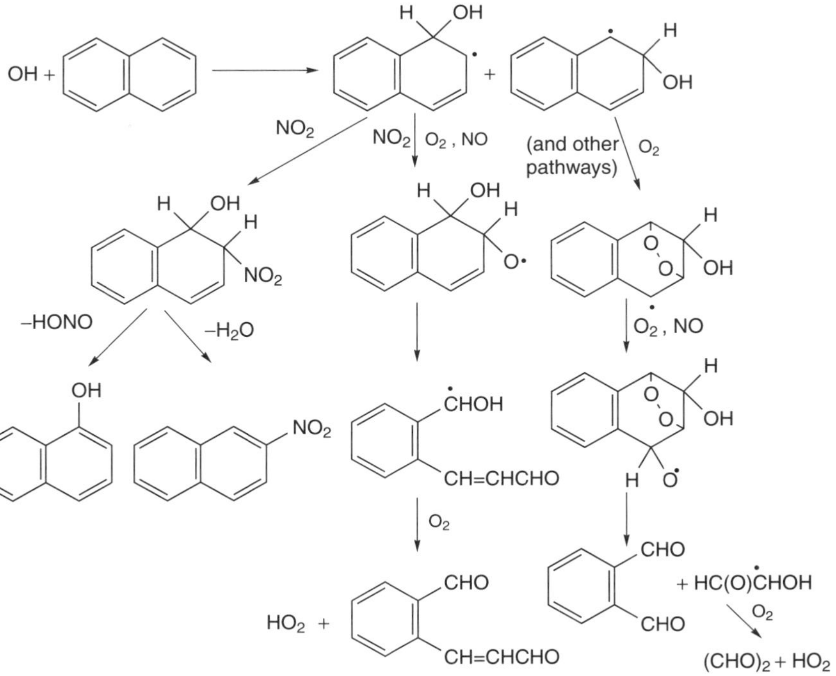

Figure 4.3 Postulated reaction scheme for naphthalene with the hydroxyl radical (in the presence of NO_x). Note that the OH-naphthalene adduct with the OH adding at the 2-position will also produce 2-naphthol, 1-nitronaphthalene and 2-formylcinnamaldehyde by reactions analogous to those shown for the adduct formed from OH addition at the 1-position

bonds (such as to the cyclopenta-fused ring of acenaphthylene). Carbonyl compounds will be products of the H-atom abstraction pathway (e.g. 1-naphthalenecarboxaldehyde from 1-methylnaphthalene). The reactions subsequent to formation of the OH–PAH adduct are shown in Figure 4.3 for naphthalene, the simplest, most abundant and most thoroughly studied PAH. The distribution of products from the naphthalene reaction with the OH radical is shown in Figure 4.4. As shown in Figure 4.3, if the OH radical adds to the 1-position of naphthalene and then adds an NO_2 to the *ortho* position, loss of water will result in 2-nitronaphthalene and loss of nitrous acid will give 1-naphthol, and analogous reactions can occur after addition of the OH radical to the 2-position of naphthalene. The full range of products has been studied for only a few PAHs (Sasaki *et al.* 1997; Reisen and Arey, 2002) and it appears that the ring-opened and decomposition (e.g. by loss of glyoxal) products dominate over the ring-retaining products containing nitro-, hydroxy-, keto- and epoxy-functional groups (see Figure 4.4).

4.4.1.2 Nitrate radical

Recent experimental data show that the reaction of naphthalene occurs by the initial addition of the NO_3 radical to the aromatic ring to form a nitrooxycyclohexadienyl-type radical (NO_3–naphthalene adduct), which then either decomposes back to reactants, reacts with NO_2, or decomposes unimolecularly (Arey 1998). For naphthalene, only ring-retaining products, including nitronaphthalenes and hydroxynitronaphthalenes, have been identified from its NO_3 radical reaction (see Figure 4.4).

4.4.2 MUTAGENIC NITRO-PAH DERIVATIVES

Bioassay-directed chemical analyses of the gas-phase reactions of 2–4-ring PAHs have resulted in the identification of mutagenic nitro-PAH (see e.g. Table 4.3) and, from phenanthrene and pyrene, nitro-PAH lactone products (Sasaki *et al.* 1995). Despite the generally low yields of nitro-PAHs ($\leq 5\%$) and nitro-PAH lactones from the OH radical-initiated PAH reactions, because of their high activity in the Ames *Salmonella typhimurium* bacterial assay (strain TA98, without microsomal activation) they accounted for a significant portion of the mutagenic activity of certain ambient air samples measured using the same assay (Sasaki *et al.* 1995). In other ambient samples, nitro-PAHs formed from NO_3 radical chemistry were determined to make a significant contribution to ambient mutagenicity (Gupta *et al.* 1996).

Because nitro-PAHs are potent mutagens and probable carcinogens (IARC 1989), the formation of nitro-PAHs from the OH and NO_3 radical reactions of at least a dozen PAHs has been studied (Atkinson and Arey 1994; Arey 1998 and references therein). Radical attack on a PAH is expected to occur mainly at

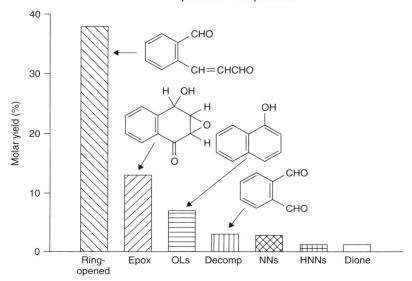

Naphthalene + OH products

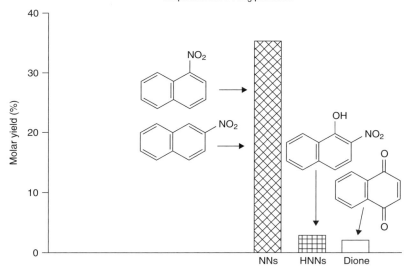

Naphthalene + NO₃ products

Figure 4.4 Molar yields of products formed from the OH radical-initiated reaction of naphthalene (top) and the NO₃ radical-initiated reaction of naphthalene (bottom). Examples of structures as identified in Sasaki *et al.* (1997) are shown in the figure. Abbreviations are: ring-opened (referring to breakage of the naphthalene ring, as in 2-formylcinnamaldehyde shown); Epox, epoxide (possible structure shown); OLs, 1- and 2-naphthol; decomp, alkoxy radical decomposition (see Figure 4.3) could lead to this product; NNs, 1- and 2-nitronaphthalene; HNNs, hydroxynitronaphthalenes; Dione, 1,4-naphoquinone

TABLE 4.3 Nitro-PAH products formed from the gas-phase reactions of selected PAHs with hydroxyl radicals and nitrate radicals (both in the presence of NO_x) and their yields (adapted from Arey 1998; Sasaki *et al.* 1997)

PAH	Nitro-PAH (yield) from PAH reaction with	
	OH	NO₃
Naphthalene	1-Nitronaphthalene (1%)	1-Nitronaphthalene (24%)
	2-Nitronaphthalene (1%)	2-Nitronaphthalene (11%)
1-Methylnaphthalene	1-Methyl-*x*-nitronaphthalenes	1-Methyl-*x*-nitronaphthalenes
	1M5NN > 1M4NN ≥ 1M6NN	1M3NN > 1M5NN ≥ 1M4NN
	(all 7 isomers ∼ 0.4%)	(all 7 isomers ∼ 30%)
2-Methylnaphthalene	2-Methyl-*x*-nitronaphthalenes	2-Methyl-*x*-nitronaphthalenes
	2M5NN > 2M6NN ∼ 2M7NN	2M4NN > 2M1NN ∼ 2M5NN
	(all 7 isomers ∼ 0.2%)	(all 7 isomers ∼ 30%)
Biphenyl	3-Nitrobiphenyl (5%)	No reaction observed
Fluoranthene	2-Nitrofluoranthene (∼ 3%)	2-Nitrofluoranthene (∼ 24%)
	7-Nitrofluoranthene (∼ 1%)	
	8-Nitrofluoranthene (∼ 0.3%)	
Pyrene	2-Nitropyrene (∼ 0.5%)	4-Nitropyrene (∼ 0.06%)
	4-Nitropyrene (∼ 0.06%)	

the ring position(s) of highest electron density and the OH–PAH or NO_3–PAH adduct formed may then add NO_2 in the *ortho* position. Thus, nitro-PAHs formed by radical-initiated atmospheric reactions are often distinct from the nitro-PAHs which are found in emission sources such as diesel exhaust, where the latter are the nitro-isomers as occur from electrophilic nitration. Furthermore, as may be deduced from Table 4.3, the presence of certain nitro-PAHs and/or nitro-PAH ratios in ambient samples can be used to distinguish air masses where OH radical reaction has dominated from air masses where NO_3 radical reaction has occurred.

4.4.3 AMBIENT NITRO-PAH MEASUREMENTS

Laboratory product studies can be verified through ambient measurements which show the importance of nitro-PAH formation from both OH and NO_3 radical-initiated gas-phase PAH reactions. For example, Figure 4.5 shows the gas chromatography–mass spectrometry (GC–MS) analysis of the nitro-PAHs of molecular weight 247, nitrofluoranthenes (NFs) and nitropyrenes (NPs), present in ambient air samples collected in Torrance and Claremont, CA, and in a diesel exhaust particle sample. The four-ring PAH fluoranthene and pyrene are generally found at levels similar to one another, both in diesel emissions and in ambient air, and the dominance of 1-NP in diesel exhaust (Figure 4.5c) is consistent with this isomer being the electrophilic nitration product of pyrene and with pyrene being more reactive toward electrophilic nitration than is

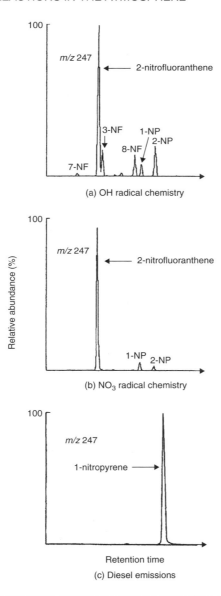

Figure 4.5 Mass chromatograms showing the molecular ion of the nitrofluoranthenes (NFs) and nitropyrenes (NPs) in (a) an ambient particle sample from Torrance, CA, in which the 2-NF, 7-NF, 8-NF and 2-NP isomers are attributed to OH radical-initiated chemistry; (b) an ambient particle sample from Claremont, CA, in which the high 2-NF/2-NP ratio suggests NO₃ radical-initiated formation of 2-NF (note that the 1-NP in both these ambient samples is assumed to be from direct emissions); and (c) a diesel exhaust particle sample in which 1-NP was the most abundant nitro-PAH observed. Note that (c) was analyzed under different GC–MS conditions, resulting in a different retention time for 1-NP

fluoranthene. Although 1-NP is present in the Torrance sample (Figure 4.5a), as is a small amount of fluoranthene's electrophilic nitration product 3-NF, the most abundant MW 247 nitro-PAH in ambient air is 2-NF. The profile of the nitro-PAHs at Torrance resembles that expected from OH radical-initiated reaction, i.e. dominated by 2-NF and 2-NP with smaller amounts of 7-NF and 8-NF; the 1-NP and 3-NF present are presumed to arise from direct emissions from combustion sources. It should be noted that while 2-NF, 1-NP and 2-NP were always observed in ambient particle samples collected in California and 2-NF was always most abundant in these samples, 3-NF was only occasionally present above the detection limit. Because of the high yield of 2-NF from the reaction of fluoranthene with the NO_3 radical (Table 4.3), an air mass in which the NO_3 radical is present would result in enhanced amounts of 2-NF relative to 2-NP. The very high 2-NF/2-NP ratio in the Claremont night-time sample then indicates the occurrence of night-time NO_3 radical chemistry (Figure 4.5b). Recently, we have confirmed a similar dominance of 2-NF in ambient particles collected at Redlands, CA, during a period of moderate photochemical air pollution, where the presence of NO_3 radicals was deduced from the measured ambient air profiles of the methylnitronaphthalenes observed (Gupta et al. 1996).

4.5 CONCLUSIONS

For PAHs present in ambient air in the gas phase, reaction with the OH radical will result in lifetimes of generally less than 1 day, limiting their range of impact but producing products which may have important health and ecological consequences. The reaction products of the PAHs are, in general, more polar than the PAHs themselves and these products, therefore, are more likely to condense and become particle-associated. The products identified include highly mutagenic nitro-PAHs and nitro-PAH lactones, together with a variety of other products, including ring-opened dicarbonyls, for which no toxicological data exist. While gas-phase reactions limit the atmospheric transport of the most volatile species, PAHs and PAH photochemical reaction products which are particle-associated or distributed between the gas and particle phase may undergo long-range atmospheric transport to remote areas of the globe.

REFERENCES

Arey J (1998) Atmospheric reactions of PAHs including formation of nitroarenes. In Neilson AH (ed.), *The Handbook of Environmental Chemistry, Vol. 3, Part I, PAHs and Related Compounds*. Springer-Verlag, Berlin, pp. 347–385.

Arey J, Zielinska B, Atkinson R and Winer AM (1987) Polycyclic aromatic hydrocarbon and nitroarene concentrations in ambient air during a wintertime high-NO_x episode in the Los Angeles basin. *Atmospheric Environment*, **21**, 1437–1444.

Atkinson R (1991) Kinetics and mechanisms of the gas-phase reactions of the NO_3 radical with organic compounds. *Journal of Physical and Chemical Reference Data*, **20**, 459–507.

Atkinson R (1994) Gas-phase tropospheric chemistry of organic compounds. *Journal of Physical and Chemical Reference Data*, Monograph **2**, 1–216.

Atkinson R and Arey J (1994) Atmospheric chemistry of gas-phase polycyclic aromatic hydrocarbons: formation of atmospheric mutagens. *Environmental Health Perspectives*, **102**(suppl 4), 117–126.

Atkinson R, Winer AM and Pitts JN Jr (1986) Estimation of night-time N_2O_5 concentrations from ambient NO_2 and NO_3 radical concentrations and the role of N_2O_5 in night-time chemistry. *Atmospheric Chemistry*, **20**, 331–339.

Atkinson R, Arey J, Winer AM and Zielinska B (1988) *A Survey of Ambient Concentrations of Selected Polycyclic Aromatic Hydrocarbons (PAHs) at Various Locations in California*. Final Report to Contract No. A5-185-32. California Air Resources Board, Sacramento, CA.

Baek SO, Field RA, Goldstone ME, Kirk PW, Lester JN and Perry R (1991) A review of atmospheric polycyclic aromatic hydrocarbons: sources, fate and behavior. *Water, Air and Soil Pollution*, **60**, 279–300.

Behymer TD and Hites RA (1988) Photolysis of polycyclic aromatic hydrocarbons adsorbed on fly ash. *Environmental Science and Technology*, **22**, 1311–1319.

Bidleman TF (1988) Atmospheric processes. *Environmental Science and Technology*, **22**, 361–367.

Calvert JG, Atkinson R, Becker KH, Kamens RM, Seinfeld JH, Wallington TJ and Yarwood G (2002) *The Mechanisms of Atmospheric Oxidation of Aromatic Hydrocarbons*. Oxford, Oxford University Press, 566 pp.

Cavalieri E and Rogan E (1998) Mechanisms of tumor initiation by polycyclic aromatic hydrocarbons in mammals. In Neilson AH (ed.), *The Handbook of Environmental Chemistry, Vol. 3, Part J. PAHs and Related Compounds*. Springer-Verlag, Berlin, pp. 81–117.

Ciccioli P, Cecinato A, Brancaleoni E, Frattoni M, Zacchei P and De Castro Vasconcellos P (1995) The ubiquitous occurrence of nitro-PAH of photochemical origin in airborne particles. *Annali di Chimica*, **85**, 455–469.

Coutant RW, Brown L, Chuang JC, Riggin RM and Lewis RG (1988) Phase distribution and artifact formation in ambient air sampling for polynuclear aromatic hydrocarbons. *Atmospheric Environment*, **22**, 403–409.

Daisey JM, Lewandowski CG and Zorz M (1982) A photoreactor for investigations of the degradation of particle-bound polycyclic aromatic hydrocarbons under simulated atmospheric conditions. *Environmental Science and Technology*, **16**, 857–861.

de Maagd PG-J and Vethaak AD (1998) Biotransformation of PAHs and their carcinogenic effects in fish. In Neilson AH (ed.), *The Handbook of Environmental Chemistry, Vol. 3, Part J. PAHs and Related Compounds*, Springer-Verlag, Berlin, pp. 265–309.

Dimashki M, Harrad S and Harrison RM (2000) Measurements of nitro-PAH in the atmospheres of two cities. *Atmospheric Environment*, **34**, 2459–2469.

Dunstan TDJ, Mauldin RF, Jinxian Z, Hipps AD, Wehry EL and Mamantov G (1989) Adsorption and photodegradation of pyrene on magnetic, carbonaceous, and mineral subfractions of coal stack ash. *Environmental Science and Technology*, **23**, 303–308.

Fan Z, Kamens RM, Hu J, Zhang J and McDow S (1996) Photostability of nitro-polycyclic aromatic hydrocarbons on combustion soot particles in sunlight. *Environmental Science and Technology*, **30**, 1358–1364.

Feilberg A, Poulsen MWB, Nielsen T and Skov H (2001) Occurrence and sources of particulate nitro-polycyclic aromatic hydrocarbons in ambient air in Denmark. *Atmospheric Environment*, **35**, 353–366.

George LA, Hard TM and O'Brien RJ (1999) Measurement of free radicals OH and HO_2 in Los Angeles smog. *Journal of Geophysical Research*, **104**, 11643–11655.

Graedel TE and Crutzen PJ (1993) *Atmospheric Change: An Earth System Perspective*. WH Freeman, New York, 446 pp.

Graedel TE and Weschler CJ (1981) Chemistry within aqueous atmospheric aerosols and raindrops. *Reviews of Geophysics and Space Physics*, **19**, 505–539.

Grosjean D, Fung K and Harrison J (1983) Interactions of polycyclic aromatic hydrocarbons with atmospheric pollutants. *Environmental Science and Technology*, **17**, 673–679.

Gupta P, Harger WP and Arey J (1996) The contribution of nitro- and methylnitronaphthalenes to the vapor-phase mutagenicity of ambient air samples. *Atmospheric Environment*, **30**, 3157–3166.

Hein R, Crutzen PJ and Heimann M (1997) An inverse modeling approach to investigate the global atmospheric methane cycle. *Global Biogeochemical Cycles*, **11**, 43–76.

Howsam M and Jones KC (1998) Sources of PAHs in the environment. In Neilson AH (ed.), *The Handbook of Environmental Chemistry, Vol. 3, Part I. PAHs and Related Compounds*. Springer-Verlag, Berlin, pp. 137–174.

IARC (1983) Polynuclear aromatic compounds. Part 1, chemical, environmental and experimental data. In *IARC Monographs on the Evaluation of the Carcinogenic Risk of Chemicals to Humans*, vol. 32, International Agency for Research on Cancer, Lyon, 453 pp.

IARC (1989) Diesel and gasoline engine exhausts and some nitroarenes. In *IARC Monographs on the Evaluation of the Carcinogenic Risk of Chemicals to Humans*, vol. 46. International Agency for Research on Cancer, Lyon, 458 pp.

Jang M and McDow SR (1995) Benz[a]anthracene photodegradation in the presence of known organic constituents of atmospheric aerosols. *Environmental Science and Technology*, **29**, 2654–2660.

Kamens RM, Guo Z, Fulcher JN and Bell DA (1988) Influence of humidity, sunlight, and temperature on the daytime decay of polyaromatic hydrocarbons on atmospheric soot particles. *Environmental Science and Technology*, **22**, 103–108.

Korfmacher WA, Wehry EL, Mamantov G and Natusch DFS (1980) Resistance to photochemical decomposition of polycyclic aromatic hydrocarbons vapor-adsorbed on coal fly ash. *Environmental Science and Technology*, **14**, 1094–1099.

Ligocki MP and Pankow JF (1989) Measurements of the gas/particle distributions of atmospheric organic compounds. *Environmental Science and Technology*, **23**, 75–83.

Logan JA (1985) Tropospheric ozone: seasonal behavior, trends, and anthropogenic influence. *Journal of Geophysical Research*, **90**, 10463–10482.

McDow SR, Sun Q, Vartiainen M, Hong Y, Yao Y, Fister T, Yao R and Kamens RM (1994) Effect of composition and state of organic components on polycyclic aromatic hydrocarbon decay in atmospheric aerosols. *Environmental Science and Technology*, **28**, 2147–2153.

Odum JR, McDow SR and Kamens RM (1994) Mechanistic and kinetic studies of the photodegradation of benz[a]anthracene in the presence of methoxyphenols. *Environmental Science and Technology*, **28**, 1285–1290.

Pankow JF (1987) Review and comparative analysis of the theories on partitioning between the gas and aerosol particulate phases in the atmosphere. *Atmospheric Environment*, **21**, 2275–2283.

Peters J and Seifert B (1980) Losses of benzo[a]pyrene under the conditions of high-volume sampling. *Atmospheric Environment*, **14**, 117–119.

Phousongphouang PT and Arey J (2002) Rate constants for the gas-phase reactions of a series of alkylnaphthalenes with the OH radical. *Environmental Science and Technology*, **36**. 1947–1952.

Phousongphouang PT and Arey J (2003) Rate constants for the gas-phase reactions of a series of alkylnaphthalenes with the nitrate radical. *Environmental Science and Technology*, in press.

Pitts JN Jr, Paur H-R, Zielinska B, Arey J, Winer AM, Ramdahl T and Mejia V (1986) Factors influencing the reactivity of polycyclic aromatic hydrocarbons adsorbed on filters and ambient POM with ozone. *Chemosphere*, **15**, 675–685.

Prinn RG, Weiss RF, Miller BR, Huang J, Alyea FN, Cunnold DM, Fraser PJ, Hartley DE and Simmonds PG (1995) Atmospheric trends and lifetime of CH_3CCl_3 and global OH concentrations. *Science*, **269**, 187–192.

Prinn RG, Huang J, Weiss RF, Cunnold DM, Fraser PJ, Simmonds PG, McCulloch A, Harth C, Salameh P, O'Doherty S, Wang RHJ, Porter L and Miller BR (2001) Evidence for substantial variations of atmospheric hydroxyl radicals in the past two decades. *Science*, **292**, 1882–1888.

Ramdahl T (1983) Retene — a molecular marker of wood combustion in ambient air. *Nature*, **306**, 580–582.

Reisen F and Arey J (2002) Reactions of hydroxyl radicals and ozone with acenaphthene and acenaphthylene. *Environmental Science and Technology*, **36**, 4302–4311.

Sasaki J, Arey J and Harger WP (1995) Formation of mutagens from the photooxidations of 2–4-ring PAH. *Environmental Science and Technology*, **29**, 1324–1335.

Sasaki J, Aschmann SM, Kwok ESC, Atkinson R and Arey J (1997) Products of the gas-phase OH and NO_3 radical-initiated reactions of naphthalene. *Environmental Science and Technology*, **31**, 3173–3179.

Truex TJ, Norbeck JM, Arey J, Kado N and Okamoto R (1998) *Evaluation of Factors that Affect Diesel Exhaust Toxicity*. Final Report to Contract No. 94–312, California Air Resources Board, Sacramento, CA.

Wania F and Mackay D (1996) Tracking the distribution of persistent organic pollutants. *Environmental Science and Technology*, **30**, 390–396A.

Wilson NK, McCurdy TR and Chuang JC (1995) Concentrations and phase distributions of nitrated and oxygenated polycyclic aromatic hydrocarbons in ambient air. *Atmospheric Environment*, **29**, 2575–2584.

Yokley RA, Garrison AA, Wehry EL and Mamantov G (1986) Photochemical transformation of pyrene and benzo[a]pyrene vapor deposited on eight coal stack ashes. *Environmental Science and Technology*, **20**, 86–90.

Metabolic Activation of PAHs: Role of DNA Adduct Formation in Induced Carcinogenesis

FARIDA AKCHA[1], THIERRY BURGEOT[1], JEAN-FRANÇOIS NARBONNE[2] AND PHILIPPE GARRIGUES[2]

[1]IFREMER, Nantes, France
[2]Laboratory of Physico- and Toxicochemistry of Natural Systems (LPTC), University of Bordeaux I, Talence, France

5.1 INTRODUCTION

Among environmental pollutants, polycyclic aromatic hydrocarbons (PAHs) act as procarcinogens once absorbed and metabolically activated by organisms, directing their reactivity towards the nucleophilic groups of cellular macromolecules. Metabolic activation of PAH to DNA adducts is considered as a crucial event in chemical carcinogenesis, involving covalent binding between the chemical carcinogen and the DNA (Miller and Miller 1981). Different pathways of metabolic activation to DNA adducts have been proposed for the PAH benzo[a]pyrene (B[a]P), a model compound in carcinogenic studies.

This chapter covers the pathways involved in B[a]P activation to DNA adducts and the biological significance of DNA adducts, both in general and in the context of environmental protection. How this knowledge is being used as indicators of a biological response to exposure of invertebrates to PAHs is covered in Chapter 10. The usefulness for the development of biomarkers is addressed in Chapter 16.

5.2 PATHWAYS INVOLVED IN B[a]P ACTIVATION TO DNA ADDUCTS

5.2.1 B[a]P ACTIVATION TO B[a]P DIOL-EPOXIDE-10-N2dG

B[a]P activation to B[a]P diol-epoxide-10-N2dG, the most frequently described pathway (Sims *et al.* 1974; Grover 1986), involves the formation of bay-region

PAHs: An Ecotoxicological Perspective. Edited by Peter E.T. Douben.
© 2003 John Wiley & Sons Ltd

diol-epoxides as ultimate carcinogenic PAH metabolites. Diol-epoxides result from two-electron oxidation of B[a]P by cytochrome P450, especially iso-forms P450IA1 and P450IA2 (Hall *et al.* 1989). Their formation is initiated by P450-mediated oxidation of B[a]P to B[a]P-arene oxides. Following hydration catalyzed by epoxide hydrolase (EH), B[a]P-arene oxides lead thus to the for-mation of B[a]P-dihydrodiols, which are oxidized by P450 to their respective B[a]P-diol-epoxides (Figure 5.1).

Among the B[a]P-dihydrodiols resulting from B[a]P metabolism by cyto-chrome P450 monooxygenases, (±)-7,8-B[a]P-diol is predominantly formed because of the stereo- and regio-selectivity of both cytochrome P450 and EH (Adams *et al.* 1995). It is also the only B[a]P-dihydrodiol that can be metabolized to a diol-epoxide to produce DNA adduct formation in vertebrates. The corre-sponding diol-epoxide exists in *syn* and *anti* forms, each presenting two *cis* and *trans* enantiomers. The *anti* form is produced in larger amounts and shows greater reactivity (especially for the *trans* enantiomer) to DNA (Hall and Grover 1990). Through reaction with DNA, this diol-epoxide leads to the formation of BPDE-10-N2dG, the predominant adduct of this activation pathway (Figure 5.1), which results from covalent binding of (+)-7R,8S-dihydroxy-9S,10R-epoxy-7,8,9,10,tetrahydrobenzo[a]pyrene (BPDE) in C10 with the N2 amino group of deoxyguanosine. BPDE-10-N2dG was isolated and identified *in vivo* for the first time by high-performance liquid chromatography (HPLC) coupled to UV and radioactivity detection and by co-chromatography with BPDE standards (Sims *et al.* 1974). Thereafter, the ^{32}P postlabeling technique has been widely used for analysis of this type of DNA adduct (Gupta 1996).

As BPDE reactivity is minimal or even null with the other DNA bases, the BPDE-10-N2dG adduct has long been considered as the only adduct associated with the diol-epoxide pathway. However, thanks to radical cation theory (Cavalieri and Rogan 1992, 1995), our understanding of the adduct profile obtained by this pathway has recently been improved.

5.2.2 STABLE AND DEPURINATING DNA ADDUCTS

Studies of the radical cation pathway have distinguished two types of DNA adducts: stable and depurinating adducts. During classical DNA extraction procedures, stable adducts remain strongly linked to DNA by a covalent bond. To the opposite, the formation of depurinating adducts weakens DNA, causing the N-glycosidic bond to break during extraction. This results in a complete loss of deoxyribose and the formation of an abasic site. Although depurinating DNA adducts have been identified essentially during investigations of the radical cation pathway, two of them (BPDE-10-N7Ade and BPDE-10-N7Gua) originating from the diol-epoxide pathway have recently been isolated both *in vitro* and *in vivo* in rodents (Figure 5.2) (Chen *et al.* 1996).

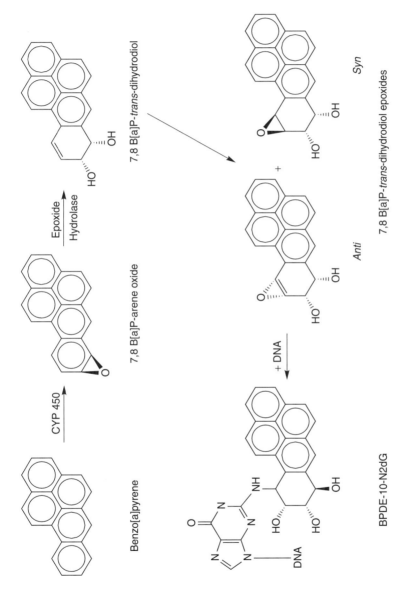

Figure 5.1 B[a]P activation pathway to diol-epoxides

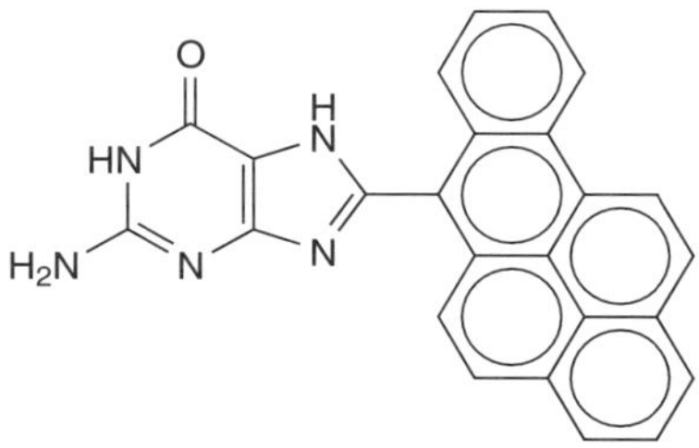

B[a]P-6-C8Gua

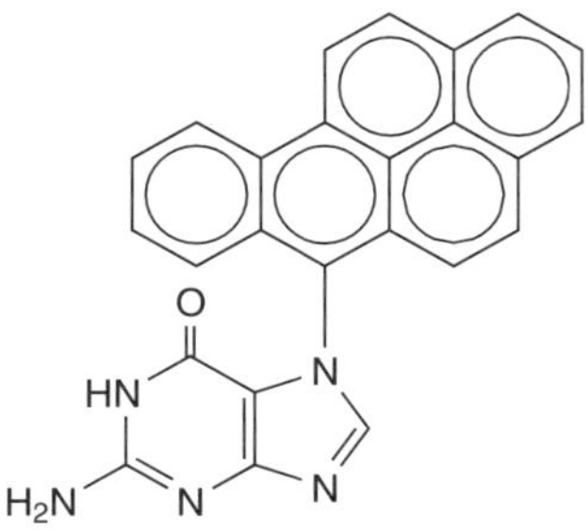

B[a]P-6-N7Gua

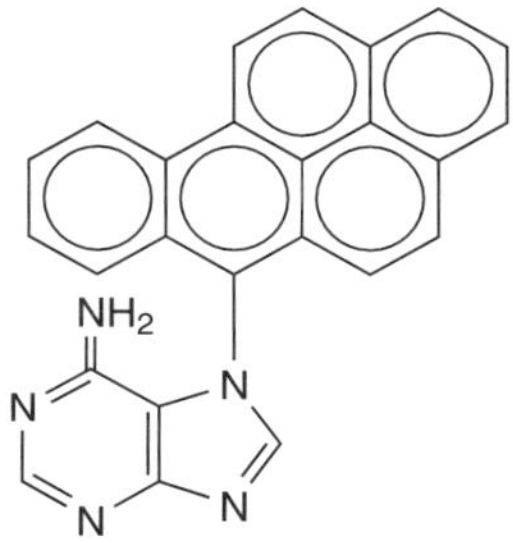

B[a]P-6-N7Ade

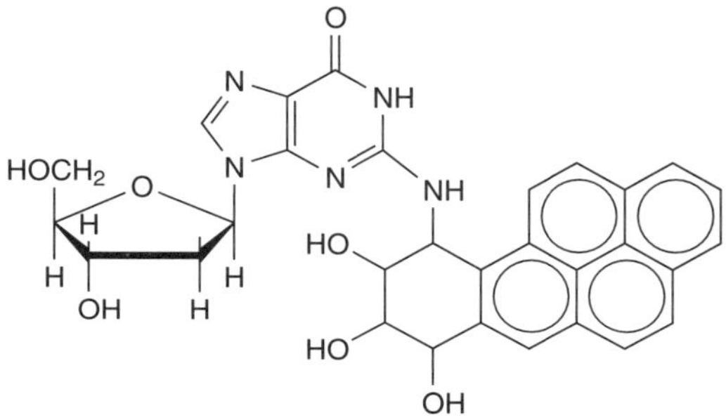

BPDE-10-N2dG

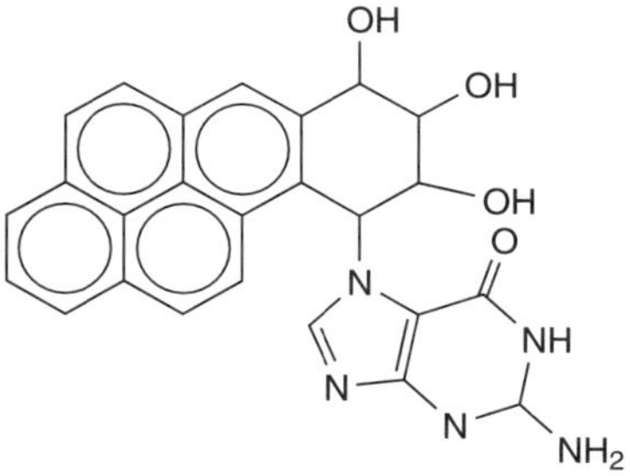

BPDE-10-N7Gua

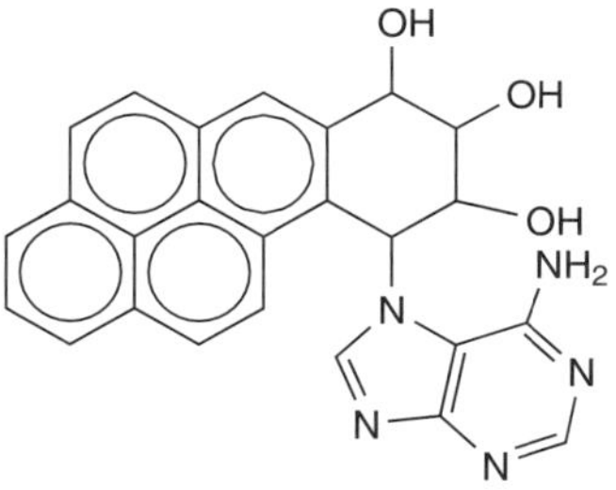

BPDE-10-N7Ade

Figure 5.2   Stable and depurinating B[a]P adducts produced *in vitro* and *in vivo* in rodents by diol-epoxide and radical cation pathways

5.2.3   THE RADICAL CATION PATHWAY

Radical cation theory is based on the formation of a PAH radical cation resulting from P450-catalyzed one-electron oxidation of PAH. Although originally contested, this theory is now considered to account for the predominant pathway of DNA adduct formation in vertebrates (Cavalieri and Rogan 1992).

5.2.3.1   Role of cytochrome P450 in the radical cation pathway: *in vitro* studies

The theory that cytochrome P450 could catalyze one-electron oxidation of B[a]P was quite original, as it had previously been regarded as catalyzing two-electron oxidation exclusively. Several studies have contributed to the notion that cytochrome P450 has 'peroxidase-like' activity.

   One-electron oxidation has been proposed as an initial step in B[a]P oxidation by cytochrome P450. It has been shown *in vitro* that B[a]P quinones originate from a radical cation (6-B[a]P$^{\bullet+}$), resulting from one-electron oxidation of B[a]P by cytochrome P450 (Cavalieri *et al.* 1988, 1990). Due to its electrophilic property, this radical has been shown to react *in vitro* with deoxyguanosine to form stable and depurinating DNA adducts (Rogan *et al.* 1988). The first depurinating DNA adduct characterized in this way was 7-(benzo[a]pyrene-6-yl)guanine (B[a]P-6-N7Gua). More extensive studies then led to identification *in vitro* in rat liver microsomes (Devanesan *et al.* 1992; Chen *et al.* 1996) and cells (Devanesan *et al.* 1996) of new B[a]P adducts from both the radical cation and the diol-epoxide pathways. Thus, the depurinating DNA adducts, B[a]P-6-N7Ade, B[a]P-6-N7Gua and B[a]P-6-C8Gua, were identified by fast atom bombardment combined with tandem mass spectrometry (FAB MS – MS) or fluorescence line narrow spectrometry (FLNS). In addition to the predominant stable adduct BPDE-10-N2dG, two depurinating adducts resulting from the diol-epoxide pathway, namely BPDE-10-N7Ade and BPDE-10-N7Gua, were also identified, although at low levels (Devanesan *et al.* 1992; Chen *et al.* 1996). In these studies, 80% of the total DNA adducts produced *in vitro* in rat liver microsomes and nuclei were depurinating DNA adducts generated mainly by the radical cation pathway: B[a]P-6-N7Ade (38 – 58%) > B[a]P-6-N7Gua (10 – 20%) > B[a]P-6-C8Gua (5 – 14%) > BPDE-10-N7Ade (0.2 – 0.5%) > BPDE-10-N7Gua (0.2%). Stable DNA adducts represented 20% of the total DNA adducts produced (BPDE-10-N2dG 22%, unidentified stable adducts 5%). Thus, the diol-epoxide pathway appears to be predominant *in vitro* for stable DNA adduct formation. In fact, the formation of stable DNA adducts by the radical cation pathway (B[a]P-6-C8dG, B[a]P-6-N3dG, B[a]P-6-N2dG) has only been reported following *in vitro* incubation of electrochemical oxidation products of B[a]P with dG (Ramakrishna *et al.* 1992).

5.2.3.2   The radical cation pathway: *in vivo* studies

The formation of DNA adducts from the radical cation pathway has been studied *in vivo* in mouse skin and rat mammary gland. Several adducts were induced after topical application of B[a]P for 4 h on mouse skin (Chen *et al.* 1996). Depurinating DNA adducts B[a]P-6-C8Gua (34%), B[a]P-6-N7Ade (22%), B[a]P-6-N7Gua (10%), BPDE-10-N7Ade (3%) and BPDE-10-N7Gua (2%) accounted for 71% of the total adducts produced, 66% of which originated from the radical cation pathway. Stable adducts accounted for only 28% of the total (BPDE-10-N2dG 22% and unidentified stable adducts 6%).

The DNA adduct profile was also determined after injection of B[a]P into rat mammary gland (Todorovic *et al.* 1997). Two depurinating adducts were predominantly formed in this tissue, B[a]P-6-C8Gua and B[a]P-6-N7Ade. The stable adducts consisted of BPDE-10-N2dG essentially (64% of stable adducts produced) and three unidentified types.

In vertebrates, the radical cation pathway appears to be the predominant means of DNA adduct formation *in vitro* and *in vivo*. Although some differences in the adduct profile have been detected between tissues and species, certain results were common to all of the studies: (a) depurinating DNA adducts are formed mainly by the radical cation pathway and constitute up to 70% of the total adducts produced; (b) stable adducts, which are produced in lower amounts, originate from the diol-epoxide pathway. In fact, it has never been demonstrated *in vivo* that stable adducts can be formed by the radical cation pathway.

5.2.4   THE B[a]P BENZYLIC ESTER PATHWAY

This PAH activation pathway proposed by Flesher and Sydnor (1973) is based on enzymatic reactions of PAH substitution that result in the production of PAH benzylic esters reactive for DNA (Figure 5.3). This pathway consists of three steps, each involving a substitution reaction of PAHs or their metabolites (Stansbury *et al.* 1994). During the first step, the PAH molecule is bioalkylated by methyl substitution at its highest electron density position. In the case of B[a]P, this substitution leads to a 6-methyl B[a]P formation (Flesher *et al.* 1983, 1990). During the second step, the methyl group is oxidized into a hydroxymethyl group. For several aromatic hydrocarbons, this hydroxymethyl metabolite can bind *in vitro* to DNA in the presence of ATP (Rogan *et al.* 1980). During the third step, esterification of the hydroxyl chain by transfer of an acetate or a sulfate group leads to the formation of a benzylic ester that can bind to DNA to form stable adducts (Okuda *et al.* 1989).

In the rat and mouse, activation of B[a]P by this pathway led to the formation of two aralkyl-B[a]P DNA adducts *in vitro* and *in vivo* (Suhr *et al.* 1989): N^2-(benzo[a]pyrene-6-ylmethyl)-deoxyguanosine (the predominant one) and N^6-(benzo[a]pyrene-6-ylmethyl)-deoxyadenosine. However, the relative share of these aralkyl adducts in the formation of stable B[a]P DNA adducts remains unclear.

5.3   BIOLOGICAL SIGNIFICANCE OF DNA ADDUCTS

5.3.1   ROLE OF DNA ADDUCT FORMATION IN PAH MUTAGENICITY

The Ames' test (Sakai *et al.* 1985) and the development of mutagenicity tests on animal cell cultures (Huberman *et al.* 1976) have allowed PAH mutagenicity to

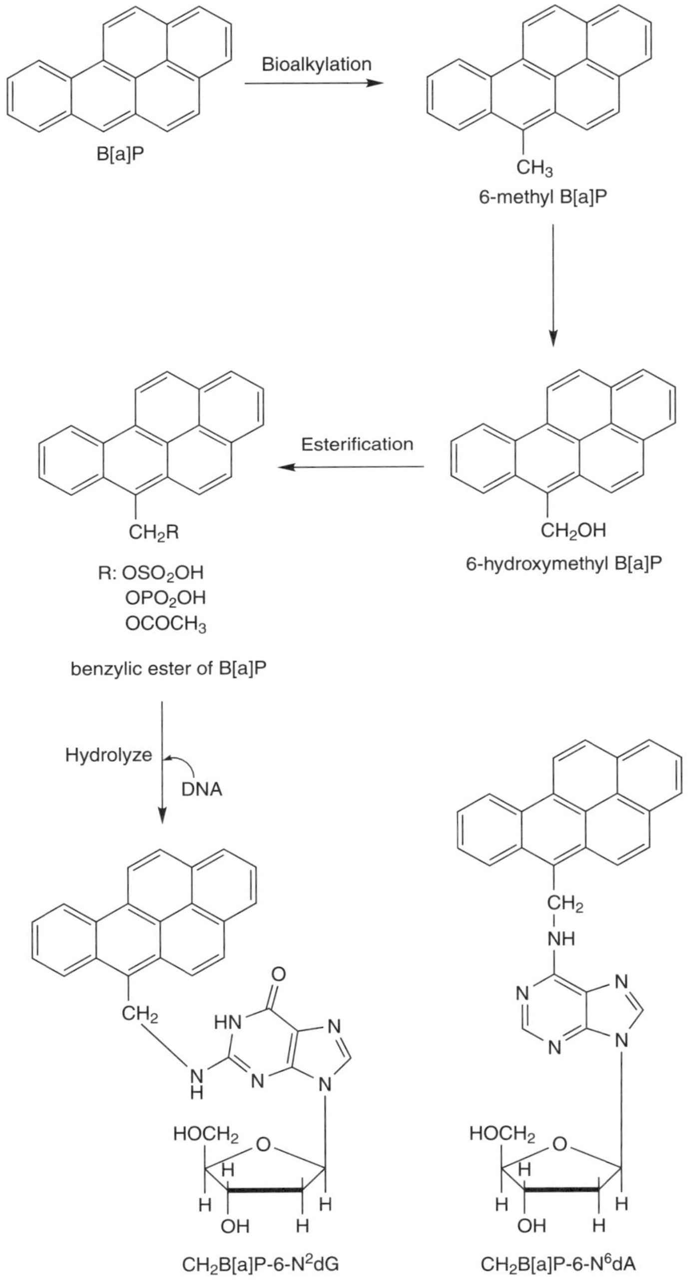

Figure 5.3 The benzylic ester pathway (adapted from Stansbury *et al*. 1994)

be studied *in vitro*. Other methods have facilitated *in vivo* study of xenobiotic-induced mutations in a rodent locus or in specific genetic markers of bacteria and transgenic animals (De Vries *et al*. 1997). The demonstration of the metabolic activation of PAH to DNA adducts has led to several studies intended to demonstrate a causal relationship between DNA adduct levels and the potent mutagenicity of this class of pollutants.

5.3.1.1   Mutagenicity of stable DNA adducts of B[a]P

Mutations induced in *Escherichia coli* by (+)-anti-BPDE have been determined *in vivo* by bacterial transformation following *in vitro* adduction of the supF plasmid gene (Rodriguez and Loechler 1993). Among the resulting mutations, 45% were by base substitution, 24% by base inversion, 23% by base insertion, and 8% by base deletion. The frequency of GC → TA (57%), CC → AT (23%) and GC → CC (20%), the most frequently observed mutations by base substitution, is directly dependent on the nucleotidic sequence, which appears to determine adduct conformation in DNA (Jelinsky *et al*. 1995).

The mechanisms involved in the appearance of these mutations have still not been clearly identified. The formation of DNA adducts, whose repair can be partially or totally inefficient, may lead to a mutation during DNA replication that is then established during DNA synthesis (Yang *et al*. 1982). *In vitro* studies have also shown that the (+)-anti-BPDE adduct could induce inhibition of the extension of DNA synthesis as well as a misreading of the DNA template by DNA polymerase (Hruszkewycz *et al*. 1992).

Several DNA repair mechanisms have been identified: the base excision repair system (BER), the nucleotide excision repair system (NER) and the DNA mismatch repair system. The exonucleasic $3' \rightarrow 5'$ activity of DNA polymerase III, which excises mismatched nucleotides, can also be added to the previous ones. NER is considered to be most important for the repair of stable BPDE DNA adducts, for which differences have been observed in the repair of *trans* and *cis* adducts (Celotti *et al*. 1993), with the *trans* adducts showing better reparability. These differences in repair can be attributed to differences in the accessibility of DNA repair enzymes, which is directly dependent on the three-dimensional structure of the adduct in the DNA molecule (Jernström and Gräslund 1994). It is noteworthy that another study showed preferential DNA repair in the coding regions of the phosphoribosyltransferase hypoxanthine (HPRT) human gene (Chen *et al*. 1991). Although this tends to reduce the role of stable DNA adducts in B[a]P mutagenicity and carcinogenesis, it has nonetheless been clearly demonstrated in rodents and humans that a deficiency in these DNA repair systems is responsible for an increase in mutational frequency and cancer risk (De Vries *et al*. 1997).

5.3.1.2   Mutagenicity of depurinating DNA adducts of B[a]P

The formation of depurinating DNA adducts is also a source of mutations. These mutations are caused by the infidelity of DNA replication at abasic sites as a result of adduct depurination. Abasic sites, when unrepaired, block the progression of DNA and RNA polymerases, resulting in a disruption of DNA replication and gene transcription (Cuniasse *et al*. 1990). In certain cases, the presence of an unrepaired apurinic site can allow damaged DNA to be read by DNA polymerases. Nevertheless, the presence of such a lesion is not controlled by DNA polymerases, which insert a base not always complementary for that lost by depurination (Gentil *et al*. 1992). Thus, abasic sites produced by adduct depurination can cause genetic mutations by substitutions of a single base. In a bacterium, adenine is generally incorporated in front of the abasic site (Shaaper *et al*. 1983). In mammalian cells, the different DNA bases are inserted according to the following order: A (33%) > C (28%) > T (26%) > G (12%) (Gentil *et al*. 1992).

Depurinating DNA adducts are repaired predominantly by BER (Frosina *et al*. 1996). During a first step, the modified base is excised by specific DNA glycosylases responsible for cutting the glycosidic bond. During a second step, the sugar–phosphate bond is cut in $3'$ by AP-endonucleases, leading to a $5'$-phosphate deoxyribose. The resulting break is repaired either by I and β DNA polymerases (when repair is limited to a single nucleotide) or by δ and/or ε DNA polymerases (when repair involves 6–13 nucleotides including the abasic site).

5.3.2 MUTATIONS INDUCED BY PAH DNA ADDUCTS IN 'CANCER GENES'

The mutagenicity of PAHs is due to their ability to form stable and depurinating DNA adducts. The predominant causes of mutations by stable and depurinating B[a]P DNA adducts are respectively GC → TA base substitutions and G → T punctual mutations. Many studies have checked the assumption that the carcinogenicity of B[a]P is related to a capacity to induce mutations in 'cancer genes' (Bartsch 1996). The mutational specificity of B[a]P DNA adducts in the p53 gene and *ras* family genes has been hence compared to the mutational profile found in these genes in individuals with cancer. These studies have been carried out in laboratory animals with cancer induced by B[a]P exposure and in persons whose exposure to PAHs may have been an etiological factor for their cancer.

5.3.2.1 p53 Tumour suppresser gene

Several studies have found a relationship between DNA adduct formation and p53 gene activity in mammals. In 1994, it was shown for the first time *in vivo* that an increase in the synthesis of p53 protein is directly dependent on the level of BPDE DNA adducts in mouse skin (Bjelogrlic *et al*. 1994). The role of B[a]P in cellular proliferation, a key cancer event, was then demonstrated in hamsters exposed to B[a]P-Fe$_2$O$_3$ particles (Wolterbeek *et al*. 1995). In this latter study, a correlation between the level of BPDE DNA adducts, cell proliferation, and expression of p53 protein was found in the tracheal cells of exposed animals. This suggested that exposure to B[a]P induces cell proliferation by producing mutations in p53 gene.

The ability of B[a]P-diol and B[a]P-diol-epoxide to induce mutations in the p53 gene has been demonstrated *in vitro* in cells of patients with xeroderma pigmentosum (a disease due to a deficiency in DNA repair mechanisms) (Quan and States 1996). Predominant induction of G → T mutations by B[a]P exposure occurs in the p53 gene, producing mutational hotspots depending on the DNA sequence, chromatin structure, and the dose applied. As this type of mutation is characteristic of those found in human lung cancer (Husgafvel-Pursiainen *et al.* 1995), the results obtained are of considerable importance to an understanding of chemical carcinogenesis.

Puisieux *et al.* (1991) have shown that 80% of exons 5, 7 and 8 of gene p53 (representing mutational hotspots in lung and liver cancers) are BPDE targets, as mutations by GC → TA base pair substitutions were associated with BPDE exposure. These p53 gene mutations represent, 46% and 92%, respectively, of all those observed in lung and liver cancers.

5.3.2.2 *ras* Family oncogenes

Mutations induced by B[a]P exposure in mice lung have been studied *in vivo* in the K-*ras* oncogene (Mass *et al.* 1993). These studies identified B[a]P as a causative agent of diverse mutations occurring predominantly in codon 12, where more than 80% involved G → T single base substitution. This kind of mutation in the K-*ras* oncogene is also well represented in lung adenocarcinoma in humans, accounting for more than 57% of total mutations of codon 12 (Husgafvel-Pursiainen *et al.* 1993). Similar observations have been made for codons 12, 13 and 61 of H-*ras* oncogene in patients with larynx cancer (Stern *et al.* 1993). As in the case of p53, the B[a]P spectrum in the mutational hotspots of K- and H-*ras* oncogenes shows many similarities with the spectra found in these genes in human epidemiological studies.

More recent studies have considered the role of depurinating DNA adducts in PAH-induced mutations in H-*ras* oncogene of mice skin papillomas (Chakravarti *et al.* 1995). Investigations of the relationship between stable and depurinating DNA adduct formation and mutations induced in H-*ras* oncogene showed a positive correlation for depurinating DNA adducts only. Moreover, a similar study *in vivo* found a correlation between tumor initiation and the level of depurinating DNA adducts (Chen *et al.* 1996). In the case of B[a]P, the depurinating DNA adducts produced, B[a]P-C8-Gua and B[a]P-N7-Gua (44% of total DNA adducts), B[a]P-N7-Ade (30%), were associated respectively with G → T and A → T mutations, representing more than 75% of those found in this type of tumor in mice.

5.4   CONCLUSION AND RELEVANCE FOR ENVIRONMENTAL PROTECTION

The activation of PAH to DNA adducts involves three different pathways in vertebrates: the diol-epoxide, the radical cation, and the benzylic ester pathway. However, only the first two have been convincingly documented. PAH–DNA adducts have been most widely measured by the ^{32}P postlabeling technique, which is limited to analysis of stable DNA adducts only. As depurinating DNA adducts represent 80% of the total adducts produced *in vitro* and *in vivo* in rodents, studies are generally concerned with only 20% of the total DNA adducts resulting from PAH exposure. Accordingly, if the biological significance of this type of DNA damage is to be determined correctly, it is essential to improve the identification of adducts by diversifying the techniques used for their measurement.

Insofar as they constitute a crucial event in cancer development, DNA adducts appear to be potential biomarkers of the genotoxicity of this class of pollutants. As the metabolic activation of PAH into DNA adducts is rapid, they can also be considered as early biomarkers of exposure to these carcinogenic compounds. Depurinating DNA adducts seem to play a significant role in PAH carcinogenesis and also represent the vast majority of DNA adducts produced in vertebrates. Although additional data on depurinating DNA adducts would be of considerable value for the validation of DNA adducts as molecular dosimeters of PAH exposure and cancer risk, their measurement is, however, still limited by difficulties in standard synthesis.

Finally it is noted that the formation of PAH–DNA adducts has also been studied in invertebrates. Nevertheless, the studies were limited to the measurement of stable DNA adducts only, for which absence of structural identification did not bring any information on the nature of the activation pathways involved.

The studies referred to in this chapter indicate the usefulness of measuring PAH–DNA adducts with regard to human cancer epidemiology and environmental biomonitoring. PAH–DNA adducts have been widely measured in fish and bivalves for aquatic biomonitoring (see Table 10.1, later in this volume). In several flatfish species, a correlation has been demonstrated between the level of DNA adducts in the liver and PAH concentration in the sediment (Van der Oost *et al*. 1994; French *et al*. 1996). Moreover, the time- and dose-dependent increase of PAH–DNA adducts in the liver led to the proposal of DNA adducts as biomarkers of PAH genotoxicity in fish (Pfau 1997). For these marine organisms, the identification of a BPDE-10-N2dG–DNA adduct validated the existence of a diol-epoxide pathway similar to that of higher vertebrates (Varanasi *et al*. 1989; Padròs and Pelletier 2001). Although no PAH activation pathway has been elucidated in bivalves, results of the studies carried out in the sentinel species *Mytilus* led to the same conclusions (Venier *et al*. 1996; Akcha *et al*. 2000a, 2000b). Despite the broad application of PAH–DNA adducts in biomonitoring programs, the interpretation of adduct data is however still limited for marine chemical risk assessment, due to the absence of adduct identification and lack of ecological context.

REFERENCES

Adams JD, Yagi H, Levin W and Jerina DM (1995) Stereo-selectivity and regio-selectivity in the metabolism of 7,8-dihydrobenzo[a]pyrene by cytochrome P450, epoxide hydrolase and hepatic microsomes from 3-methylcholanthrene-treated rats. *Chemical–Biological Interactions*, **95**, 57–77.
Akcha F, Izuel C, Venier P, Budzinski H, Burgeot T and Narbonne J-F (2000a) Enzymatic biomarker measurement and study of DNA adduct formation in B[a]P-contaminated mussel, *Mytilus galloprovincialis*. *Aquatic Toxicology*, **49**, 269–287.
Akcha F, Ruiz S, Zamperon C, Venier P, Burgeot T, Cadet J and Narbonne J-F (2000b) Benzo-[a]-pyrene-induced DNA damage in *Mytilus galloprovincialis*. Measurement of bulky DNA adducts and DNA oxidative damage in terms of 8-oxo-7,8-dihydro-2′-deoxyguanosine. *Biomarkers*, **5**, 355–367.
Bartsch H (1996) DNA adducts in human carcinogenesis: etiological relevance and structure–activity relationship. *Mutation Research*, **340**, 67–79.
Bjelogrlic NM, Mäkinen M, Stenbäck F and Vähäkangas K (1994) Benzo[a]pyrene-7,8-diol-9,10-epoxide-DNA adducts and increased p53 protein in mouse skin. *Carcinogenesis*, **15**, 771–774.
Cavalieri EL and Rogan EG (1992) The approach to understanding aromatic hydrocarbon carcinogenesis. The central role of radical cations in metabolic activation. *Pharmacological Therapy*, **55**, 183–199.
Cavalieri EL and Rogan EG (1995) Central role of radical cations in metabolic activation of polycyclic aromatic hydrocarbons. *Xenobiotica*, **25**, 677–688.
Cavalieri EL, Rogan EG, Cremonesi P and Devanesan PD (1988) Radical cations as precursor in the metabolic formation of quinones from benzo[a]pyrene and 6-fluorobenzo[a]pyrene. Fluorosubstitution as a probe for one-electron oxidation in aromatic substrates. *Biochemical Pharmacology*, **37**, 2173–2182.
Cavalieri EL, Rogan EG, Devanesan PD, Cremonesi P, Cerny RL, Gross ML and Bodell WJ (1990) Binding of benzo[a]pyrene to DNA by cytochrome P-450-catalyzed one-electron oxidation in rat liver microsomes and nuclei. *Biochemistry*, **29**, 4820–4827.
Celotti L, Ferraro P, Furlan D, Zanesi N and Pavanello S (1993) DNA repair in human lymphocytes treated *in vitro* with (±)-*anti* and (±)-*syn*-benzo[a]pyrene diol-epoxide. *Mutation Research*, **294**, 117–126.
Chakravarti D, Pelling JC, Cavalieri EL and Rogan EG (1995) Relating aromatic hydrocarbon-induced DNA adducts and c-Harvey-*ras* mutations in mouse skin papillomas: the role of apurinic sites. *Proceedings of the National Academy of Science of the United States of America*, **92**, 10422–10426.
Chen L, Devanesan PD, Higginbotham S, Ariese F, Jankowiak R, Small GJ, Rogan EG and Cavalieri EL (1996) Expanded analysis of benzo[a]pyrene-DNA adducts formed *in vitro* and in mouse skin: their significance in tumor initiation. *Chemical Research in Toxicology*, **9**, 897–903.
Chen RH, Maher VM and McCormick JJ (1991) Lack of a cell cycle-dependent stand bias for mutations induced in the HPRT gene by (±)-7β, 8α-dihydroxy-9α, 10α-epoxy-7,8,9,10-tetrahydrobenzo[a]pyrene in excision repair deficient cells. *Cancer Research*, **51**, 2587–2592.
Cuniasse P, Fazakerley GV, Guschlbauer V, Kaplan BE and Sowers LC (1990) The abasic site as a challenge to DNA polymerase. A nuclear magnetic resonance study of G, C and T opposite a model abasic site. *Journal of Molecular Biology*, **213**, 303–314.

De Vries A, Dollé MET, Broekhof JLM, Muller JJA, Dinant Kroese E, Van Kreijl CF, Capel PJA, Vijg J and Van Steeg H (1997) Induction of DNA adducts and mutations in spleen, liver and lung of *XPA*-deficient/*lacZ* transgenic mice after oral treatment with benzo[a]pyrene: correlation with tumor development. *Carcinogenesis*, **18**, 2327–2332.

Devanesan PD, Ramakrishna NVS, Todorovic R, Rogan EG, Cavalieri EL, Jeong H, Jankowiak R and Small GJ (1992) Identification and quantitation of benzo[a]pyrene-DNA adducts formed by rat liver microsomes *in vitro*. *Chemical Research in Toxicology*, **5**, 302–309.

Devanesan PD, Higginbotham S, Ariese F, Jankowiak R, Suh M., Small GJ, Cavalieri EL and Rogan EG (1996) Depurinating and stable Benzo[a]pyrene–DNA adducts formed in isolated rat liver nuclei. *Chemical Research in Toxicology*, **9**, 1113–1116.

Flesher JW and Sydnor KL (1973) Possible role of 6-hydroxymethylbenzo[a]pyrene as a proximate carcinogen of benzo[a]pyrene and 6-methylbenzo[a]pyrene. *International Journal of Cancer*, **11**, 433–437.

Flesher JW, Stansbury KH, Kadry AM and Myers SR (1983) Bio-alkylation of benzo[a]pyrene in rat lung and liver. In Rydstrom J, Montelius J, Bengtsson M (eds), *Extrahepatic Drug Metabolism and Chemical Carcinogenesis*, Elsevier Science, Amsterdam, pp. 237–238.

Flesher JW, Myers SR and Stansbury KH (1990) The site of substitution of the methyl group in the bioalkylation of benzo[a]pyrene. *Carcinogenesis*, **11**, 493–496.

French BL, Reichert WL, Hom T, Nishimoto M, Sanborn HR and Stein JE (1996) Accumulation and dose-response of hepatic DNA adducts in English sole (*Pleuronectes vetulus*) exposed to a gradient of contaminated sediments. *Aquatic Toxicology*, **36**, 1–16.

Frosina G, Fortini P, Rossi O, Carrozzino F, Raspaglio G, Cox LS, Lane DP, Abbondandolo A and Dogliotti E (1996) Two pathways for base excision repair in mammalian cells. *Journal of Biological Chemistry*, **271**, 9573–9578.

Gentil A, Cabral-Neto JB, Mariage-Samson R, Margot A, Imbach JL, Rayner B and Sarasin A (1992) Mutagenicity of a unique apurinic/apyrimidinic site in mammalian cells. *Journal of Molecular Biology*, **227**, 981–984.

Grover PL (1986) Pathways involved in the metabolism and activation of polycyclic hydrocarbons. *Xenobiotica*, **16**, 915–931.

Gupta RC (1996) ^{32}P-postlabeling for detection of DNA adducts. In Pfeifer GP (ed.), *Technologies for Detection of DNA Damage and Mutations*. Plenum, New York, pp. 45–61.

Hall M, Forrester LM, Parker DK, Grover PL and Wolf CR (1989) Relative contribution of various forms of cytochrome P450 to the metabolism of benzo[a]pyrene by human liver microsomes. *Carcinogenesis*, **10**, 1815–1821.

Hall M and Grover PL (1990) Polycyclic aromatic hydrocarbons, metabolism, activation and tumor initiation. In Cooper CS, Grover PL (eds), *Chemical Carcinogenesis and Mutagenesis. Handbook of Experimental Pharmacology*. Springer-Verlag, London, pp. 327–372.

Hruszkewycz AM, Canella KA, Peltonen K, Kotrappa L and Dipple A (1992) DNA polymerase action on benzo[a]pyrene-DNA adducts. *Carcinogenesis*, **13**, 2347–2352.

Huberman JG, Sachs L, Yang SK and Gelboin HV (1976) Identification of mutagenic metabolites of benzo[a]pyrene in mammalian cells. *Proceedings of the National Academy of Science of the United States of America*, **73**, 607–611.

Husgafvel-Pursiainen K, Hackman P, Radanpää M, Anttila S, Karjalainen A, Partanen T, Taikinia-Aho O, Heikkilä L and Vainio H (1993) K-*ras* mutations in human adenocarcinoma of the lung: association with smoking and occupational exposure to asbestos. *International Journal of Cancer*, **53**, 250–256.

Husgafvel-Pursiainen K, Ridanpaa M, Anttila S and Vainio H (1995) p53 and *ras* gene mutations in lung cancer: implications for smoking and occupational exposures. *Journal of Occupational and Environmental Medicine*, **37**, 69–76.

Jelinsky SA, Liu T, Geacintov NE and Loechler EL (1995) The major N^2-Gua adduct of the (+)-*anti* diol epoxide is capable of inducing G → A and G → C, in addition to G → T, mutations. *Biochemistry*, **34**, 13545–13553.

Jernström B and Gräslund A (1994) Covalent binding of benzo[a]pyrene 7,8-dihydrodiol 9,10 epoxides to DNA: molecular structures, induced mutations and biological consequences. *Biophysical Chemistry*, **49**, 185–199.

Mass MJ, Jeffers AJ, Ross JA, Nelson G, Galati AJ, Stoner GD and Nesnow S (1993) Ki-*ras* oncogene mutations in tumors and DNA adducts formed by benz[j]acetanthrylene and benzo[a]pyrene in the lungs of strain A/J mice. *Molecular Carcinogenesis*, **8**, 186–192.

Miller EC and Miller JA (1981) Searches for ultimate chemical carcinogens and their reactions with cellular macromolecules. *Cancer*, **47**, 2327–2345.

Okuda H, Nojima H, Miwa K, Watanabe N and Watabe T (1989) Selective covalent binding of the active sulfate ester of the carcinogen 5-(hydroxymethyl)chrysene to the adenine residue of calf thymus DNA. *Chemical Research in Toxicology*, **2**, 15–22.

Padròs J and Pelletier E (2001) Subpicogram determination of (+)-anti-benzo[a]pyrene diol-epoxide adducts in fish albumin and globin by high-performance liquid chromatography with fluorescence detection. *Analytica Chimica Acta*, **426**, 71–77.

Pfau W (1997) DNA adducts in marine and freshwater fish as biomarkers of environmental contamination. *Biomarkers*, **2**, 145–151.

Puisieux A, Lim S, Groopman J and Ozturk M (1991) Selective targeting of p53 gene mutational hotspots in human cancers by etiologically defined carcinogens. *Cancer Research*, **51**, 6185–6189.

Quan T and States JC (1996) Preferential DNA damage in the p53 gene by benzo[a]pyrene metabolites in cytochrome P4501A1-expressing xeroderma pigmentosum group A cells. *Molecular Carcinogenesis*, **16**, 32–43.

Ramakrishna NVS, Gao F, Padmavathi NS, Cavalieri EL, Rogan EG, Cerny RL and Gross ML (1992) Model adducts of benzo[a]pyrene and nucleosides formed from its radical cation and diol epoxide. *Chemical Research in Toxicology*, **5**, 293–302.

Rodriguez R and Loechler EL (1993) Mutational specificity of (+)-anti-diol epoxide of benzo[a]pyrene in a *supF* gene of an *Escherichia coli* plasmid: DNA sequence context influences hotspots, mutagenic specificity and mutagenesis. *Carcinogenesis*, **14**, 373–383.

Rogan EG, Roth RW, Katomski-Beck PA, Laubscher JR and Cavalieri EL (1980) Non-enzymatic ATP-mediated binding of hydroxymethyl derivatives of aromatic hydrocarbons to DNA. *Chemical-Biological Interactions*, **31**, 51–63.

Rogan E, Cavalieri E, Tibbels S, Cremonesi P, Warner C, Nagel D, Tomer K, Cerny R and Gross M (1988) Synthesis and identification of benzo[a]pyrene–guanine nucleoside adducts formed by electrochemical oxidation and by horseradish peroxidase-catalyzed reaction of benzo[a]pyrene with DNA. *Journal of the American Chemical Society*, **110**, 4023–4029.

Sakai M, Yoshida D and Mizusaki S (1985) Mutagenicity of polycyclic aromatic hydrocarbons and quinones on *Salmonella typhimurium* TA97. *Mutation Research*, **156**, 61–67.

Shaaper RM, Kunkel TA and Loeb LA (1983) Infidelity of DNA synthesis associated with bypass of apurinic sites. *Proceedings of the National Academy of Science of the United States of America*, **80**, 487–491.

Sims P, Grover PL, Swaisland A, Pal K and Hewer A (1974) Metabolic activation of benzo[a]pyrene proceeds by a diol epoxide. *Nature*, **252**, 326–328.

Stansbury KH, Flesher JW and Gupta RC (1994) Mechanism of aralkyl–DNA adduct formation from benzo[a]pyrene *in vivo*. *Chemical Research in Toxicology*, **7**, 254–259.

Stern SJ, Degawa M, Martin MV, Guengericb FP, Kaderlik RK, Ilett KF, Breau R, McGhee M, Montague D, Lyn-Cook B *et al*. (1993) Metabolic activation, DNA adducts, and H-*ras* mutations in human neoplastic and non-neoplastic laryngeal tissue. *Journal of Cellular Biochemistry*, **17F**, 129–137.

Suhr Y-J, Liem A, Miller EC and Miller JA (1989) Metabolic activation of the carcinogen 6-hydroxymethylbenzo[a]pyrene: formation of an electrophilic sulfuric acid ester and benzylic DNA adducts in rat liver *in vivo* and in reactions *in vitro*. *Carcinogenesis*, **10**, 1519–1528.

Todorovic R, Ariese F, Devanesan P, Jankowiak R, Small GJ, Rogan E and Cavalieri E (1997) Determination of benzo[a]pyrene- and 7,12-dimethylbenz[a]anthracene–DNA adducts formed in rat mammary glands. *Chemical Research in Toxicology*, **10**, 941–947.

Van der Oost R, Van Schooten F-J, Ariese F, Heida H, Satumalay K and Vermeulen NPE (1994) Bioaccumulation, biotransformation and DNA binding of PAHs in feral eel (*Anguilla anguilla*) exposed to polluted sediments: a field survey. *Environmental Toxicology and Chemistry*, **13**, 859–870.

Varanasi U, Reichert WL, Le Eberhart B-T and Stein JE (1989) Formation and persistence of benzo[a]pyrene-diolepoxide–DNA adducts in liver of English sole (*Parophrys vetulus*). *Chemical-Biological Interactions*, **69**, 203–216.

Venier P, Canova S and Levis AG (1996). DNA adducts in *Mytilus galloprovincialis* and *Zosterisessor ophiocephalus* collected from PAC-polluted and references sites of the Venice Lagoon. *Polymers and Aromatic Compounds*, **11**, 67–73.

Wolterbeek APM, Roggeband R, Baan RA, Feron VJ and Rutten AAJJL (1995) Relation between benzo[a]pyrene–DNA adducts, cell proliferation and p53 expression in tracheal epithelium of hamsters fed a high β-carotene diet. *Carcinogenesis*, **16**, 1617–1622.

Yang LL, Maher VM and McCormick JJ (1982) Relationship between excision repair and the cytotoxic and mutagenic effect of the 'anti' 7,8-diol-9,10-epoxide of benzo[a]pyrene in human cells. *Mutation Research*, **94**, 435–447.

6

Biodegradation and General Aspects of Bioavailability

FRANK VOLKERING[1] AND ANTON M. BREURE[2]

[1]Tauw bv, Research & Development Department, Deventer, The Netherlands
[2]National Institute for Public Health and the Environment, Bilthoven, The Netherlands

6.1 INTRODUCTION

Biodegradation of polycyclic aromatic hydrocarbons is the process in which PAHs are transformed as a result of biological activity. The most important mechanism for PAH biodegradation is metabolic dissimilation, which can cause the complete degradation to biomass, CO_2 and H_2O. This is also referred to as mineralization. Mineralization is the result of the process by which chemo-organotrophic microorganisms obtain energy from the oxidation of reduced organic compounds such as PAHs. Via complex biochemical pathways, hydrogen and electrons are transferred from the reduced substrate onto an oxidized compound named the terminal electron acceptor. Other degradation mechanisms include partial transformation by aspecific oxidative (often fungal) enzymes, co-metabolic transformation, and detoxification.

For biodegradation to occur, it is essential that two factors are present: (a) sufficient microbial degrading capacity (microorganisms); and (b) bioavailable substrate (PAH). For mineralization, the presence of an appropriate electron acceptor is also required. In the first sections of this chapter, the role of these factors in the biodegradation of PAHs will be addressed. The final sections deal with the degradation of PAHs in soil and groundwater and with the impact of bioavailability on the biodegradation processes for the clean-up of PAH-polluted soil and groundwater.

6.2 MICROBIAL DEGRADING CAPACITY

As PAHs are naturally occurring compounds, it is not surprising that many different PAH-degrading microorganisms can be found in pristine environments (Sims

PAHs: An Ecotoxicological Perspective. Edited by Peter E.T. Douben.
© 2003 John Wiley & Sons Ltd

and Overcash 1983). Studies on the microbial ecology of PAH-contaminated soils have shown, however, that the numbers of PAH-degrading microorganisms, as well as the degrading capacity, are much higher in PAH-contaminated soils than in pristine soils (Carmichael and Pfaender 1997; Herbes and Schwall 1978). Thus, the indigenous microbial population changes towards PAH-tolerant and PAH-degrading microorganisms due to the presence of the contamination. Consequently, most PAH-degrading soil organisms isolated originate from PAH-contaminated sites.

Many different species of bacteria (both Gram-negative and Gram-positive), fungi, yeasts, and algae are known to degrade PAHs. Extensive overviews of organisms capable of degrading the different PAHs under aerobic conditions are given elsewhere (Cerniglia 1992; Cerniglia *et al.* 1992; Juhasz and Naidu 2000). Bacteria are generally assumed to be the most important group of soil microorganisms involved in the (natural) biodegradation of PAHs in soils and sediments (Kastner *et al.* 1994; McGillivray and Shiaris 1994). Additionally, fungi may play a significant role in PAH degradation in the top soil (Cerniglia *et al.* 1992). PAH degradation by yeast, cyanobacteria and algae is of little importance for the fate of PAHs in soil and will not be addressed here.

6.2.1   AEROBIC TRANSFORMATION

Aerobic microbial PAH transformations have been studied extensively. Many bacterial and fungal species are known to have the enzymatic capacity to oxidize PAHs, ranging in size from naphthalene (two aromatic rings) to coronene (seven aromatic rings). The aerobic transformation of PAHs always involves the incorporation of oxygen into the PAH molecule; in most cases molecular oxygen is used. There are, however, differences in the hydroxylation mechanisms of eukaryotic and prokaryotic organisms. The dominating initial steps in the bacterial and fungal PAH catabolism are shown in Figure 6.1.

6.2.1.1   Bacterial metabolism

The initial bacterial attack of the aromatic ring is usually performed by a dioxygenase, forming a *cis*-dihydrodiol. This is then converted by a hydrogenase into a dihydroxylated derivative (Figure 6.1). The next step in the degradation is ring fission, which proceeds analogously to the well-known ring fission of catechol. It can take place via the *ortho* pathway, in which ring fission occurs between the two hydroxylated carbon atoms, or via the *meta* pathway, which involves cleavage of the bond adjacent to the hydroxyl groups. After this ring fission a number of reactions can occur. For the lower molecular weight PAHs, the most common route involves the fission into a C3 compound and a hydroxy aromatic acid compound. The aromatic ring can thereafter either undergo direct fission or can be subjected to a decarboxylation,

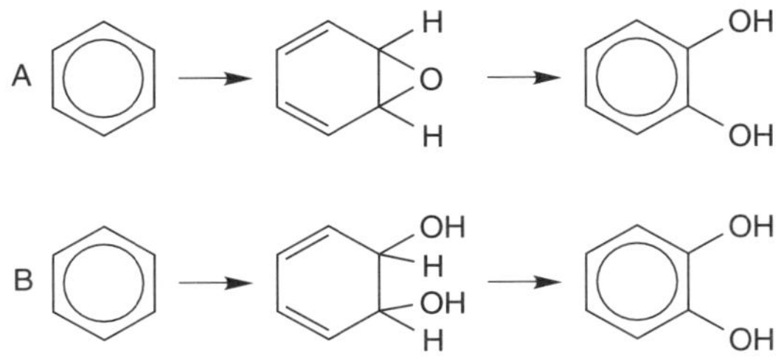

Figure 6.1   Initial steps in the microbial catabolism of homocyclic aromatic compounds. (A) fungi; (B) bacteria

leading to the formation of a dihydroxylated compound. This compound can then be dissimilated as described above. When degraded via these pathways, the low molecular weight PAHs can be completely mineralized to CO_2 and H_2O. An overview of the most common mineralization pathways of naphthalene, phenanthrene and anthracene, showing the similarities, is presented in Figure 6.2. The complete degradation of other low molecular weight PAHs, such as acenaphthene and acenaphthylene, has also been described (e.g. Komatsu *et al.* 1993). Partial degradation of low molecular weight PAHs may also occur, the products usually being hydroxy-aromatic acids or hydroxy-aromatics (Cerniglia 1992).

For the high molecular weight PAHs, the use as sole source of carbon and energy has been reported for fluoranthene, pyrene, chrysene, benzo[a]pyrene, dibenzo[ah]anthracene, and coronene (e.g. Boldrin *et al.* 1993; Caldini *et al.* 1995; Juhasz *et al.* 1997; Weissenfels *et al.* 1991). Partial degradation has been shown for many of these PAHs, and often hydroxylated polycyclic aromatic acids are formed as end products (e.g. Gibson *et al.* 1975, Kelley *et al.* 1993; Mahaffey *et al.* 1988; Roper and Pfaender 2001).

6.2.1.2   Bacterial co-metabolic transformation

Bacterial oxygenases are known to be rather non-specific enzymes. Organisms growing on one PAH (usually a low molecular weight compound) may therefore gratuitously oxidize other PAHs (e.g. Bouchez *et al.* 1995); e.g. growth on naphthalene was found to support degradation of fluoranthene (Beckles *et al.* 1998). An overview of co-metabolic transformation of benzo[a]pyrene is given elsewhere (Juhasz and Naidu 2000), the primary substrates being other PAH compounds such as naphthalene, phenanthrene, fluoranthene and pyrene, or metabolites in their degradation routes such as salicylate. Kanaly and Bartha (1999) report the co-metabolic degradation of benzo[a]pyrene in the presence of different types of mineral oil that act as primary substrate and as a means to increase the bioavailability of the five-ring PAHs.

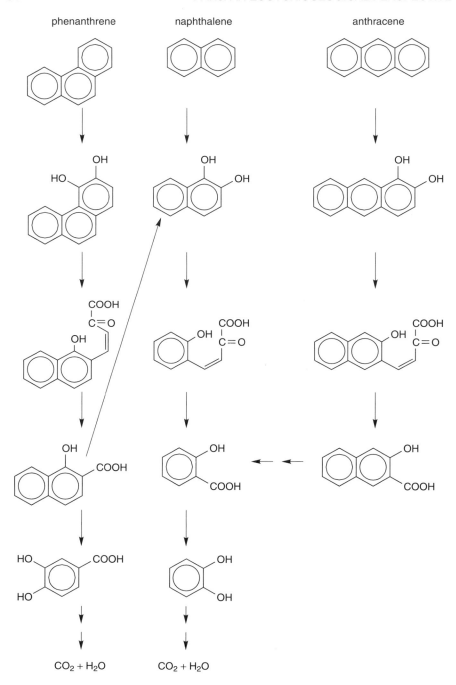

Figure 6.2   Aerobic bacterial mineralization of phenanthrene, naphthalene and anthracene

6.2.1.3 Fungal metabolism

White-rot fungi, such as *Phaenerochaete chrysosporium, Trametes versicolor* and several *Bjerkandera* species, produce highly aspecific extracellular lignin-degrading enzymes. These lignolytic enzymes have the capability of oxidizing a wide range of PAHs with three or more aromatic rings, including anthracene, phenanthrene, fluoranthene, pyrene, chrysene and benzo[a]pyrene (Cerniglia *et al*. 1992; Müncnerová and Augustin 1994). PAH metabolites of the white rot fungus attack are usually quinones, as is illustrated for anthracene in Figure 6.3. White-rot fungi are specialized in degrading lignins, which are mainly present in rotting wood. Consequently, their natural role in PAH degradation in soil may be limited (Cerniglia *et al*. 1992).

A large variety of soil fungi uses the cytochrome P450 enzyme system to oxidize PAHs as a detoxification mechanism. The direct oxidation product is an arene oxide, which can isomerize to phenols or undergo enzymatic hydration to yield *trans*-dihydrodiols (Figure 6.1). The phenols are further detoxified by conjugation with molecules such as sulfate, glucose or xylose. The *trans*-dihydrodiols are further detoxified by transformation to tetrahydrotetraols (Cerniglia *et al*. 1992; Müncnerová and Augustin 1994).

6.2.2 ANAEROBIC DEGRADATION

Under anaerobic conditions, chemo-organotrophic bacteria are able to obtain energy from the transport of electrons from a reduced organic substrate to an inorganic electron acceptor other than oxygen, such as nitrate, iron(III), manganese(IV), sulfate, or carbonate. Compared to the aerobic process, the anaerobic biodegradation of PAHs is less well documented. For many years unsubstituted aromatic hydrocarbons were considered to be persistent under anaerobic conditions, and studies in which anaerobic transformation of PAHs was observed were scarce. It was not until the last decade that conclusive evidence for the occurrence of anaerobic degradation of PAHs under anaerobic conditions has been obtained (see Table 6.1). Transformation of benzene and two- and three-ring PAHs by enriched cultures under methanogenic, sulfate-reducing, and iron-reducing conditions has been confirmed (Kazumi *et al*. 1997; Langenhoff *et al*. 1996; Nales *et al*. 1998; Zhang and Young 1997).

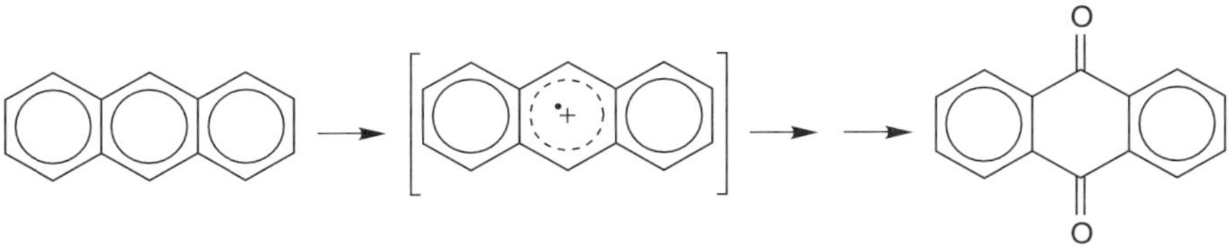

Figure 6.3 Oxidation of anthracene to anthraquinone by white rot fungi

TABLE 6.1 Potential degradability and occurrence of biodegradation of benzene and PAHs under different redox conditions

Redox condition	Benzene		Naphthalene		3- and 4-Ring PAH	
	Potential	Field	Potential	Field	Potential	Field
Aerobic	++	ooo	++	ooo	+/±	oo
Nitrate-reducing	+/?	o/?	+	o/?	+/?	?
Fe(III)-reducing	+	o	?	o	?	?
Sulfate-reducing	+	o	+	o	+/?	o
Methanogenic	+	o	±	o/?	?	?

Laboratory studies (potential):		Field studies (occurrence):	
++	Fast complete degradation	ooo	(Almost) Always degradation
+	Complete degradation	oo	Mostly
±	Partial degradation	o	Degradation demonstrated
−	No degradation	?	Insufficient or ambiguous data
?	Insufficient or ambiguous data		

The possibility of PAH-degradation under nitrate-reducing conditions has long been a very controversial subject. Although in some studies PAHs were found to be more or less readily degradable under denitrifying conditions, others obtained little or no transformation of PAHs by denitrifying organisms. Recently, however, evidence for benzene and PAH degradation coupled to nitrate-reduction is accumulating (McNally *et al.* 1998; Nales *et al.* 1998; Rockne and Strand 2001).

The relative persistence of PAHs under anaerobic conditions is caused by the high thermodynamic stability of the unsubstituted aromatic ring. As discussed above, the first step in the aerobic microbial metabolism of PAHs is the introduction of molecular oxygen into the ring, making it more susceptible to cleavage. Anaerobic organisms, however, cannot benefit from the strong oxidizing properties of molecular oxygen and have to rely on other mechanisms for destabilization of the aromatic ring. The biochemistry and physiology of the bacteria involved in anaerobic PAH degradation remain largely unknown as yet.

Some evidence exists that under conditions with a low oxygen concentration (< 2 mg/L), nitrate-reducing microorganisms are able to use the available oxygen solely for ring-cleavage (e.g. Wilson and Bouwer 1997). Since nitrate is the terminal electron acceptor, this so-called microaerophilic degradation can be seen as an anaerobic degradation process. Therefore, the presence of small amounts of oxygen may be essential for the anaerobic (microaerophilic) degradation of PAHs to occur.

6.3 PAH BIOAVAILABILITY: ITS IMPORTANCE FOR BIODEGRADATION

Generally, biodegradation rates of many readily degradable organic compounds (e.g. phenol under aerobic conditions) are much lower in soil than in liquid cultures. It is commonly accepted that this effect is mainly caused by the substrate not being present in the aqueous phase (Mihelcic *et al.* 1993; see also Chapter 3). This is especially true for hydrophobic organic compounds such as PAHs. Studies have revealed that most organic compounds can only be taken up by bacteria when they are present in an aqueous phase and therefore the mass transfer of the pollutant to the bulk of the aqueous phase is a prerequisite for biodegradation. When the mass transfer limits the biodegradation process, this is termed limited bioavailability. The term 'bioavailability' is also used with a somewhat different meaning by ecotoxicologists and, to avoid confusion, several researchers have tried to give a definition for limited bioavailability. The definition to be used in this chapter is:

A pollutant has a limited bioavailability when a physicochemical barrier between the pollutant and the organism limits its uptake rate by organisms (Volkering 1996).

For a better insight into the processes that play a role in the bioavailability of hydrophobic compounds in soil, it is essential to understand the interactions between the soil matrix, the pollutant and the microorganisms. These interactions are dependent on:

1.   The type and physicochemical characteristics of the pollutant.
2.   The type and physicochemical characteristics of the soil.
3.   The type and state of the microorganisms.
4.   Other factors, such as the temperature, groundwater current, etc.

The first two factors determine the form in which the contaminants occur in soil. The different physical forms possible for organic contaminants are illustrated in Figure 6.4. Compounds such as PAHs can be dissolved in porewater, adsorbed onto soil particles, absorbed into soil particles, or be present as a separate phase, which can be a liquid (e.g. creosote) or a solid phase (e.g. coal tar). As bioavailability is a concept based on mass transfer rates, transport of the

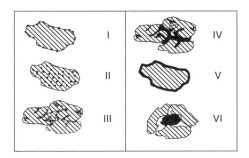

Figure 6.4 Different physical forms of organic pollutants with a limited bioavailability in soil I: adsolbed, II: absorbed, III: in porewater, IV: liquid or solid in pores, V: liquid film, VI: solid particles

pollutant to the aqueous bulk phase (dissolution or desorption) is the key process governing the bioavailability of the contaminant (e.g. Volkering *et al.* 1992; Mulder *et al.* 2001)

An important concept in the bioavailability of PAHs is the aging process. When the residence time of PAH in the soil increases the desorption rate of the pollutants decreases. This is caused by slow sorption into micropores and organic matter, and polymerization or covalent binding to the organic fraction of the soil. Additionally, degradation of the lower molecular weight compounds will result in a 'heavy' residual pollution with a lower bioavailability. One consequence of aging is a decreased biodegradability and a decreased toxicity of the compounds. Due to the sorption of PAH compounds, degrading organisms can get into contact with the pollutant to a lesser extent and therefore biodegradation is inhibited and the residual concentration increases (Burgos *et al.* 1999; White *et al.* 1999).

6.4   BIODEGRADATION OF PAHs IN SOIL AND SEDIMENTS

6.4.1   TRANSFORMATION PROCESSES

6.4.1.1   Aerobic conditions

As PAHs are naturally occurring compounds and many soil microorganisms have the capability of aerobic PAH transformation, it is not surprising that PAH degradation in soil is a ubiquitous process. In most studies on the degradation of PAHs in natural systems, the results are expressed as the removal efficiency, as determined by measurement of the residual PAH concentrations. However, since the formation of aromatic metabolites is known to be a common phenomenon in the aerobic bacterial PAH metabolism, the biodegradation of PAHs in soil can be expected to result in the accumulation of these metabolites. Table 6.2 provides an overview of frequently detected metabolites of selected PAHs.

Only a few studies have considered the fate of PAHs in natural systems. Herbes and Schwall (1978) found that after incubation of ^{14}C-labeled PAHs in an acclimated soil, 20–60% of the amount of PAHs that had disappeared could be detected as soluble or bound ^{14}C compounds. Although the toxicity of the metabolites is often lower than the toxicity of the parent compound, their bioavailability may be higher because of their higher aqueous solubilities as a consequence of their polar character. Measuring the change in toxicity of the soil or groundwater before and after treatment provides a useful way of evaluating the results of bioremediation of contaminated soil, including any information on the impact of metabolite formation. Wang *et al.* (1990) and Baudgrasset *et al.* (1993) have used this method for PAHs. In both cases it was found that the toxicity of the contaminated soil decreased along with PAH biodegradation. Haeseler *et al.* (2001) showed stimulation of PAH degradation in polluted soil from several former manufactured gas plants to result in a short,

TABLE 6.2 Frequently observed metabolites from selected PAHs

PAH	Intermediates	
	Bacterial (aerobic catabolism, co-metabolism)	Fungal
Naphthalene	Salicylic acid	Naphthalene-1,2-quinone; naphthalene-1,4-quinone; 1-naphthol; 2-naphthol
Phenanthrene	1-Hydroxy-2-naphthoic acid	3-, 4- and 9-Hydroxyphenanthrene; phenanthrene 9,10-quinone; phenanthrene *trans*-dihydrodiols
Anthracene	2-Hydroxy-3-naphthoic acid	Anthraquinone; 1-anthrol; anthracene *trans*-1,2-dihydrodiol 9,10-anthracenedione
Pyrene	1-Hydroxypyrene; 4-hydroxyperinaphthenone; pyrene dihydrodiols	Pyrene-1,6-quinone; pyrene-1,8-quinone; 1-hydroxypyrene
Benzo[a]pyrene (B[a]P)	B[a]P dihydrodiols	B[a]P-1,6-quinone; B[a]P-3,6-quinone; B[a]P-6,12-quinone; B[a]P dihydrodiols

temporary increase of toxicity of leachate from the soils, due to the formation of intermediates with higher mobility compared to the parent compound. At the end of the experiments the removal of PAHs was incomplete, but the toxicity of the leachate after remediation was negligible.

Thus, although the formation of metabolites is likely to occur, this is usually not reflected in toxicity. A possible explanation for this phenomenon is that the intermediates, which are more reactive than saturated PAHs, undergo polymerization reactions or chemical reactions with the soil organic matter. Recent investigations have shown that under natural conditions, intermediates of PAH degradation are irreversibly incorporated in the humic soil fraction (e.g. Burgos *et al*. 1999; Kastner *et al*. 1999; Nieman *et al*. 1999).

6.4.1.2 Anaerobic conditions

The possibility of anaerobic PAH degradation is no guarantee that it will play a significant role in the actual fate of PAHs in anaerobic sediments. Because marine sediments and heavily polluted saturated soil systems are often anaerobic, many data on the behavior of PAHs under anaerobic conditions are available. Since direct proof for the occurrence of anaerobic PAH degradation in sediments is difficult to obtain, usually indirect lines of evidence are presented, such as degradation in anaerobic microcosms or *in situ* mesocosms, calculation of degradation rates from concentration profiles along the contaminated plume, etc. Using one or more of these approaches, degradation of benzene and two- or three-ring PAHs was observed under iron-reducing, sulfate-reducing, and methanogenic conditions (Coates *et al.* 1997; Thierrin *et al.* 1993; Weiner and Lovley 1998b). At most sites, however, anaerobic degradation of these compounds was found not to occur. Although in a number of cases this may be explained by the detection methods used not being sufficiently sensitive, there are also many sites at which careful study showed no anaerobic degradation potential to be present in the soil (e.g. Weiner and Lovley 1998a).

High molecular weight PAHs are usually considered to be persistent under anaerobic conditions (e.g. Sharak-Genthner *et al.* 1997), although some indications for anaerobic transformation of four- and five-ring PAHs have been obtained. Concluding, it can be said that, contrary to aerobic degradation, the anaerobic degradation of PAH is not an ubiquitous soil process.

6.4.1.3 Biodegradation of PAH mixtures

Soil pollution with PAHs almost always consists of mixtures of different compounds. The effect of the presence of several different PAHs on degradation kinetics is only partly understood. Low molecular weight PAHs may stimulate degradation of high molecular weight PAHs by acting as a co-substrate. On the other hand, competitive inhibition has also been observed (Beckles *et al.* 1998; Bouchez *et al.* 1995, 1999). In a recent study, Bouchez *et al.* (1999) showed that an indigenous bacterial population was more effective in degrading the PAH mixture present in the soil than especially selected strains with high degrading capacity. This indicates that the role of competitive inhibition in polluted soils in the field, which usually contain adapted consortia, will be smaller than in laboratory studies with pure and mixed bacterial cultures.

6.4.2 BIOAVAILABILITY

Although the biodegradation pathways of the different PAHs are very similar, their biodegradation rates in soil differ considerably. Generally, first-order degradation kinetics are observed and the degradation rate is found to decrease with an increasing number of rings. An overview of degradation rates of PAHs in contaminated soils is given by Wilson and Jones (1993). As an example, laboratory studies with creosote-polluted soil showed that two-ring PAHs exhibited half-lives < 10 days, three-ring PAHs showed half-lives < 100 days, and four- and five-ring PAHs were found to have half-lives of > 100 days (McGinnis *et al.* 1988). In another study (Herbes and Schwall 1978), the same differences in degradation rates of the different PAHs were observed, although slightly higher degradation rates were found.

Two factors are thought to be responsible for these different degradation rates. First, the bacterial uptake rates of the compounds with higher molecular weights have been shown to be lower than the uptake rates of low molecular weight PAHs. The second and most important factor is the bioavailability of

PAHs. Because of their low aqueous solubilities, PAHs will occur in soil mainly in association with soil matter or as a separate phase. Several studies have shown a good correlation between the octanol–water partition coefficient, the aqueous solubility, and the soil–water partition coefficient of a pollutant (e.g. Dzombak and Luthy 1984; Karickhoff *et al.* 1979). The fact that high molecular weight PAHs have the highest octanol–water coefficients explains the difference in the observed half-lives of PAHs in soil. An illustrative example of this phenomenon is given in Figure 6.5, in which the first-order degradation constants for several different PAHs in estuarine sediment, as reported by Durant *et al.* (1995), are plotted as a function of their octanol–water coefficients. When PAHs are present as a separate phase, such as tar particles, their bioavailability is even lower than that of sorbed PAHs and degradation rates of separate phase PAHs are often negligible (Weissenfels *et al.* 1992).

Besides the low degradation rates, the low bioavailability is also an important cause of another problem in the bioremediation of PAH-contaminated soil: the high residual PAH concentrations after clean-up. In several studies, the limited bioavailability of the residual PAHs was found to prevent further biodegradation (Breedveld and Karlsen 2000; Erickson *et al.* 1993; Weissenfels *et al.* 1992). Bioavailability testing seems to be a good method for predicting residual PAH levels after bioremediation (e.g. Tang and Alexander 1999). Generally, the residual PAH concentrations in soils after biological clean-up are found to be too high to allow unrestricted use of the soils according to most governmental guidelines (Wilson and Jones 1993).

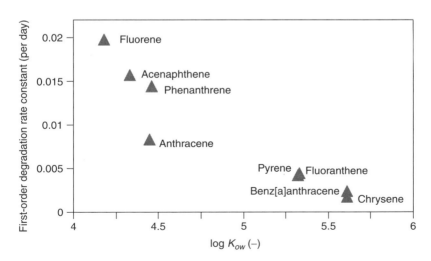

Figure 6.5 PAH degradation rates observed in soil microcosm studies (Durant *et al.* 1995) as a function of log K_{ow}

6.5 ROLE OF BIOAVAILABILITY IN BIOREMEDIATION
OF PAH-POLLUTED SOIL AND GROUNDWATER

Biological soil remediation is based on creating the proper conditions for (indigenous) microorganisms to degrade hazardous organic pollutants such as PAHs. It has several advantages over the other remediation techniques in terms of cost and soil functionality. Two different bioremediation approaches can be distinguished:

1. *Ex situ treatment*, in which the contaminated soil is excavated and transferred to a place where it can be treated. Techniques such as landfarming, composting in biopiles, and treatment in a bioreactor have been used successfully to treat PAH-polluted soil.
2. *In situ treatment* in which the polluted soil is left in place and remains essentially undisturbed. The goal of the treatment is to create the right conditions for biodegradation in the soil. It is a cost-effective method, but has the disadvantage that it is difficult to control the processes in the soil.

Landfarming is a low-intensity type of *ex situ* treatment that may involve nutrient addition (nitrogen, phosphorus), pH adjustment, and occasional tilling and wetting of the soil. Since these treatments are not applied to increase bioavailability, the residual concentration after remediation is mainly determined by the bioavailability of the PAHs. Intensive *ex situ* remediation techniques, such as treatment in bioreactors, may involve processes to increase PAH bioavailability, such as mixing, increasing temperature, the use of extractants, etc. In many cases, the soil will be cleaned in a shorter time due to increased mass transfer rates, resulting from the physicochemical treatments (e.g. Mulder *et al.* 1998; Volkering *et al.* 1998; Bonten 2001) but the ecological and physicochemical structure of the soil will be severely altered, facilitating only limited use of the cleaned soil.

The applicability of *in situ* bioremediation for PAH pollutions is limited due to the low bioavailability of the PAHs. Consequently, *in situ* treatment is restricted to the mobile PAHs present in groundwater or porewater and to PAHs that can desorb or dissolve from the solid phase in soil to the water phase within the period of the treatment. Aeration of groundwater via techniques such as air sparging has been shown to be a reasonably effective method for these pollutions. Some attempts have been made to stimulate PAH degradation by enhancing PAH bioavailability using intensive *in situ* techniques. Even with the most promising approaches, usually a combination of physicochemical and biological techniques, the result is likely to be a situation in which a substantial residual pollution will remain. However, this residual concentration will consist of PAH compounds with a low bioavailability. Under these conditions the fluxes towards target organisms will be reduced, resulting in a relatively low ecotoxicological risk (see e.g. Alexander 1995).

A special *in situ* approach in which risk-based considerations play an important role is plume management by monitored natural attenuation (MNA). MNA is defined as:

> The use of natural processes within the context of a carefully controlled and monitored site cleanup approach that will reduce contaminant concentrations to levels that are protective of human health and the environment within a reasonable time frame (USEPA 1997).

MNA is usually applied for mobile contaminants, such as monoaromatic hydrocarbons or chlorinated aliphatic hydrocarbons. However, MNA has also been used successfully for PAHs, either alone or as a post-remediation treatment. When the (residual) immobile contamination source causes no toxicological risks, natural biodegradation processes are sufficient to prevent the spreading of contaminants from the site, e.g. into the ground and surface waters, and the bioaccumulation in organisms.

The applicability of all of these bioremediation techniques would be greatly enhanced when risk-based strategy is followed for the remediation of a polluted site. Recently, the Dutch government formulated a so-called site-specific approach in which multifunctional use of the cleaned soil is no longer the main target, and the future function of the polluted site is considered prior to the remediation as well as the risks that the site imposes (VROM 1999). When, after a biotechnological treatment, a reduction in toxicity is shown and can be made operational, this would open new possibilities for the use of bioremediation techniques. In that case the success of a remediation strategy will be determined by the degree in which the toxicity of the polluted site and the risk of spreading into groundwater and surface waters are reduced, instead of the degree in which the total concentration of the pollutant is reduced.

REFERENCES

Alexander M (1995) How toxic are toxic chemicals in soil? *Environmental Science and Technology*, **29**, 2713–2717.

Baudgrasset F, Baudgrasset S and Safferman SI (1993) Evaluation of the bioremediation of a contaminated soil with phytotoxicity tests. *Chemosphere*, **26**, 1365–1374.

Beckles DM, Ward CH and Hughes JB (1998) Effect of mixtures of polycyclic aromatic hydrocarbons and sediments on fluoranthene biodegradation patterns. *Environmental Toxicology and Chemistry*, **17**, 1246–1251.

Boldrin B, Tiehm A and Fritzsche C (1993) Degradation of phenanthrene, fluorene, fluoranthene, and pyrene by a *Mycobacterium* sp. *Applied and Environmental Microbiology*, **59**, 1927–1930.

Bonten LTC (2001) Improving Bioremediation of PAH-contaminated Soils by Thermal Pretreatment. PhD Thesis, Wageningen University, The Netherlands.

Bouchez M, Blanchet D and Vandecasteele JP (1995) Degradation of polycyclic aromatic hydrocarbons by pure strains and by defined strain associations — inhibition phenomena and co-metabolism. *Applied Microbiology and Biotechnology*, **43**, 156–164.

Bouchez M, Blanchet D, Bardin V, Haeseler F and Vandecasteele JP (1999) Efficiency of defined strains and of soil consortia in the biodegradation of polycyclic aromatic hydrocarbon (PAH) mixtures. *Biodegradation*, **6**, 429–435.

Breedveld GD and Karlsen DA (2000) Estimating the availability of polycyclic aromatic hydrocarbons for bioremediation of creosote contaminated soils. *Applied Microbiology and Biotechnology*, **54**, 255–261.

Burgos WD, Berry DF, Bhandari A and Novak JT (1999) Impact of soil–chemical interactions on the bioavailability of naphthalene and 1-naphthol. *Water Research*, **33**(18), 3789–3795.

Caldini G, Cenci G, Manenti R and Morozzi G (1995) The ability of an environmental isolate of *Pseudomonas* fluorescences to utilize chrysene and other four-ring polynuclear aromatic hydrocarbons. *Applied Microbiology and Biotechnology*, **44**, 225–229.

Carmichael LM and Pfaender FK (1997) Polynuclear aromatic hydrocarbon metabolism in soils — relationship to soil characteristics and preexposure. *Environmental Toxicology and Chemistry*, **16**, 666–675.

Cerniglia CE (1992) Biodegradation of polycyclic aromatic hydrocarbons. *Biodegradation*, **3**, 351–368.

Cerniglia CE, Sutherland JB and Crow SA (1992) Fungal metabolism of aromatic hydrocarbons. In Winkelman G (ed.), *Microbial Degradation of Natural Products*. VCH, Weinheim, pp. 193–217.

Coates JD, Woodward J, Allen J, Philp P and Lovley DR (1997) Anaerobic degradation of polycyclic aromatic hydrocarbons and alkanes in petroleum-contaminated marine harbor sediments. *Applied and Environmental Microbiology*, **63**, 3589–3593.

Dzombak DA and Luthy RG (1984) Estimating adsorption of polycyclic aromatic hydrocarbons on soils. *Soil Science*, **137**, 292–308.

Durant ND, Wilson LP and Bouwer EJ (1995) Microcosm studies of subsurface PAH-degrading bacteria from a former manufactured gas plant. *Journal of Contaminant Hydrology*, **17**, 213–223.

Erickson DC, Loehr RC and Neuhauser EF (1993) PAH loss during bioremediation of manufactured gas plant site soils. *Water Research*, **27**, 911–919.

Gibson DT, Mahadevan V, Jerina DM, Yagi H and Yeh HJC (1975) Oxidation of the carcinogens benzo[a]pyrene and benzo[a]anthracene to dihydrodiols by a bacterium. *Science*, **189**, 295–297.

Haeseler F, Blanchet D, Werner P and Vandecasteele JP (2001) Ecotoxicological characterization of metabolites produced during PAH biodegradation in contaminated soils. In Magar VS, Johnson G, Ong SK, Leeson A (eds), *Bioremediation of Energetics, Phenolics, and Polycyclic Aromatic Hydrocarbons*. Battelle Press, Columbus, OH, USA.

Herbes SE and Schwall LR (1978) Microbial transformation of polycyclic aromatic hydrocarbons in pristine and petroleum contaminated sediments. *Applied and Environmental Microbiology*, **35**, 306–316.

Juhasz AL, Britz ML and Stanley GA (1997) Degradation of benzo[a]pyrene, dibenzo[ah]anthracene and coronene by *Burkholderia cepacia*. *Water Science and Technology*, **36**, 45–51.

Juhasz AL and Naidu R (2000) Bioremediation of high molecular weight polycyclic aromatic hydrocarbons: a review of the microbial degradation of benzo[a]pyrene. *International Biodeterioration and Biodegradation*, **45**, 57–88.

Kanaly RA and Bartha R (1999) Cometabolic mineralization of benzo[a]pyrene caused by hydrocarbon additions to soil. *Environmental Toxicology and Chemistry*, **18**, 2186–2190.

Karickhoff SW, Brown DS and Scott TA (1979) Sorption of hydrophobic pollutants on natural sediments. *Water Research*, **13**, 241–248.

Kastner M, Breuerjammali M and Mahro B (1994) Enumeration and characterization of the soil microflora from hydrocarbon-contaminated soil sites able to mineralize polycyclic aromatic hydrocarbons (PAH). *Applied Microbiology and Biotechnology*, **41**, 267–273.

Kastner M, Streibich S, Beyrer M, Richnow HH and Fritsche W (1999) Formation of bound residues during microbial degradation of [C-14]anthracene in soil. *Applied and Environmental Microbiology*, **65**, 1834–1842.

Kazumi J, Caldwell ME, Suflita JM, Lovley DR and Young LY (1997) Anaerobic degradation of benzene in diverse anoxic environments. *Environmental Science and Technology*, **31**, 813–818.

Kelley I, Freeman JP, Evans FE and Cerniglia CE (1993) Identification of metabolites from the degradation of fluoranthene by *Mycobacterium* sp. strain PYR-1. *Applied and Environmental Microbiology*, **59**, 800–806.

Komatsu T, Omori T and Kodama T (1993) Microbial degradation of the polycyclic aromatic hydrocarbons acenaphthene and acenaphthylene by a pure bacterial culture. *Bioscience, Biotechnology and Biochemistry*, **57**, 864–865.

Langenhoff AAM, Zehnder AJB and Schraa G (1996) Behaviour of toluene, benzene and naphthalene under anaerobic conditions in sediment columns. *Biodegradation*, **7**, 267–274.

Mahaffey WR, Gibson DT and Cerniglia CE (1988) Bacterial oxidation of chemical carcinogens: formation of polycyclic aromatic acids from benzo[a]anthracene. *Applied and Environmental Microbiology*, **54**, 2415–2423.

McGillivray AR and Shiaris MP (1994) Relative role of eukaryotic and prokaryotic microorganisms in phenanthrene transformation in coastal sediments. *Applied and Environmental Microbiology*, **60**, 1154–1159.

McGinnis GD, Borazjani H, McFarland LK, Pope DF and Strobel DA (1988) Characterization and laboratory testing soil treatability studies for creosote and pentachlorophenol sludges and contaminated soil. *USEPA Report No. 600/2–88/055*, R.S. Kerr Environmental Laboratory, Ada, OK, USA.

McNally DL, Mihelcic JR and Lueking DR (1998) Biodegradation of three- and four-ring polycyclic aromatic hydrocarbons under aerobic and denitrifying conditions. *Environmental Science and Technology*, **32**, 2633–2639.

Mihelcic JR, Lueking DR, Mitzell RJ and Stapleton JM (1993) Bioavailability of sorbed- and separate-phase chemicals. *Biodegradation*, **4**, 141–153.

Mulder H, Breure AM, Andel JG van, Grotenhuis JTC and Rulkens WH (1998) The influence of hydrodynamic conditions on naphthalene dissolution and subsequent biodegradation. *Biotechnology and Bioengineering*, **57**, 145–154.

Mulder H, Breure AM and Rulkens WH (2001) Application of a mechanistic desorption–biodegradation model to describe the behavior of polycyclic aromatic hydrocarbons in peat soil aggregates. *Chemosphere*, **42**, 285–299.

Müncnerová D and Augustin J (1994) Fungal metabolism and detoxification of polycyclic aromatic hydrocarbons — a review. *Bioresource Technology*, **48**, 97–106.

Nales M, Butler BJ and Edwards EA (1998) Anaerobic benzene biodegradation: a microcosm survey. *Bioremediation Journal*, **2**, 125–144.

Nieman JKC, Sims RC, Sims JL, Sorensen DL, McLean JE and Rice JA (1999) [C-14]pyrene-bound residue evaluation using MIBK fractionation method for creosote-contaminated soil. *Environmental Science and Technology*, **33**, 776–781.

Rockne KJ and Strand SE (2001) Anaerobic biodegradation of naphthalene, phenanthrene, and biphenyl by a denitrifying enrichment culture. *Water Research*, **35**, 291–299.

Roper JC and Pfaender FK (2001) Pyrene and chrysene fate in surface soil and microcosms. *Environmental Toxicology and Chemistry*, **20**, 223–230.

Sharak-Genthner BR, Townsend GT, Lantz SE and Mueller JG (1997) Persistence of poly-cyclic aromatic hydrocarbon components of creosote under anaerobic enrichment conditions. *Archives of Environmental Contamination and Toxicology*, **32**, 99-105.

Sims RM and Overcash MR (1983) Fate of polynuclear aromatic compounds (PNAs) in soil-plant systems. *Residue Reviews*, **88**, 1-68.

Tang JX and Alexander M. (1999) Mild extractability and bioavailability of polycyclic aromatic hydrocarbons in soil. *Environmental Toxicology and Chemistry*, **18**, 2711-2714.

Thierrin J, Davis GB, Barber C, Patterson BM, Pribac F, Power TR and Lambert M (1993) Natural degradation rates of BTEX compounds and naphthalene in a sulphate reducing groundwater environment. *Hydrological Sciences Journal*, **38**, 309-322.

USEPA, United States Environmental Protection Agency (1997) *Use of Monitored Natural Attenuation at Superfund, RCRA Corrective Action, and Underground Storage Tank Sites*. OSWER (Office of Solid Waste Emergency Response) Directive 9200.4-17.

Volkering F (1996) Bioavailability and Biodegradation of Polycyclic Aromatic Hydrocar-bons. PhD Thesis, Wageningen University, The Netherlands.

Volkering F, Breure AM and Rulkens WH (1998) Microbial aspects of the use of surfac-tants in biological soil remediation. *Biodegradation*, **8**, 401-417.

Volkering F, Breure AM, Sterkenburg A and Andel JG van (1992) Microbial degradation of polycyclic aromatic hydrocarbons. Effect of substrate availability on bacterial growth kinetics. *Applied Microbiology and Biotechnology*, **36**, 548-552.

VROM, The Netherlands Ministry of Housing, Spatial Planning, and the Environment (1999) Governmental view on the soil-use specific and cost-effective approach to soil contamination. Lower House, Dutch Parliament, 1999-2000, 25411, No. 7 (in Dutch).

Wang X, Yu X. and Bartha R (1990) Effect of bioremediation on polycyclic aromatic hydrocarbon residues in soil. *Environmental Science and Technology*, **24**, 1086-1089.

Weiner JM and Lovley DR (1998a) Anaerobic benzene degradation in petroleum-contami-nated aquifer sediments after inoculation with a benzene-oxidizing enrichment. *Applied and Environmental Microbiology*, **64**, 775-778.

Weiner JM and Lovley DR (1998b) Rapid benzene degradation in methanogenic sedi-ments from petroleum-contaminated aquifer. *Applied and Environmental Microbiol-ogy*, **64**, 1937-1939.

Weissenfels WD, Klewer HJ and Langhoff J (1992) Adsorption of polycyclic aromatic hydrocarbons (PAHs) by soil particles: influence on biodegradability and biotoxicity. *Applied Microbiology and Biotechnology*, **36**, 689-696.

Weissenfels WD, Beyer M, Klein J and Rehm H-J (1991) Microbial metabolism of fluoran-thene: isolation and identification of ring fission products. *Applied Microbiology and Biotechnology*, **34**, 528-535.

White JC, Alexander M and Pignatello JJ (1999) Enhancing the bioavailability of organic compounds sequestered in soil and aquifer solids. *Environmental Toxicology and Chemistry*, **18**, 182-187.

Wilson SC and Jones KC (1993) Bioremediation of soil contaminated with polynuclear aromatic hydrocarbons (PAHs): a review. *Environmental Pollution*, **81**, 229-249.

Wilson LP and Bouwer EJ (1997) Biodegradation of aromatic compounds under mixed oxygen/denitrifying conditions: a review. *Journal of Industrial Microbiology and Biotechnology*, **18**, 116-130.

Zhang XM and Young LY (1997) Carboxylation as an initial reaction in the anaerobic metabolism of naphthalene and phenanthrene by sulfidogenic consortia. *Applied and Environmental Microbiology*, **63**, 4759-4764.

PART III

Bioavailability, Exposure and Effects in Environmental Compartments

7

An Overview of the Partitioning and Bioavailability of PAHs in Sediments and Soils*

ROBERT M. BURGESS[1], MICHAEL J. AHRENS[2], CHRISTOPHER W. HICKEY[2], PIETER J. DEN BESTEN[3], DORIEN TEN HULSCHER[3], BERT VAN HATTUM[4], JAMES P. MEADOR[5] AND PETER E. T. DOUBEN[6]

[1]US EPA, ORD/NHEERL Atlantic Ecology Division, Narragansett, RI, USA
[2] National Institute of Water and Atmospheric Research (NIWA), Hamilton, New Zealand
[3]Institute for Inland Water Management and Waste Water Treatment (RIZA), Lelystad, The Netherlands
[4]Institute for Environmental Studies, Vrije Universiteit, Amsterdam, The Netherlands
[5]Northwest Fisheries Science Center, National Oceanic and Atmospheric Administration, Seattle, WA, USA
[6] Unilever R&D Colworth, Safety and Environmental Assurance Centre, Sharnbrook, UK

7.1 INTRODUCTION

Understanding and predicting any adverse effects of PAHs depends on generating a reliable measure or estimate of how much PAH is available for uptake. Simply knowing the total amount of PAH in soil, water or sediment is insufficient for determining whether or not these compounds are bioavailable. For example, in sediments it has been observed that bioaccumulation and toxicity to organisms correlate much more strongly with interstitial water contaminant

---

* Mention of trade names or commercial products does not constitute endorsement or recommendation for use. Although the research described in this article has been funded wholly by the US Environmental Protection Agency it has not been subjected to Agency-level review. Therefore, it does not necessarily reflect the views of the Agency. This is Contribution No. AED-02-084 of the US EPA Office of Research and Development National Health and Environmental Effects Research Laboratory's Atlantic Ecology Division. The views expressed in this chapter do not represent those of Unilever or its companies.

PAHs: An Ecotoxicological Perspective. Edited by Peter E.T. Douben.
© 2003 John Wiley & Sons Ltd

concentrations, specifically the freely dissolved concentrations, rather than with bulk sediment concentrations (Adams *et al.* 1985; DiToro *et al.* 1991). In addition to the partitioning of PAHs between water and sediment phases, organism tissues also seek equilibrium or near-equilibrium with PAH concentrations in the surrounding system. The PAHs most available to equilibrate are those that are freely dissolved, since these are capable of transferring from one phase to another and passing through biological membranes (Wang and Fisher 1999). Unlike the freely dissolved PAHs, PAHs associated with sediment organic carbon, soot (or black) carbon or any other phase are not readily bioavailable (McElroy *et al.* 1989; DiToro *et al.* 1991; Gustafsson *et al.* 1997; Accardi-Dey and Gschwend 2002). However, because partitioning is a dynamic process, PAHs associated with a particulate or colloidal phase can eventually desorb into the freely dissolved phase and consequently increase in bioavailability. Conditions which affect the desorption of PAHs from the various phases are dependent on the organism, the environment, and the PAH molecule. A final complicating factor to understanding PAH bioavailability is appreciating that organisms are exposed to PAHs from several sources: overlying and interstitial waters, ingested food and sediments.

This chapter provides an overview of PAH partitioning, including the different phases involved, and bioavailability in relation principally to bioaccumulation. It also covers how to predict bioaccumulation with the aid of basic environmental information. The chapter concludes with a brief discussion of methods for assessing bioavailability. In doing so, the focus is on PAH partitioning and bioavailability in sediments; however, when relevant the behavior of PAHs in soils will also be discussed.

7.2   PARTITIONING

7.2.1   PARTITIONING BETWEEN TWO PHASES

The mechanism by which PAHs are dispersed in aquatic environments was originally described as via partitioning between an aqueous and a particulate phase (Chiou *et al.* 1979; Karickhoff *et al.* 1979; Brown and Flagg 1981; Chiou *et al.* 1983), with the aqueous phase being represented by interstitial water and the particulate phase being sediment. The approach of assuming an equilibrium or near-equilibrium of PAH concentrations between the aqueous and the particulate phase is generally known as equilibrium partitioning (EqP) theory. Equilibrium partitioning theory evolved from the physical–chemical concept that all substances in a physical system (such as the environment) distribute until equilibrium or near-equilibrium is achieved between all relevant phases as a function of affinity (Weber and Gould 1966; Hamelink *et al.* 1971). Supported by the work of Mackay (1979) and others on fugacity, EqP was used to explain the relative distributions of organic pollutants, including PAHs, in environmental matrices such as sediments and soils.

As a PAH molecule establishes a dynamic equilibrium or near-equilibrium between the particulate and aqueous phases, it is affected by factors intrinsic to these phases. These factors cause PAHs to undergo sorption to the particulate phase and then desorption back into the aqueous phase if the equilibrium changes. Because these factors ultimately determine the freely dissolved concentrations of PAHs, understanding these factors is also critical for understanding bioavailability (Kenaga and Goring 1980; Adams *et al.* 1985; McElroy *et al.* 1989; Lake *et al.* 1990; Landrum and Robbins 1990; DiToro *et al.* 1991; Hamelink *et al.* 1994; Meador *et al.* 1995a).

The concept of EqP can be quantified in the following way starting with the principles of fugacity:

$$C_{PAH} = f_{PAH} Z_{PAH} \qquad (7.1)$$

where C_{PAH} is the concentration (mol/m^3) of a PAH and f_{PAH} and Z_{PAH} are the fugacity (Pa) and fugacity capacity (mol/m^3/Pa), respectively, of the PAH at equilibrium or near-equilibrium in a given phase. Partitioning between two phases (the superscripts 1 and 2) on a fugacity basis is expressed as:

$$\frac{C_{PAH^1}}{C_{PAH^2}} = \frac{f_{PAH^1} Z_{PAH^1}}{f_{PAH^2} Z_{PAH^2}} \qquad (7.2)$$

If phases 1 and 2 are identified as equilibrium concentrations of a PAH in the particulate (C_p) and dissolved (C_d) phases, and the units are changed to μg/kg and μg/L, respectively, the partition coefficient (K_p) (L/kg) of a PAH between sediment and interstitial water can be generated:

$$K_p = \frac{C_p}{C_d} \qquad (7.3)$$

In general, the data used in these calculations are expressed on a dry weight basis. Organic matter coating most inorganic particles is the primary phase with which hydrophobic PAHs are associated. Normalization of K_p by the particulate organic carbon concentration (f_{oc}) (kg organic carbon/kg sediment) results in the organic carbon normalized partition coefficient K_{oc} (in L/kg organic carbon) for the PAH in question:

$$K_{oc} = \frac{K_p}{f_{oc}} \qquad (7.4)$$

Several laboratory studies using spiked sediments generated PAH K_ps and K_{oc}s based on this understanding of partitioning (e.g. Herbes 1977; Karickhoff *et al.* 1979; Means *et al.* 1979, 1980). The relationship between C_p and C_d in Equation 7.3 is assumed to be linear (e.g. Figure 7.1a) which allows for the generic normalization by f_{oc} in Equation 7.4. In practice, generating experimentally based K_{oc}s is relatively difficult and expensive; consequently an estimation technique based on the linear relationship between K_{oc} and the relatively easier to generate octanol–water partition coefficient (K_{ow}) was developed. Log K_{ow}s for PAHs range from about 2.00 to 7.00 (Table 7.1) (Mackay *et al.* 1992;

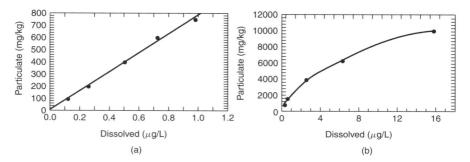

Figure 7.1   Example isotherms of (a) linear partitioning with pyrene and (b) non-linear partitioning with phenanthrene to soils and sediments (Means *et al.* 1980; Young and Weber 1995)

TABLE 7.1   Molecular weight (MW), water solubility (S), octanol–water (K_{ow}), organic carbon (K_{oc}) normalized and soot carbon (K_{sc}) normalized partition coefficients for several PAHs

PAH	MW	log S (mol/L)[1]	Log (K_{ow})[1]	Log (K_{oc})[2]	Log (K_{sc})[3]
Fluorene	166	−4.94	4.21	3.8	6.2
Phenanthrene	178	−5.21	4.57	4.2	6.4
Anthracene	178	−6.60	4.53	4.1	6.7
Pyrene	202	−6.19	4.92	4.8	7.0
Fluoranthene	202	−5.93	5.08	4.8	6.9
Chrysene	228	−8.06	5.71	5.4	8.2
Benzo[a]anthracene	228	−7.32	5.67	5.5	7.8
Benzo[a]pyrene	252	−7.82	6.11	6.1	8.4
Dibenzo[ah]anthracene	278	−8.67	6.71	6.8	9.3

[1] DiToro and McGrath (2000).
[2] Gustafsson and Gschwend (1997a) and Karickhoff (1981).
[3] Gustafsson and Gschwend (1997a).

Boethling and Mackay 2000). Based on this linear relationship, the K_{oc} of many PAHs can be predicted (Karickhoff *et al.* 1979; Means *et al.* 1980; DiToro *et al.* 1991) (Table 7.1).

7.2.2   PARTITIONING INCORPORATING SOOT CARBON

In the early days of EqP development, particulate and aqueous phases were assumed to be compositionally uniform and the only phases involved in the partitioning process. By the early 1980s, actual measurements of the partitioning of PAHs in marine sediments from Washington State (USA) began to cast doubt on the universal applicability of the two-phase EqP model (Socha and Carpenter 1987). One explanation for this deviation was the presence of a third sedimentary phase, into which PAHs preferentially partitioned more than 'normal' or diagenetic organic carbon. They concluded, as did Readman *et al.* (1984, 1987) with sediment from the Tamar Estuary (UK), that petrogenic PAHs followed two-phase partitioning, while PAHs from pyrogenic sources did not. The candidate third phase was the soot or black carbon formed during the pyrolysis of fossil fuels (see Chapter 3). Subsequently, several studies from around the globe have corroborated PAHs to be occluded in or partitioned to soot carbon or soot-like particles (e.g. coal) and not to be partitioning to 'normal' organic carbon, as predicted by EqP (McGroddy and Farrington 1995; Meador *et al.* 1995a; Maruya *et al.* 1996; McGroddy *et al.* 1996; Gustafsson *et al.* 1997; Jonker and Smedes 2000; Accardi-Dey and Gschwend 2002). These studies suggest that soot carbon serves to 'trap' PAHs early in the formation process, while simultaneously acting as a partitioning phase for petrogenic PAHs.

The shortcomings of the original, simple two-phase partitioning model has sparked current scientific debate of the appropriate model for describing PAH partitioning in the environment. Gustafsson and Gschwend (1997a, 1999) and Bucheli and Gustafsson (2000) provided mechanistic explanations for the interaction of soot carbon with PAHs and proposed the following expansion of the partitioning model:

$$K_p = K_{oc}f_{oc} + K_{sc}f_{sc} \qquad (7.5)$$

in which K_{sc} (L/kg soot carbon) and f_{sc} (kg soot/kg) are the soot carbon partition coefficient and sediment soot carbon content, respectively. More recently, Bucheli and Gustafsson (2000) and Accardi-Dey and Gschwend (2002) further expanded this equation to include nonlinear partitioning. The potential for soot carbon and PAH interactions can be further appreciated by considering the amount of the soot found in the environment (Goldberg 1985; Gustafsson *et al.* 1997, Gustafsson and Gschwend 1998; Middelburg *et al.* 1999). Soot carbon has been found to be approximately 30% of organic carbon in some sediments (Middelburg *et al.* 1999), although considerable uncertainty and controversy continues to exist regarding the optimal method of quantifying soot carbon. K_{sc} often exceeds K_{oc} by two orders of magnitude (Table 7.1), according to the few studies available (e.g. Gustafsson and Gschwend 1997a; Accardi-Dey and Gschwend 2002). Consequently, only a very small amount of freely dissolved PAH will be present in interstitial waters, and thus bioavailable, if soot carbon is present in the sediment. Thus, soot carbon can be instrumental in limiting the bioavailability of PAHs in sediment and soil systems. (Paine *et al.* 1996; Lamoureux and Brownawell 1999; Accardi-Dey and Gschwend 2002)

7.2.3   FURTHER DEVELOPMENTS IN PARTITIONING

The nature of diagenetic or 'normal' organic carbon has also come under scrutiny: rather than an amorphous organic carbon phase, several researchers propose that sedimentary organic carbon consists of two domains, 'rubbery' and 'glassy' (Xing and Pignatello 1996; Xing *et al.* 1996; LeBoeuf and Weber 1999;

Weber *et al.* 1999). The rubbery domain is similar to the original interpretation of sediment organic carbon, being homogeneous and amorphous, while the glassy domain is more condensed. PAHs interact differently with the two forms of carbon: dissolving into the rubbery carbon (absorption) and associating primarily with the surface of glassy carbon (adsorption). Another significant aspect of these domains are their kinetics. The rubbery domain is generally described as having relatively rapid sorption and desorption kinetics, while the glassy domain is reported as showing extremely slow sorption/desorption behavior (Pignatello and Xing 1996). This description of partitioning helps to explain the non-linear behavior between C_p and C_d observed in some studies (Figure 7.1b), with the rubbery domain showing linear partitioning and the glassy domain non-linear partitioning (Weber *et al.* 1999). Mathematically, the relationship between C_p and C_d becomes more complicated and frequently is expressed using a Freundlich relationship or isotherm:

$$C_p = K_p C_d^{\frac{1}{n}} \tag{7.6}$$

where the exponent $1/n$ fits the non-linear portion of the C_p and C_d relationship. This identification of two carbon domains with their intrinsic differences in sorption and desorption kinetics has led to speculation that the bioavailability of PAHs and other organic contaminants shows both linear and non-linear attributes. As discussed in Section 7.3, this means that the more readily desorbed linear PAHs are most bioavailable, because they equilibrate with the interstitial waters far more rapidly than the non-linear PAHs.

However, the various types of sedimentary phases with which PAHs interact do not stop with soot, rubbery and glassy carbon. Luthy *et al.* (1997) proposed the existence of numerous particulate and interstitial water phases with which organic contaminants associate. Specifically for PAHs, these phases include aqueous phase colloids and non-aqueous phase liquids (NAPLs), as well as soot carbon and the rubbery and glassy forms of organic carbon (Figure 7.2). Colloidal carbon is composed of 'humic' substances primarily consisting of humic and fulvic acids dissolved in the interstitial waters (Sigleo and Means 1990; Gustafsson and Gschwend 1997b). Several studies have demonstrated a strong affinity of PAHs for colloidal carbon (Wijayaratne and Means 1984; Gauthier *et al.* 1987; Chin and Gschwend 1992; Mitra and Dickhut 1999; MacKay and Gschwend 2001). NAPLs, like coal tars, are often present in the vicinity of manufactured gas plants and consist of a plethora of liquid organic phases with a high affinity for PAHs (Lane and Loehr 1992; Luthy *et al.* 1994; Mahjoub *et al.* 2000). Each of these phases has potentially unique partitioning characteristics affecting PAH fate and bioavailability. Other researchers have emphasized the importance of yet other aspects of particulate organic carbon, such as polarity (i.e. the presence of moieties in organic matter including functional groups containing nitrogen and oxygen) and aromaticity to explain PAH partitioning (Gauthier *et al.* 1987; Grathwohl 1990; Kile *et al.* 1995).

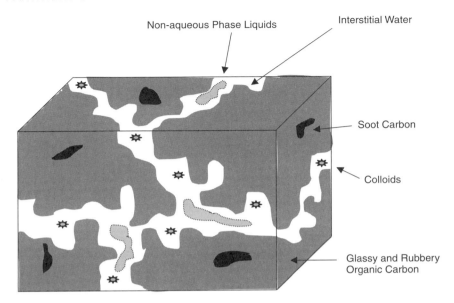

Figure 7.2   Sedimentary and interstitial water binding phases known to affect the fate and bioavailability of PAHs in an imaginary cube of sediment

7.3   BIOAVAILABILITY AND BIOACCUMULATION

It has been argued on the basis of, amongst other factors, the partitioning discussed above, that the total concentration of a pollutant in a sediment or soil is not indicative of its bioavailability. Rather, because bioavailability may be affected by the type of partitioning and rate of desorption of contaminants from sediments and soils, a commonly accepted view is that bioavailability (measured as biodegradability, bioaccumulation or toxicity), is most strongly related to the contaminants that can exchange rapidly between phases in their surroundings. As discussed above, the contaminants already dissolved in the interstitial waters (a very small amount) and those that are sorbed to the linear domain carbon are frequently considered the most bioavailable (Figure 7.3). The bioavailability of the non-linear domain contaminants is not known; presumably, because of their very slow desorption kinetics, they are not bioavailable but this needs to be investigated (Alexander 2000; Bosma *et al.* 1997; Carmichael *et al.* 1997; Cornelissen 1999; Fuchsman and Barber 2000; Kraaij 2001; Loehr and Webster 1996; Reid *et al.* 2000a; Sijm *et al.* 2000; White and Alexander 1996; White *et al.* 1999; Williamson *et al.* 1998, Cuypers *et al.* 2002, Gevao *et al.* 2001, Harkey *et al.* 1995, Scribner *et al.* 1992). This pool of dissolved and rapidly exchanging contaminants can vary between sediments and soils. In Section 7.2, it was shown how our ability to measure or estimate the concentrations of dissolved contaminants with EqP has developed to a rather sophisticated level using

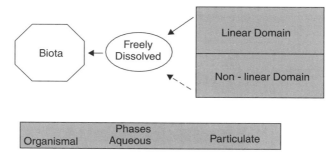

Figure 7.3 Conceptual illustration of the partitioning of PAHs between the aqueous and particulate phases important to bioavailability. Dashed arrow indicates the non-linear domain is not known to contribute bioavailable PAHs

organic carbon and other sediment characteristics. At present we are unable to accurately predict or relate the amounts of rapidly exchanging contaminants to any specific sediment characteristics. Given this level of understanding of bioavailability, it is therefore timely to review the relevant factors that determine the bioaccumulation of PAHs by organisms.

7.3.1   BIOCONCENTRATION AND BIOACCUMULATION FACTORS: BACKGROUND

Two general approaches have been developed for modeling contaminant accumulation by organisms: EqP-based (Di Toro *et al.* 1991) and kinetics-based (Thomann 1989). The EqP-based approach, just as with the aqueous and particulate phases, assumes equilibrium or near-equilibrium conditions between the organisms and the aquatic environment. Conversely, the kinetic approaches describe bioaccumulation as the net effect of the rate processes — uptake and elimination (Opperhuizen 1991; Lee 1992). For the following discussion, we will focus on the EqP-based approach because of its wide usage.

Bioconcentration (BCF) and bioaccumulation factors (BAF and BSAF) are useful coefficients that can indicate equilibrium or near-equilibrium exposure and expected tissue residues based on environmental concentrations. In the literature, there are inconsistencies in the actual calculations used to derive the various tissue-based coefficients (e.g. Thomann and Komlas 1999; Gobas 2000; Mackay and Fraser 2000). For the purposes of the discussion here, the equations are:

$$BCF = \frac{C_t}{C_d} \tag{7.7}$$

$$BAF = \frac{C_t}{C_p} \tag{7.8}$$

$$\text{BSAF} = \left(\frac{C_t}{f_{lip}} \right) \bigg/ \left(\frac{C_p}{f_{oc}} \right) \qquad (7.9)$$

where C_t is the concentration in the tissues and f_{lip} is the fraction of tissue lipids. Units for BCF, BAF and BSAF are L/kg tissue, kg sediment/kg tissue, and kg organic carbon/kg lipid, respectively. As with the earlier particulate phase calculations, all concentrations should be dry weights. If known (or reasonably approximated), all bioaccumulation factors should be specified in terms of exposure time (e.g. day 10 BAF) because of the large variability that occurs from initial exposure to equilibrium or near-equilibrium. Whereas BCFs tend to increase with hydrophobicity, reflecting the greater affinity of more hydrophobic compounds for organism tissue relative to water, BAFs show no correlation with hydrophobicity as a result of similar partitioning of hydrophobic compounds between tissue and sediment (Meador *et al.* 1995b). Because the BSAF relates organic carbon-normalized sediment concentrations to lipid-normalized tissue concentrations, variability between tissue and sediment PAH concentrations is even further reduced. The BSAF approach also assumes that PAHs and other organic contaminants have similar affinities for sediment organic carbon and tissue lipid and thus should be equivalent to about one (also see Section 7.3.3).

In addition to calculations based on measured values (e.g. C_p and C_t), expected BCF and BAF values based on uptake and elimination rate constants determined in the laboratory can be useful for predicting equilibrium or near-equilibrium tissue residues in field-collected organisms because metabolic conversion is included in their derivation. The equilibrium or near-equilibrium BCF and BAF can be determined as follows:

$$BCF = \frac{k_1}{k_2} \qquad (7.10)$$

$$BAF = \frac{k_s}{k_2} \qquad (7.11)$$

where, k_1, k_2 and k_s are the uptake clearance rate constant for water exposures (L/kg tissue/h), the elimination rate constant (h^{-1}), and uptake clearance rate constant for sediment exposures (kg sediment/kg tissue/h), respectively (Landrum *et al.* 1992).

7.3.2   SEDIMENT PROPERTIES

As discussed, bioavailability is affected directly by sediment properties such as particulate organic carbon content and composition. However, the EqP approach for predicting PAH bioavailability (DiToro *et al.* 1991) evolved before our appreciation of 'new' particulate phases such as soot carbon and their very different partitioning characteristics. It has been found that the presence of soot carbon in sediments alters partitioning and reduces the bioavailability of PAHs (e.g. Hickey $et\ al.$ 1995; Meador $et\ al.$ 1995a,b; Paine $et\ al.$ 1996; Lamoureux and Brownawell 1999). Thus, in areas where sediments are known to contain soot carbon, use of the simple K_{oc} to K_{ow} relationship to predict new K_{oc}s may not be appropriate. This practice is likely to overestimate PAH bioavailability, since contaminants are actually more strongly associated with the soot in sediments than assumed (i.e. $K_{sc} \gg K_{oc}$) (Table 7.1). Obviously, this further complicates attempts at predicting bioavailability. To address this uncertainty, the US EPA sediment benchmarks for PAH mixtures, which estimate PAH bioavailability in contaminated sediments, use a site-specific correction and/or direct measurement of interstitial water PAHs in instances where soot carbon may be present and affecting bioavailability (US EPA 2002 a,b).

Another important sediment variable affecting PAH bioavailability is grain size, especially for organisms that are particle size-selective (Lydy and Landrum 1993; Harkey $et\ al.$ 1994; Kukkonen and Landrum 1994; Wang $et\ al.$ 2001). For example, Harkey $et\ al.$ (1994) measured PAH accumulation by amphipods exposed to a range of sediment particle sizes ($< 63\ \mu m$) and found contaminants with similar K_{ow}s partitioned differently among the various fractions resulting in varying bioavailability. Similarly, the effects of other phases illustrated in Figure 7.2 need to be better understood with regard to their effects on bioavailability.

The dynamic nature of partitioning means that PAH bioavailability will also vary as a function of where in the dissolved–particulate equilibration an organism is exposed. It is well known that contaminants from freshly PAH-spiked sediments are more readily desorbed and thus more bioavailable than PAHs from aged sediments (Kukkonen and Landrum 1995, 1998; Chung and Alexander 1998) and that prolonged sediment–contaminant contact time tends to decrease bioavailability (Leppänen and Kukkonen 2000). Similarly, aging of PAH-spiked soil reduces bioavailability to earthworms, plants such as wheat, and microorganisms capable of contaminant biodegradation (Alexander 1995; Hatzinger and Alexander 1995; see also Chapters 6 and 10). The reasons for this behavior are not well understood. Possible explanations include mechanisms discussed earlier as well as irreversible (i.e. non-desorbing) chemical bonding between PAHs and sediments, and increased contact along with molecular diffusion allows PAHs to penetrate deeper into mineral micropores or surface-coatings, resulting in PAHs increasingly resistant (or retarded) to desorption.

For example, bioavailability of phenanthrene to earthworms was further decreased when soils were subjected to wetting and drying during the aging period (White $et\ al.$ 1997, 1998). Similarly, uptake of anthracene by the worm $Eisenia\ fetida$ was reduced from 32.2% in unaged spiked soil to 13.7% after 203 days (Tang $et\ al.$ 1998; Tang and Alexander 1999). Further, uptake rates, BSAFs and leachability declined in the following order: spiked soil > source soil > treated soil > aged soil.

7.3.3 ORGANISM PROPERTIES

Because PAHs are hydrophobic and lipophilic, the lipid content of an organism will have a strong effect on its tissue PAH concentrations (Stegeman and Teal 1973; Mackay 1982). PAHs are assumed to partition to a similar extent between organism lipid and sediment organic carbon (i.e. $K_{ow} \approx K_{oc} \approx$ BCF), such that the maximum value for BSAF is predicted to be close to unity and constant with regard to hydrophobicity. However, some investigations have reported that the dependency of BSAF on contaminant hydrobicity is tenuous (e.g. Connor 1984; Markwell *et al.* 1989; Connell and Markwell 1990; Lee *et al.* 1993). Furthermore, departure from unity occurs due to the slightly greater PAH affinity of biological lipid relative to sediment organic carbon. Thus, the theoretical bioaccumulation potential (TBP) is somewhat higher and ranges between 1 and 4 (McFarland and Clarke 1986; Di Toro *et al.* 1991; Lee 1992; Boese *et al.* 1995; Clarke and McFarland 2000). Actual measurements of BSAFs vary by several orders of magnitude (Van Brummelen *et al.* 1998; Van Hattum *et al.* 1998; Wong *et al.* 2001), although for benthic invertebrates their median typically lies between 0.3 and 3 (Tracey and Hansen 1996). For example, Lee *et al.* (1993) derived a generic BSAF of 1.7, while Thomann *et al.* (1992) predicted field BSAFs of approximately 0.8–1.0 for persistent contaminants with log K_{ow}s of 2–5. As well as BSAFs of 1–10 for contaminants with log K_{ow}s between 5 and 8 (Thomann *et al.* 1992). Thus, by using this range of BSAFs and knowing the sediment PAH concentrations and the respective amounts of organic carbon in sediment and lipid in tissue, the equilibrium or near-equilibrium PAH concentration of a given organism can often be estimated to within an order of magnitude.

Whenever BCFs or BSAFs are below the theoretical maximum, factors such as biotransformation of parent compound, changing exposure concentrations, reduced bioavailability, or insufficient time for accumulation may be the cause. If metabolic capacity can be accounted for, then spatial or temporal factors may be explored. The resultant EqP-based value for both the predicted BCF and equilibrium or near-equilibrium BSAF values would be for species that do not metabolize the compound of interest.

One study has examined the uncertainty associated with estimating the TBP based on equilibrium partitioning predictions (Clarke and McFarland 2000). These authors performed a sensitivity analysis on the various parameters of the BSAF equation to determine which have the greatest influence on the result. Their study examined bioaccumulation of PAHs for several invertebrate species and concluded that the highest variability came from lipid content and tissue concentration, which were due, in part, to values close to the limit of detection and the influence of outliers. Uncertainty, as determined by total error (= systematic method error + propagated measurement error) was generally consistent for most PAH compounds; however, two (pyrene and fluoranthene) exhibited substantially more total error. In general, bioaccumulation was underestimated 10% of the time but was overestimated 41% of the time, leading to the conclusion that TBP was a reasonably accurate, if somewhat excessive, predictor of bioaccumulation.

While lipid reservoirs ultimately set the upper limit to how much PAH may be accumulated by an organism (i.e. the maximum body-burden), other factors determine the rate of uptake. The rate and extent of PAH uptake from food strongly depends on lifestyle and feeding strategy; i.e. whether an organism is a suspension feeder, a deposit feeder, a herbivore browsing on particle surfaces, or a predator (Leppanen 1995; Kaag *et al*. 1997). Owing to their greater ingestion rates, deposit feeders tend to accumulate more PAHs than suspension feeders. For example, Hickey *et al*. (1995) measured higher PAH body burdens for deposit-feeding tellinid clams (*Macomona liliana*) compared to suspension-feeding cockles (*Austrovenus stutchburyi*) and oysters (*Crassostrea gigas*) (Figure 7.4). BSAFs deviated by one to three orders of magnitude from the predicted equilibrium BSAF value of 1.7 (1–4). Moreover, a markedly different pattern in bioaccumulation was observed between *M. liliana* and the cockles and oysters, with BSAF values decreasing with increasing log K_{ow} for the suspension feeders, while remaining more or less constant for *M. liliana*. Such differences could be described by differences in either bioavailability or elimination rates (Meador *et al*. 1995b). Grazers and epipsammic browsers, such as many snails and meiobenthic copepods, also commonly have high PAH body burdens due to preferentially feeding on organic films or food particles that are concentrated with contaminants (Carman *et al*. 1997; Forbes and Forbes 1997; Bennett *et al*. 1999).

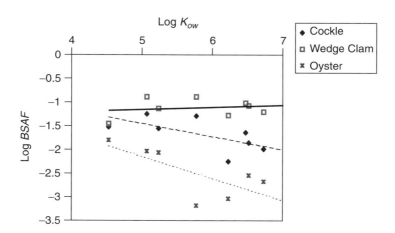

Figure 7.4   Field measurements of PAH BSAFs for several species with differing feeding strategies and digestive chemistries including cockles (*Austrovenus stutchburyi*), wedge clams (*Macomona liliana*) and oysters (*Crassostrea gigas*; unpublished data) from Manukau Harbour (New Zealand) (Hickey *et al*. 1995). PAH log K_{ow}s were approximately 4.50–7.00

In recent years, an important area of research regarding PAH bioavailability has developed around characterizing the gut chemistry of benthic organisms. Deposit feeders ingest and process several times their body weight in sediment per day (Cammen 1980; Lopez and Levinton 1987). The physicochemical conditions that sediment particles are subjected to during gut passage can be quite different than those in ambient sediments. Deposit-feeder gut fluids typically have elevated enzymatic activities, high organic colloidal concentrations (e.g. proteins), and display strong surfactant properties (Mayer *et al.* 1997; Weston and Mayer 1998a,b). Furthermore, deposit feeder guts show pronounced microbial activity and can often be slightly reducing, while gut pH typically varies between 5 and 8 (Plante and Jumars 1992; Ahrens and Lopez 2001). Combined with vigorous peristalsis and fluid reflux mechanisms, guts therefore represent an intense sediment microenvironment. Under these conditions it is likely that PAHs will desorb from the sediments, increasing the freely dissolved concentrations and thus the bioavailability. For example, Mayer *et al.* (1996) and Voparil and Mayer (2000) found PAH solubilization by gut fluids of a polychaete (*Arenicola marina*) and a holothuroid (*Parastichopus californicus*) to be several orders of magnitudes greater than predicted by EqP. Since species differ in gut fluid strength (e.g. surfactancy, colloidal content, enzymatic activity and other physicochemical parameters; Mayer *et al.* 1997), contaminant desorption rates and thus bioavailability from food and sediments may be expected to differ among species. Mayer *et al.* (2001) recently compared benzo[a]pyrene solubilization from spiked sediments among gut fluids from 18 benthic invertebrates and found nearly 10-fold differences in extraction efficiency compared to an organic solvent extraction (Figure 7.5). In practice, the amount of PAH accumulated is called the absorption or assimilation or accumulation efficiency (α or AE). AE may vary greatly between different species and individuals, the reasons for which are still poorly understood, but, in part, are probably due to differences in digestive chemistry. It is important to realize that, while AE is not necessarily a constant for an individual or species, different AEs can, nonetheless, result in similar equilibrium or near-equilibrium body burdens. Consider the following scenario (Figure 7.6). Species 1 and 2 differ in the efficiency of their digestion, while being exposed to the same sediment PAH pool. Consequently, species 2 has a higher instantaneous absorption efficiency and uptake rate than species 1. Nonetheless, due to similar lipid content, species 1 and 2 will eventually accumulate similar (lipid-normalized) PAH concentrations, and the PAH of interest will appear equally bioavailable if measured at equilibrium or near-equilibrium. Species 3, on the other hand, is exposed to an overall larger pool of PAHs because of inherently different physiological characteristics. For example, it could have higher gut fluid surfactancy, enabling it to solubilize PAHs that are inaccessible to the other two species.

Because contaminant uptake is a time-dependent or kinetic process, longer gut passage times are likely to result in more extensive desorption and accumulation than shorter ones. Since ingestion rate is inversely related to gut passage

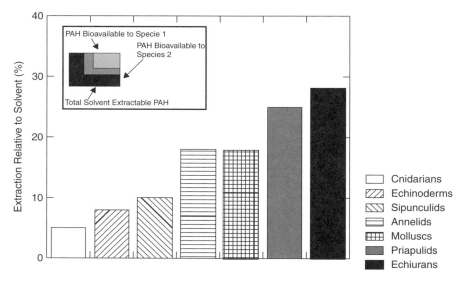

Figure 7.5   Extraction of benzo[a]pyrene from sediment by digestive fluids of several types of marine invertebrates compared to solvent extraction (Mayer *et al*. 2001). On the *y* axis, a value of 100% is equivalent to the solvent extraction. Inset schematic shows how bioavailability can vary between species depending on the extraction efficiency of their gut chemistry

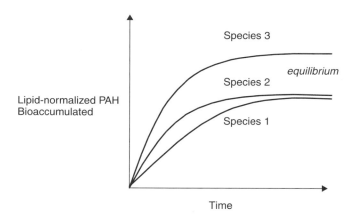

Figure 7.6   Idealized uptake kinetics for a PAH in three sediment-ingesting species

time, organisms with high ingestion rates, and consequently short gut passage times, will probably solubilize and accumulate a smaller amount of PAHs per gut filling than species with low ingestion rates (all other factors being equal). Ahrens *et al*. (2001b) have shown for the polychaete *Nereis succinea* that adult animals commonly have higher absorption efficiencies for food as well as PAHs and PCBs than do juveniles. However, despite the fact that gut passage times may differ by an order of magnitude between adults and juveniles, absorption efficiencies generally vary by less than a factor of two. This may be explained by the finding that organic contaminants tend to desorb very rapidly in gut fluids. For example, desorption of a tetrachlorobiphenyl and hexachlorobenzene by gut fluids of two polychaete worm species showed two-thirds of the bioavailable contaminant fraction to be solubilized within the first minute of gut fluid contact while the remainder desorbed within 6 h (Ahrens *et al.* 2001a). While accumulation of PAHs by ingestion necessarily requires passage of food through the intestine, elimination also appears to depend on gut passage. Discontinuous feeding decreases elimination of contaminants (Landrum and Scavia 1983; Frank *et al.* 1986; Leppänen 1995), suggesting that sediments can both deliver and remove contaminants. Further, there have been concerns expressed that some high molecular weight PAHs would have inadequate time to establish equilibrium or near-equilibrium between the sediments and organisms because the equilibration duration approaches organismal lifespans (Connell 1988). These concerns are probably not important, given gut chemistry effects on increasing desorption kinetics.

According to EqP theory, even though the BSAF is derived from sediment concentrations, there is no implicit assumption of sediment ingestion. The BSAF value integrates exposure from all sources (i.e. food, water, and sediment ingestion) because it is assumed that the concentrations of chemicals in the different matrices occur in predictable proportions. At equilibrium or near-equilibrium, the organism would receive equal exposure from all phases, hence it does not matter which one is measured. It must be appreciated that for a given sediment, very little of the frequently large amount of contaminant present will ever be bioavailable, just as the amount of contaminant present in the aqueous phase is often minute for hydrophobic chemicals like PAHs. In general, sediment concentrations are used to estimate bioavailability because they are the easiest to determine, they are less variable than water or food concentrations, and they are among the focus for regulatory action (see Chapter 18). Additionally, it may not matter if the main source of a hydrophobic compound (e.g. PAH) to the organism varies between ventilation of water or ingestion of sediment or prey; the sediment concentration can still be used to represent accumulation from all sources at equilibrium or near-equilibrium.

7.4 ASSESSMENT OF BIOAVAILABILITY IN SEDIMENTS AND SOILS

Although a lot of effort has been put into understanding and trying to predict the partitioning phenomena and the distribution of contaminants between phases, our estimates are still only good to within an order of magnitude. As discussed earlier, several factors influence PAH distributions, including the amount and quality of natural organic material present in a soil or sediment, the presence of soot carbon, and the contact time of a chemical with the sediment or soil (Landrum *et al.* 1992; McGroddy and Farrington 1995; Pignatello and Xing 1996; Gustafsson *et al.* 1997; Huang and Weber 1997; Luthy *et al.* 1997; Chung and Alexander 1999; Cornelissen 1999; Ghosh *et al.* 2000). Of course, this lack of ultimately certainty in our ability to predict PAH distributions directly affects confidence in our estimates of bioavailability (as shown in our discussion of bioaccumulation, above). The following section describes several direct approaches for assessing bioavailable concentrations of PAHs. By directly measuring bioavailability, the uncertainties acknowledged to exist with partitioning are avoided. However, direct measures frequently incur greater costs and more labor and have their own technical artifacts. Approaches for directly measuring the bioavailability of PAHs can be grouped into two categories: (a) chemical; and (b) biological or toxicological (Table 7.2).

TABLE 7.2   Different chemical and biological/toxicological methods to estimate or measure the bioavailable fraction of contaminants in sediment

Method	Phase sampled
Chemical	
SPMDs and related methods	Freely dissolved in interstitial water
Extraction with solvents of variable strength	Variable portion of linear domain (sometimes also part of the non-linear sorbed fractions)
Resin-based desorption	Known part of linear domain
Biological/toxicological	
Bioaccumulation in soft-bodied sediment-dwelling organisms or in organisms feeding on surface water[1]	Freely dissolved in interstitial water and unknown portion of linear domain
Toxicity testing with soft-bodied sediment-dwelling organisms or in organisms feeding on surface water[1]	Freely dissolved in interstitial water and unknown portion of linear domain

[1] In the case that exposure via the ingestion of sediment is the dominant route, comparisons between biological and chemical techniques can be biased by differences between species in their ability to accumulate (part of) the non-linear sorbed fraction.

7.4.1   CHEMICAL ASSESSMENTS

Chemical methods for measuring bioavailability described in the recent literature can be broken into three types (Table 7.2). These include:

- Methods which mimic uptake by organisms from the water phase (biomimetric approach). Examples include: extraction using SPMDs (semi-permeable membrane devices) (Awata *et al.* 1999; Leppänen and Kukkonen 2000; SPMEs (solid-phase microextraction fibers) (Mayer *et al.* 2000; Leslie *et al.* 2002), or PTD (polyethylene tube dialyses) (MacRae and Hall 1998). The use of membranes such as polyethylene are thought to mimic the passage of contaminants through cell membranes like those in the gut.

- Selective (mild) extraction procedures that only remove an operationally defined fraction of the total amount of contaminant present in a sediment or soil. This is performed by using a combination of solvents and comparing extracted amounts with accumulation measurements. Another approach is the use of gut fluids (Voparil and Mayer 2000) or prepared solutions containing enzymes and amino acids (Mayer *et al.* 1996; Weston and Mayer 1998a; Lamoureux and Brownawell 1999). The choice of solvents depends on the organisms that are used for comparison (Kelsey *et al.* 1997; Reid *et al.* 2000b; Tang *et al.* 1999; Tang and Alexander 1999). A similar approach is taken when applying supercritical fluid extraction (Björklund *et al.* 2000; Hawthorne and Grabanski 2000).

- Use of resins like Tenax, cyclodextrin XAD and C-18 along with desorption kinetics to selectively remove contaminants from sediments during a predetermined time period. It is believed that this approach captures part or all of the linearly sorbed fraction because of the high desorption rate constants associated with this fraction (Lake *et al.* 1996; ten Hulscher *et al.* 1997; Cornelissen *et al.* 1998, 2001; MacRae and Hall 1998; Williamson *et al.* 1998; Lamoureux and Brownawell 1999; Reid *et al.* 2000a; Kraaij 2001; Cuypers *et al.* 2002).

All of the above approaches assume that a desorption step to an aqueous phase must take place before a compound can be taken up by organisms (Figure 7.3). The results obtained with the various (chemical) methods described above are sometimes highly variable but this may be attributed to the different approaches that are used. It should also be appreciated that the concept of a bioavailable fraction, while intuitively attractive, may not be meaningful when organisms of very different physiology, ecology and exposure routes are compared. Thus, PAH bioavailability to microorganisms that are exposed exclusively via interstitial waters or sediment surfaces may be quite different from bioavailability to sediment-ingesting macrofauna. For example, when bioavailability is measured using microorganisms (e.g. degradation studies), to a great extent the desorption kinetics control how much chemical can be degraded by determining how much is present in the interstitial water. Chemical methods for measuring bioavailability that act in the same way (e.g. resins like XAD, Tenax and cyclodextrin) have been shown to correlate well with biodegradability, because in both cases the same rate-limiting step is studied (Bosma *et al.* 1997; Carmichael *et al.* 1997; Cornelissen *et al.* 1998; Cuypers *et al.* 2000, 2002; Reid *et al.* 2000b). Studies to measure bioaccumulation in organisms are usually designed in such a way that the contaminant source (in the sediment or soil) is not exhausted. The concentrations in organisms are usually measured after a prolonged exposure time during which equilibrium or near-equilibrium a situation has been reached. In this case, chemical methods measuring exposure concentrations in the interstitial water (e.g. SPMDs, PTD, or SPMEs) would be expected to behave as good surrogates of organism accumulation. Recent evidence is now available that both types of measurements are related. From the knowledge gained in desorption kinetic studies, it is highly probable that interstitial water concentrations are determined by the fast desorbing pool of compounds present in sediment or soil (Ten Hulscher *et al.* 1997; Cornelissen *et al.* 1998; Kraaij 2001). For example, Kraaij (2001) established the relationship between concentrations in the linear fraction, concentrations in porewater, and uptake by oligochaetes. This prediction of bioavailability can still be described with equilibrium partitioning theory. The only difference from the earlier models of partitioning and bioavailability is the use of the linear domain rather than the entire amount of particulate phase organic carbon.

7.4.2 BIOLOGICAL AND TOXICOLOGICAL ASSESSMENTS

The alternative approach for the assessment of the bioavailable fraction of contaminants is by performing laboratory exposure studies with test organisms. In these studies, either bioaccumulation or toxicity can be used to determine bioavailability. The bioaccumulation approach relates measured tissue concentrations to those predicted to occur using EqP. Similarly, the toxicological approach relates concentrations known to cause adverse effects (e.g. mortality, reduced growth) to observed effects at the EqP predicted concentrations in sediment exposures.

In the bioaccumulation approach, the bioavailable fraction in the sediment can be estimated by comparing the measured accumulation levels in the test organisms with EqP predicted concentrations (Van der Kooij *et al.* 1991; Den Besten in press). In these studies, exposure levels should be low enough to allow normal activity of the organisms and acceptable survival to yield sufficient body mass for reliable chemical measurements.

The choice of test organism should also be made dependent on the feeding strategy of the test organism (Kraaij, 2001). For a comparison with the chemical techniques (e.g. SPMDs, PTD, SPMEs or resins), ideally a test organism should be chosen that is exposed to contaminants only via the interstitial water phase. For example, bioaccumulation studies with commonly used aquatic oligochaetes measure the uptake of contaminants from two possible routes: interstitial water and ingestion of food. As a general rule, for organic chemicals with log $K_{ow} < 5$ the sediment interstitial water is the dominant source of exposure, while for contaminants with log $K_{ow} > 5$, uptake from ingested sediment particles can contribute significantly to accumulation (Belfroid *et al.* 1996). Regardless of the exposure route, it may be possible to predict accumulation on the basis of the chemical methods that measure the bioavailability when there are no species-specific differences in the digestive chemistry that result in additional sources for the uptake of contaminants (such as the non-linear domain, as indicated in Figure 7.3).

In the toxicity-testing approach, the known toxic concentrations of PAHs from water-only exposures (i.e. dissolved phase) are compared to sediment exposures for which either the interstitial water concentrations are known or the EqP estimated concentrations have been determined (DiToro *et al.* 1991). Agreement between the known and predicted measures of bioavailability provide significant support to the partitioning model and our ability to estimate bioavailability (Table 7.2).

As in the case of choosing organisms for comparing bioaccumulation between chemical assessments techniques and actual animals, it is also critical to select toxicity-testing species carefully. To avoid equivocal results, it is desirable to conduct toxicity tests with species which have a good database of sensitivity information as well as standardized methods. Over the last 25 years, several toxicity-testing species have achieved this criterion. Unfortunately, the pool of test species is biased towards amphipods and limited with regard to other representative organisms. Development of sediment toxicity tests using new species is therefore needed.

7.5 SUMMARY

In this chapter, the evolution of our understanding of the partitioning of PAHs in sediment and soil environments was discussed. Understanding PAH distributions, using equilibrium partitioning theory (EqP), allows for estimating the bioavailability of PAHs in the environment. Total measures of PAH concentrations in the particulate phase of a soil or sediment do not equate to the bioavailable concentrations. Rather, bioavailability is best understood, at this time, as PAH dissolved in interstitial water and the fraction of PAH associated with the rapidly desorbing linear domain portion of the organic carbon present in the particulate phase. Other phases, including soot carbon, colloids and NAPLs, further complicate our interpretation of PAH partitioning and bioavailability. The bioavailability of PAHs, expressed as accumulation and toxicity, can also be predicted using EqP or measured directly with chemical and biological/toxicological tools. Several approaches are available for depicting accumulation including BCFs, BAFs and BSAFs. These methods are complemented by chemical techniques for measuring bioaccumulation, including the use of SPMDs. Toxicity testing offers another way to directly measure PAH bioavailability. With our evolving understanding of PAH partitioning and bioavailability, the ability to determine the risk to benthic organisms associated with PAH-contaminated soils and sediments is improving.

REFERENCES

Accardi-Dey A and Gschwend PM (2002) Assessing the combined roles of natural organic matter and black carbon as sorbents in sediments. *Environmental Science and Technology*, **36**, 21–29.

Adams WJ, Kimerle RA and Mosher RG (1985) Aquatic safety assessment of chemicals sorbed to sediments. In *Aquatic Toxicology and Hazard Assessment: Seventh Symposium* (ASTM STP 854). Cardwell RD, Purdy R and Bahner RC (eds), American Society for Testing and Materials, Philadelphia, PA, pp. 429–453.

Ahrens MJ and Lopez GR (2001) *In vivo* characterization of the gut chemistry of small deposit feeding polychaetes. In Aller JY, Woodin SA, and Aller RC (eds), *Organisms–Sediment Interactions* University of South Carolina Press, Columbia, SC, pp. 349–368.

Ahrens MJ, Hertz J, Lamoureux EM, Lopez GR, McElroy AE and Brownawell BJ (2001a) The role of digestive surfactants in determining bioavailability of sediment-bound hydrophobic organic contaminants to two deposit-feeding polychaetes. *Marine Ecology Progress Series*, **212**, 145–157.

Ahrens MJ, Hertz J, Lamoureux EM, Lopez GR, McElroy AE and Brownawell BJ (2001b) The effect of body size on digestive chemistry and absorption efficiencies of food and sediment-bound organic contaminants in *Nereis succinea* (Polychaeta). *Journal of Experimental Marine Biology and Ecology*, **263**, 185–209.

Alexander M (1995) How toxic are toxic chemicals in soil? *Environmental Science and Technology*, **29**, 2713–2717.

Alexander M (2000) Aging, bioavailability, and overestimation of risk from environmental pollutants. *Environmental Science and Technology*, **34**, 4259–4265.

Awata H, Johnson KA and Anderson TA (1999) Passive sampling devices as surrogates for evaluating bioavailability of aged chemicals in soil. *Toxicological and Environmental Chemistry*, **73**, 25–42.

Belfroid AC, Sijm DTHM and van Gestel CAM (1996) Bioavailability and toxicokinetics of hydrophobic aromatic compounds in benthic and terrestrial invertebrates. *Environmental Reviews*, **4**, 276–299.

Bennett A, Bianchi TS, Means JC and Carman KR (1999) The effects of polycyclic aromatic hydrocarbon contamination and grazing on the abundance and composition of microphytobenthos in salt marsh sediments (Pass Fourchon, LA) I. A microcosm experiment. *Journal of Experimental Marine Biology and Ecology*, **242**, 1–20.

Björklund E, Nilsson T, Bøwadt S, Pilorz K, Mathiasson L and Hawthorne SB (2000) Introducing selective supercritical fluid extraction as a new tool for determining sorption/desorption behavior and bioavailability of persistent organic pollutants in sediment. *Journal of Biochemical and Biophysical Methods*, **43**, 295–311.

Boese BL, Winsor M, Lee H II, Echols S, Pelletier J and Randall R (1995) PCB congeners and hexachlorobenzene biota sediment accumulation factors for *Macoma nasuta* exposed to sediments with different total organic carbon contents. *Environmental Toxicology and Chemistry*, **14**, 303–310.

Boethling RS and Mackay D (2000) *Handbook of Property Estimation Methods for Chemicals: Environmental and Health Sciences*. Lewis, CRC Press, Boca Raton, FL, p. 481.

Bosma T, Middeldorp P, Schraa G and Zehnder A (1997) Mass transfer limitation of biotransformation: quantifying bioavailability. *Environmental Science and Technology*, **31**, 248–252.

Brown DS and Flagg EW (1981) Empirical prediction of organic pollutant sorption in natural sediments. *Journal of Environmental Quality*, **10**, 382–386.

Bucheli TD and Gustafsson O (2000) Quantification of the soot–water distribution coefficient of PAHs provides mechanistic basis for enhanced sorption observations. *Environmental Science and Technology*, **34**, 5144–5151.

Cammen LM (1980) Ingestion rate: an empirical model for aquatic deposit feeders and detritivores. *Oecologia*, **44**, 303–310.

Carman KR, Fleeger JW and Pomarico SM (1997) Response of a benthic food web to hydrocarbon contamination. *Limnology and Oceanography*, **42**, 561–571.

Carmichael L, Chrisman R and Pfaender F (1997) Desorption and mineralization kinetics of phenanthrene and chrysene in contaminated soils. *Environmental Science and Technology*, **31**, 126–132.

Chin, Y-P and Gschwend PM (1992) Partitioning of polycyclic aromatic hydrocarbons to marine porewater organic colloids. *Environmental Science and Technology*, **26**, 1621–1626.

Chiou CT, Peters LJ and Freed VH (1979) A physical concept of soil–water equilibria for non-ionic organic compounds. *Science*, **206**, 831–832.

Chiou CT, Porter PE and Schmedding DW (1983) Partition equilibria of non-ionic organic compounds between soil organic matter and water. *Environmental Science and Technology*, **17**, 227–231.

Chung N and Alexander M (1998) Differences in sequestration and bioavailability of organic compounds in dissimilar soils. *Environmental Science and Technology*, **32**, 855–860.

Chung N and Alexander M (1999) Effect of concentration on sequestration and bioavailability of two polycyclic aromatic hydrocarbons. *Environmental Science and Technology*, **33**, 2605–3608.

Clarke JU and McFarland VA (2000) Uncertainty analysis for an equilibrium partitioning-based estimator of polynuclear aromatic hydrocarbon bioaccumulation potential in sediments. *Environmental Toxicology and Chemistry*, **19**, 360–367.

Connell DW (1988) Bioaccumulation behaviour of persistent organic chemicals with aquatic organisms. *Reviews of Environmental Contamination and Toxicology*, **101**, 117–154.

Connell DW and Markwell RD (1990) Bioaccumulation in the soil to earthworm system. *Chemosphere*, **20**, 91–100.

Connor MS (1984) Fish/sediment concentration ratios for organic compounds. *Environmental Science and Technology*, **18**, 31–35.

Cornelissen G (1999) Mechanism and Consequences of Slow Desorption of Organic Compounds from Sediments. PhD Thesis, University of Amsterdam, the Netherlands, p. 297.

Cornelissen G, Rigterink H, Ferdinandy MMA and van Noort PCM (1998) Rapidly desorbing fractions of PAHs in contaminated sediments as a predictor of the extent of bioremediation. *Environmental Science and Technology*, **32**, 966–970.

Cornelissen G, Rigterink H, ten Hulscher DEM, Vrind BA and van Noort PCM (2001) A simple Tenax extraction method to determine the availability of sediment-sorbed organic compounds. *Environmental Toxicology and Chemistry*, **20**, 706–711.

Cuypers CT, Grotenhuis T, Joziasse J and Rulkens W (2000) Rapid persulfate oxidation predicts PAH bioavailability in soils and sediments. *Environmental Science and Technology*, **34**, 2057–2063.

Cuypers C, Pancras T, Grotenhuis T and Rulkens W (2002) The estimation of PAH bioavailability in contaminated sediments using hydroxypropyl-β-cyclodextrin and Triton X-100 extraction techniques. *Chemosphere*, **46**, 1235–1245.

Den Besten PJ (2002) Contamination in intertidal areas: risks of increased bioavailability. *Aquatic Ecosystem Health and Management.* (in press).

DiToro DM and McGrath JA (2000) Technical basis for narcotic chemicals and polycyclic aromatic hydrocarbon criteria. II. Mixtures and sediments. *Environmental Toxicology and Chemistry*, **19**, 1971–1982.

DiToro DM, Zarba CS, Hansen DJ, Berry WJ, Swartz RC, Cowan CE, Pavlou SP, Allen HE, Thomas NA and Paquin PR (1991) Technical basis for establishing sediment quality criteria for non-ionic organic chemicals using equilibrium partitioning. *Environmental Toxicology and Chemistry*, **10**, 1541–1583.

Forbes VE and Forbes TL (1997) Dietary absorption of sediment-bound fluoranthene by a deposit-feeding gastropod using the ^{14}C:^{51}Cr dual-labeling method. *Environmental Toxicology and Chemistry*, **16**, 1002–1009.

Frank AP, Landrum PF and Eadie BJ (1986) Polycyclic aromatic hydrocarbon rates of uptake, depuration, and biotransformation by Lake Michigan *Stylodrilus heringianus*. *Chemosphere*, **15**, 317–330.

Fuchsman PC and Barber TR (2000) Spiked sediment toxicity testing of hydrophobic organic chemicals: bioavailability, technical considerations, and applications. *Soil and Sediment Contamination*, **9**, 197–218.

Gauthier TD, Seitz WR and Grant CL (1987) Effects of structural and compositional variations of dissolved humic materials on pyrene K_{oc} values. *Environmental Science and Technology*, **21**, 234–248.

Gevao B, Mordaunt C, Semple KT, Piearce TG and Jones KC (2001) Bioavailability of non-extractable (bound) pesticide residues to earthworms. *Environmental Science and Technology*, **35**, 501–507.

Ghosh U, Gillette JS, Luthy RG and Zare RN (2000) Microscale location, characterization, and association of polycyclic aromatic hydrocarbons on harbor sediment particles. *Environmental Science and Technology*, **34**, 1729–1736.

Gobas F (2000) Bioconcentration and biomagnification in the aquatic environment. In Boethling RS and Mackay D (eds), *Handbook of Property Estimation Methods for Chemicals*. Lewis, New York, pp. 191–227.

Goldberg ED (1985) *Black Carbon in the Environment: Properties and Distribution*. Wiley, New York, p. 198.

Grathwohl P (1990) Influence of organic matter from soils and sediments from various origins on the sorption of some chlorinated aliphatic hydrocarbons: implications on K_{OC} correlations. *Environmental Science and Technology*, **24**, 1687–1693.

Gustafsson O and Gschwend PM (1997a) Soot as a strong partition medium for polycyclic aromatic hydrocarbons in aquatic systems. In Eganhouse RP (ed.), *Molecular Markers in Environmental Geochemistry*. American Chemical Society, Washington, DC, pp. 365–381.

Gustafsson O and Gschwend PM (1997b) Aquatic colloids: concepts, definitions, and current challenges. *Limnology and Oceanography*, **42**, 519–528.

Gustafsson O and Gschwend PM (1998) The flux of black carbon to surface sediments on the New England continental shelf. *Geochimica et Cosmochimica Acta*, **62**, 465–472.

Gustafsson O and Gschwend PM (1999) Phase distribution of hydrophobic chemicals in the aquatic environment. In Baveye *et al.* (eds), *Bioavailability of Organic Xenobiotics in the Environment*. Kluwer Academic, Amsterdam, pp. 327–348.

Gustafsson O, Haghseta F, Chan C, MacFarlane J and Gschwend PM (1997) Quantification of the dilute sedimentary soot phase: implications for PAH speciation and bioavailability. *Environmental Science and Technology*, **31**, 203–209.

Hamelink JL, Waybrant RC and Ball RC (1971) A proposal: exchange equilibria control the degree chlorinated hydrocarbons are biologically magnified in lentic environments. *Transactions of the American Fisheries Society*, **100**, 207–214.

Hamelink JL, Landrum PF, Bergman HL and Benson WH (1994) *Bioavailability: Physical, Chemical and Biological Interactions*. Lewis, CRC Press, Boca Raton, FL, p. 239.

Harkey GA, Lydy MJ, Kukkonen J and Landrum PF (1994) Feeding selectivity and assimilation of PAH and PCB in *Diporeia* spp. *Environmental Toxicology and Chemistry*, **13**, 1445–1455.

Harkey G, van Hoof P and Landrum P (1995) Bioavailability of polycyclic aromatic hydrocarbons from a historically contaminated sediment core. *Environmental Toxicology and Chemistry*, **14**, 1551–1560.

Hatzinger P and Alexander M (1995) Effect of aging of chemicals in soil on their biodegradability and extractability. *Environmental Science and Technology*, **29**, 537–545.

Hawthorne S and Grabanski C (2000) Correlating selective supercritical fluid extraction with bioremediation behavior of PAHs in a field treatment plot. *Environmental Science and Technology*, **34**, 4103–4110.

Herbes SE (1977) Partitioning of polycyclic aromatic hydrocarbons between dissolved and particulate phases in natural waters. *Water Research*, **11**, 493–496.

Hickey CW, Roper DS, Holland PT and Trower TM (1995) Accumulation of organic contaminants in two sediment-dwelling shellfish with contrasting feeding modes: deposit- (*Macomona liliana*) and filter-feeding (*Austrovenus stutchburyi*). *Archives of Environmental Contamination and Toxicology*, **29**, 221–231.

Huang W and Weber WJ (1997) A distributed reactivity model for sorption by soils and sediments. 10. Relationships between desorption, hysteresis, and the chemical characteristics of organic domains. *Environmental Science and Technology*, **31**, 2562–2569.

Jonker MTO and Smedes F (2000) Preferential sorption of planar contaminants in sediments from Lake Ketelmeer, The Netherlands. *Environmental Science and Technology*, **34**, 1620–1626.

Kaag NHBM, Foekema EM, Scholten MCT and vanStraalen NM (1997) Comparison of contaminant accumulation in three species of marine invertebrates with different feeding habits. *Environmental Toxicology and Chemistry*, **16**, 837–842.

Karickhoff SW (1981) Semi-empirical estimation of sorption of hydrophobic pollutants on natural sediments and soils. *Chemosphere*, **10**, 833–846.

Karickhoff SW, Brown DS and Scott TA (1979) Sorption of hydrophobic pollutants on natural sediments. *Water Research*, **13**, 241–248.

Kelsey JW, Kottler BD and Alexander M (1997) Selective chemical extractants to predict bioavailability of soil-aged organic chemicals. *Environmental Science and Technology*, **31**, 214–217.

Kenaga EE and Goring CAI (1980) Relationship between water solubility, soil sorption, octanol–water partitioning, and concentration of chemicals in biota. In Eaton JG, Parrish PR and Hendricks AC (eds), *Aquatic Toxicology*. ASTM STP 707. American Society for Testing and Materials, Philadelphia, PA, pp. 78–115.

Kile DE, Chiou CT, Zhou H, Li H and Xu O (1995) Partition of non-polar organic pollutants from water to soil and sediment organic matters. *Environmental Science and Technology*, **29**, 1401–1406.

Kraaij RH (2001) Sequestration and Bioavailability of Hydrophobic Chemicals in Sediment. Thesis, Institute for Risk Assessment Sciences Utrecht University, Utrecht, The Netherlands.

Kukkonen J and Landrum PF (1994) Toxicokinetics and toxicity of sediment-associated pyrene to *Lumbriculus variegatus* (Oligochaeta). *Environmental Toxicology and Chemistry*, **13**, 1457–1468.

Kukkonen J and Landrum PF (1995) Measuring assimilation efficiencies for sediment-bound PAH and PCB congeners by benthic organisms. *Aquatic Toxicology*, **32**, 75–92.

Kukkonen JVK and Landrum PF (1998) Effect of particle–xenobiotic contact time on bioavailability of sediment-associated benzo[a]pyrene to benthic amphipods, *Diporeia* spp. *Aquatic Toxicology*, **42**, 229–242.

Lake JL, Rubinstein NI, Lee H, Lake CA, Heltshe J and Pavignano S (1990) Equilibrium partitioning and bioaccumulation of sediment-associated contaminants by infaunal organisms. *Environmental Toxicology and Chemistry*, **9**, 1095–1106.

Lake JL, McKinney R, Osterman FA and Lake CA (1996) C18-coated silica particles as a surrogate for benthic uptake of hydrophobic compounds from bedded sediment. *Environmental Toxicology and Chemistry*, **15**, 2284–2289.

Lamoureux EM and Brownawell BJ (1999) Chemical and biological availability of sediment-sorbed hydrophobic organic contaminants. *Environmental Toxicology and Chemistry*, **18**, 1733–1741.

Landrum P, Eadie B and Faust W (1992) Variation in the bioavailability of polycyclic aromatic hydrocarbons to the amphipods *Diporeia* (spp.) with sediment aging. *Environmental Toxicology and Chemistry*, **11**, 1197–1208.

Landrum PF and Scavia D (1983) Influence of sediment on anthracene uptake, depuration, and biotransformation by the amphipod *Hyalella azteca*. *Canadian Journal of Fisheries and Aquatic Science*, **40**, 298–305.

Landrum PF and Robbins JA (1990) Bioavailability of sediment-associated contaminants to benthic invertebrates. In Baudo R, Giesy JP and Muntau H (eds), *Sediments: Chemistry and Toxicity of In-place Pollutants*. Lewis, Chelsea, MI, pp. 237–263.

Lane WF and Loehr RC (1992) Estimating the equilibrium aqueous concentrations of polynuclear aromatic hydrocarbons in complex mixtures. *Environmental Science and Technology*, **26**, 983–990.

LeBoeuf EJ and Weber JJ (1999) Reevaluation of general partitioning model for sorption of hydrophobic organic contaminants by soil and sediment organic matter. *Environmental Toxicology and Chemistry*, **18**, 1617–1626.

Lee H (1992) Models, muddles, and mud: predicting bioaccumulation of sediment-associated pollutants. In Burton GA (ed.), *Sediment Toxicity Assessment*. Lewis, CRC Press, Boca Raton, FL, pp. 267–293.

Lee H, Boese BL, Pelletier J, Winsor M, Specht DT and Randall RC (1993) *Bedded Sediment Bioaccumulation Tests: Guidance Manual*. EPA 600/R-93/183, United States Environmental Protection Agency, Office of Research and Development, Washington, DC.

Leppänen M (1995) The role of feeding behaviour in bioaccumulation of organic chemicals in benthic organisms. *Annales Zoologici Fennici*, **32**, 247–255.

Leppänen MT and Kukkonen JVK (2000) Effect of sediment-chemical contact time on availability of sediment-associated pyrene and benzo[a]pyrene to oligochaete worms and semi-permeable membrane devices. *Aquatic Toxicology*, **49**, 227–241.

Leslie HA, Oosthoek AJP, Busser FJM, Kraak MHS and Hermens JLM (2002) Biomimetic solid-phase microextraction to predict body residues an toxicity of chemicals that act by narcosis. *Environmental Toxicology and Chemistry*, **21**, 229–234.

Loehr R and Webster M (1996) Behavior of fresh vs. aged chemicals in soil. *Journal of Contaminant Hydrology*, **5**, 361–383.

Lopez GR and Levinton JS (1987) Ecology of deposit-feeding animals in marine sediments. *Quarterly Review of Biology*, **62**, 235–260.

Luthy RG, Dzombak DA, Peters CA, Roy SB, Ramaswami A, Nakles DV and Nott BR (1994) Remediating tar-contaminated soils at manufactured gas plant sites. *Environmental Science and Technology*, **28**, 266–276A.

Luthy RG, Aiken GR, Brusseau ML, Cunningham, SD, Gschwend PM, Pignatello JJ, Reinhard M, Traina SJ, Weber WJ and Westall JC (1997) Sequestration of hydrophobic organic contaminants by geosorbents. *Environmental Science and Technology*, **31**, 3341–3347.

Lydy MJ and Landrum PF (1993) Assimilation efficiency for sediment-sorbed benzo[a]pyrene by *Diporeia* spp. *Aquatic Toxicology*, **26**, 209–224.

Mackay D (1979) Finding fugacity feasible. *Environmental Science and Technology*, **13**, 1218–1223.

Mackay D (1982) Correlation of bioconcentration factors. *Environmental Science and Technology*, **16**, 274–278.

MacKay AA and Gschwend PM (2001) Enhanced concentrations of PAHs in groundwater at a coal tar site. *Environmental Science and Technology*, **35**, 1320–1328.

Mackay D, Shiu WY and Ma KC (1992) *Illustrated Handbook of Physical–Chemical Properties and Environmental Fate for Organic Chemicals. Volume II, Polynuclear Aromatic Hydrocarbons, Polychlorinated Dioxins and Dibenzofurans*. Lewis, CRC Press, Boca Raton, FL, p. 597.

MacKay D and Fraser A (2000) Bioaccumulation of persistent organic chemicals: mechanisms and models. *Environmental Pollution*, **110**, 375–391.

MacRae JD and Hall KJ (1998) Comparison of methods used to determine the availability of polycyclic aromatic hydrocarbons in marine sediment. *Environmental Science and Technology*, **32**, 3809–3815.

Mahjoub B, Jayr E, Bayard R and Gourdon R (2000) Phase partition of organic pollutants between coal tar and water under variable experimental conditions. *Water Research*, **34**, 3551–3560.

Markwell RD, Connell DW and Gabric AJ (1989) Bioaccumulation of lipophilic compounds from sediments by oligochaetes. *Water Research*, **23**, 1443–1450.

Maruya KA, Risebrough RW and Horne AJ (1996) Partitioning of polynuclear aromatic hydrocarbons between sediments from San Francisco Bay and their porewaters. *Environmental Science and Technology*, **30**, 2942–2947.

Mayer LM, Chen Z, Findlay RH, Fang JS, Sampson S, Self RFL, Jumars PA, Quetel C and Donard OFX (1996) Bioavailability of sedimentary contaminants subject to deposit-feeder digestion. *Environmental Science and Technology*, **30**, 2641–2645.

Mayer LM, Schick LL, Self RFL, Jumars PA, Findlay RH, Chen Z and Sampson S (1997) Digestive environments of benthic macroinvertebrate guts: enzymes, surfactants and dissolved organic matter. *Journal of Marine Research*, **55**, 785–812.

Mayer P, Vaes W, Wijnker F, Legierse K, Kraaij R, Tolls J and Hermens JLM (2000) Sensing dissolved sediment porewater concentrations of persistent and bioaccumulative pollutants using disposable solid phase microextraction fibers. *Environmental Science and Technology*, **34**, 5177–5183.

Mayer LM, Weston DP and Bock MJ (2001) Benzo[a]pyrene and zinc solubilization by digestive fluids of benthic invertebrates — a cross-phyletic study. *Environmental Toxicology and Chemistry*, **20**, 1890–1900.

McFarland VA and Clarke JU (1986) Testing bioavailability of polychlorinated biphenyls using a two-level approach. In Willey RG (ed.), *Water Quality R&D: Successfully Bridging between Theory and Application*. Hydrologic Engineering Center, Davis, CA, pp. 220–229.

McElroy AE, Farrington JW and Teal JM (1989) Bioavailability of polycyclic aromatic hydrocarbons in the aquatic environment. In Varanasi U (ed.), *Metabolism of Polycyclic Aromatic Hydrocarbons in the Aquatic Environment*. CRC Press, Boca Raton, FL, pp. 1–39.

McGroddy SE and Farrington JW (1995) Sediment porewater partitioning of polycyclic aromatic hydrocarbons in three cores from Boston Harbor, Massachusetts. *Environmental Science and Technology*, **29**, 1542–1550.

McGroddy SE, Farrington JW and Gschwend PM (1996) Comparison of the *in situ* and desorption sediment-water partitioning of polycyclic aromatic hydrocarbons and polychlorinated biphenyls. *Environmental Science and Technology*, **30**, 172–177.

Meador JP, Stein JE, Reichert WL and Varanasi U (1995a) Bioaccumulation of polycyclic aromatic hydrocarbons by marine organisms. *Reviews of Environmental Contamination and Toxicology*, **143**, 79–165.

Meador JP, Casillas E, Sloan CA and Varanasi U (1995b) Comparative bioaccumulation of polycyclic aromatic hydrocarbons from sediment by two infaunal invertebrates. *Marine Ecology Progress Series*, **123**, 107–124.

Means JC, Hassett JJ, Wood SG and Banwart WL (1979) Sorption properties of energy-related pollutants and sediments. In Jones PW and Leber P (eds), *Polynuclear Aromatic Hydrocarbons*. Ann Arbor Science, Ann Arbor, MI, pp. 327–340.

Means JC, Wood SG, Hassett JJ and Banwart WL (1980) Sorption of polynuclear aromatic hydrocarbons by sediments and soils. *Environmental Science and Technology*, **14**, 1524–1528.

Middelburg JJ, Nieuwenhuize J and Breugel PV (1999) Black carbon in marine sediments. *Marine Chemistry*, **65**, 245–252.

Mitra S and Dickhut RM (1999) Three-phase modeling of polycyclic aromatic hydrocarbon association with pore-water-dissolved organic carbon. *Environmental Toxicology and Chemistry*, **18**, 1144–1148.

Opperhuizen A (1991) Bioaccumulation kinetics: experimental data and modeling. In *Proceedings of the Sixth European Symposium on Organic Micropollutants in the Aquatic Environment*. Kluwer Academic, Amsterdam, pp. 61–70.

Paine MD, Chapman PM, Allard PJ, Murdoch MH and Minifie D (1996) Limited bioavailability of sediment PAH near an aluminum smelter: contamination does not equal effects. *Environmental Toxicology and Chemistry*, **15**, 2003–2018.

Pignatello JJ and Xing B (1996) Mechanisms of slow sorption of organic chemicals to natural particles. *Environmental Science and Technology*, **30**, 1–11.

Plante CJ and Jumars PA (1992) The microbial environment of marine deposit-feeder guts characterized via microelectrodes. *Microbial Ecology*, **23**, 257–277.

Readman JW, Mantoura RFC and Rhead MM (1984) The physico-chemical speciation of polycyclic aromatic hydrocarbons (PAH) in aquatic systems. *Fresenius Journal of Analytical Chemistry*, **319**, 126–131.

Readman JW, Mantoura RFC and Rhead MM (1987) A record of polycyclic aromatic hydrocarbon (PAH) pollution obtained from accreting sediments of the Tamar estuary, UK: evidence for non-equilibrium behavior in PAH. *Science of the Total Environment*, **66**, 73–94.

Reid VJ, Jones KC and Semple KT (2000a) Bioavailability of persistent organic pollutants in soils and sediments — a perspective on mechanisms, consequences and assessment. *Environmental Pollution*, **108**, 103–112.

Reid BJ, Stokes JD, Jones KC and Semple KT (2000b) Non-exhaustive cyclodextrin-based extraction technique for the evaluation of PAH bioavailability. *Environmental Science and Technology*, **34**, 3174–3179.

Scribner SL, Benzing TR, Sun S and Byd SA (1992) Desorption and bioavailability of aged simazine residues in soil from a continuous corn field. *Journal of Environmental Quality*, **21**, 115–120.

Sigleo AC and Means JC (1990) Organic and inorganic components in estuarine colloids: implications for sorption and transport of pollutants. *Reviews of Environmental Contamination and Toxicology*, **112**, 123–147.

Sijm D, Kraaij R and Belfroid A (2000) Bioavailability in soil or sediment: exposure of different organisms and approaches to study it. *Environmental Pollution*, **108**, 113–119.

Socha SB and Carpenter R (1987) Factors affecting pore water hydrocarbon concentrations in Puget Sound sediments. *Geochimica et Cosmochimica Acta*, **51**, 1273–1284.

Stegeman JJ and Teal JM (1973) Accumulation, release and retention of petroleum hydrocarbons by the oyster, *Crassostrea virginica. Marine Biology*, **22**, 37–44.

Tang J, Robbertson B and Alexander M (1999) Chemical extraction methods to estimate bioavailability of DDT, DDE and DDD in soil. *Environmental Science and Technology*, **33**, 4346–4351.

Tang J, Carroquino MJ, Robertson BK and Alexander M (1998) Combined effect of sequestration and bioremediation in reducing the bioavailability of polycyclic aromatic hydrocarbons in soil. *Environmental Science and Technology*, **32**, 3586–3590.

Tang J and Alexander M (1999) Mild extractability and bioavailability of polycyclic aromatic hydrocarbons in soil. *Environmental Toxicology and Chemistry*, **18**, 2711–2714.

Ten Hulscher TEM, van Noort P and van der Velde L (1997) Equilibrium partitioning theory overestimates chlorobenzene concentrations in sediment-porewater from Lake Ketelmeer, The Netherlands. *Chemosphere*, **35**, 2331–2344.

Thomann RV (1989) Bioaccumulation model of organic chemical distribution in aquatic food chains. *Environmental Science and Technology*, **23**, 699–707.

Thomann RV, Connolly JP and Parkerton TF (1992) An equilibrium model of organic chemical accumulation in aquatic food webs with sediment interaction. *Environmental Toxicology and Chemistry*, **11**, 615–630.

Thomann RV and Komlos J (1999) Model of biota-sediment accumulation factor for polycyclic aromatic hydrocarbons. *Environmental Toxicology and Chemistry*, **18**, 1060–1068.

Tracey GA and Hansen DJ (1996) Use of biota–sediment accumulation factors to assess similarity of non-ionic organic chemical exposure to benthically-coupled organisms of differing trophic mode. *Archives of Environmental Contamination and Toxicology*, **30**, 467–475.

USEPA (Environmental Protection Agency) (2002a) *Procedures for the Derivation of Site-specific Equilibrium Partitioning Sediment Benchmarks (ESBs) for the Protection of Benthic Organisms: Non-ionic Organics*. EPA-600-R-02-012. Office of Research and Development, Washington, DC, 20460 (draft).

USEPA (Environmental Protection Agency) (2002b) *Procedures for the Derivation of Equilibrium Partitioning Sediment Benchmarks (ESBs) for the Protection of Benthic Organisms: PAH Mixtures*. EPA-600-R-02-013. Office of Research and Development, Washington, DC, 20460 (draft).

Van Brummelen TA, van Hattum B, Crommentuijn T and Kalf DF (1998). Bioavailability and ecotoxicity of PAHs. In Neilson A and Hutzinger O (eds), *The Handbook of Environmental Chemistry*. Volume 3, Part J. Springer Verlag, Berlin, pp. 203–263.

Van der Kooij LA, van de Meent D, van Leeuwen CJ and Bruggeman WA (1991) Deriving quality criteria for water and sediment from the results of aquatic toxicity tests and product standards: application of the equilibrium partitioning method. *Water Research*, **25**, 697–705.

Van Hattum B, Curto Pons MJ and Cid Montañés JF (1998) Polycyclic aromatic hydrocarbons in freshwater isopods and field-partitioning between abiotic phases. *Archives of Environmental Contamination and Toxicology*, **35**, 257–267.

Voparil IM and Mayer LM (2000) Dissolution of sedimentary polycyclic aromatic hydrocarbons into the lugworm's (*Arenicola marina*) digestive fluids. *Environmental Science and Technology*, **34**, 1221–1228.

Wang W-X and Fisher NS (1999) Assimilation efficiencies of chemical contaminants in aquatic invertebrates: a synthesis. *Environmental Toxicology and Chemistry*, **18**, 2034–2045.

Wang X-C, Zhang Y-X and Chen RF (2001) Distribution and partitioning of polycyclic aromatic hydrocarbons (PAHs) in different size fractions in sediments from Boston Harbor, United States. *Marine Pollution Bulletin*, **42**, 1139–1149.

Weber WJ and Gould JP (1966) Sorption of organic pesticides from aqueous solution. In Rosen AA and Kraybill HF (eds), *Organic Pesticides in the Environment*. Advances in Chemistry Series 60, American Chemistry Society, Washington, DC, pp. 280–304.

Weber JJ, Huang W and LeBoeuf EJ (1999) Geosorbent organic matter and its relationship to the binding and sequestration of organic contaminants. *Colloids and Surfaces A: Physicochemical and Engineering Aspects*, **151**, 167–179.

Weston DP and Mayer LM (1998a) Comparison of *in vitro* digestive fluid extraction and traditional *in vivo* approaches as measures of polycyclic aromatic hydrocarbon bioavailability from sediments. *Environmental Toxicology and Chemistry*, **17**, 830–840.

Weston DP and Mayer LM (1998b) *In vitro* digestive fluid extraction as a measure of the bioavailability of sediment-associated polycyclic aromatic hydrocarbons: sources of variation and implications for partitioning models. *Environmental Toxicology and Chemistry*, **17**, 820–829.

White J and Alexander M (1996) Reduced biodegradability of desorption-resistant fractions of polycyclic aromatic hydrocarbons in soil and aquifer solids. *Environmental Toxicology and Chemistry*, **15**, 1973–1978.

White JC, Kelsey JW, Hatzinger PB and Alexander M (1997) Factors affecting sequestration and bioavailability of phenanthrene in soils. *Environmental Toxicology and Chemistry*, **16**, 2040–2045.

White JC, Quinones-Rivera A and Alexander M (1998) Effect of wetting and drying on the bioavailability of organic compounds sequestered in soil. *Environmental Toxicology and Chemistry*, **17**, 2378–2382.

White J, Hunter M, Pignatello J and Alexander M (1999) Increase in bioavailability of aged phenanthrene in soils by competitive displacement with pyrene. *Environmental Toxicology and Chemistry*, **18**, 1728–1732.

Wijayaratne RD and Means JC (1984) Sorption of polycyclic aromatic hydrocarbons by natural estuarine colloids. *Marine Environmental Research*, **11**, 77–89.

Williamson DG, Loehr RC and Kimura Y (1998) Release of chemicals from contaminated soils. *Soil Contamination*, **7**, 543–558.

Wong CS, Capel PD and Nowell LH (2001) National-scale, field-based evaluation of the biota–sediment accumulation factor model. *Environmental Science and Technology*, **35**, 1709–1715.

Xing B and Pignatello JJ (1996) Time-dependent isotherm shape of organic compounds in soil organic matter: implications for sorption mechanism. *Environmental Toxicology and Chemistry*, **15**, 1282–1288.

Xing B, Pignatello JJ and Gigliotti B (1996) Competitive sorption between atrazine and other organic compounds in soils and model sorbents. *Environmental Science and Technology*, **30**, 2432–2440.

Young TM and Weber WJ (1995) A distributed reactivity model for sorption by soils and sediments. 3. Effects of diagenetic processes on sorption energies. *Environmental Science and Technology*, **29**, 92–97.

Bioavailability, Uptake and Effects of PAHs in Aquatic Invertebrates in Field Studies

PIETER J. DEN BESTEN[1], DORIEN TEN HULSCHER[1]
AND BERT VAN HATTUM[2]

[1]*Institute for Inland Water Management and Waste Water Treatment (RIZA),
Lelystad, The Netherlands*
[2]*Institute for Environmental Studies, Vrije Universiteit, Amsterdam, The Netherlands*

8.1 INTRODUCTION

This chapter will focus on the uptake and effects of PAHs in aquatic invertebrates. First, data will be presented of PAH levels in foodchains obtained from field studies in the delta of the rivers Rhine and Meuse. Special attention will be given to different approaches that can be used to assess the bioavailability of PAHs. Finally, the contribution of PAHs in effects on aquatic invertebrates is discussed.

8.2 PAH LEVELS IN SEDIMENTS AND AQUATIC FOODCHAINS

Polycyclic aromatic hydrocarbons (PAHs) are a widely distributed group of organic pollutants originating from petrogenic, pyrogenic and natural sources (see Chapters 1, 2 and 3). With the successful abatement of point sources in the past decades, non-point sources, such as atmospheric deposition and surface runoff, nowadays constitute major inputs into the environment. They have been of environmental concern because of the mutagenic and carcinogenic properties of the metabolites of several compounds, and are included in most environmental monitoring programs described in various monographs (Neff 1979; Meador *et al.* 1995; Neilson and Hutzinger 1998).

Due to the partitioning behavior of PAHs, sediments and particulate matter are the main compartments in the aquatic system, where PAHs can be found. In sedimentation areas, such as estuaries of large rivers and harbours, increased

PAHs: An Ecotoxicological Perspective. Edited by Peter E.T. Douben.
© 2003 John Wiley & Sons Ltd

benthic exposure to PAHs can be expected. This chapter includes an example of accumulation in foodchains in the delta of the rivers Rhine and Meuse.

Time series data for the rivers Rhine and the Meuse indicate that PAH levels in sediments and particulate matter decreased during the 1980s, but that no or only small further decreases were noted in the 1990s (Gandrass and Salomons 2001; Den Besten 1997). Results for two locations in one of the main sedimentation areas (Hollandsch Diep and Haringvliet) are given in Figure 8.1, derived from a national database on water and sediment quality parameters (RIZA/RIKZ 2000). The spatial distribution of PAHs in sieved (< 63 μm) sediment fractions along the Dutch coast (Figure 8.2, derived from Laane *et al.* 1999) reveals that the highest concentrations are found adjacent to the coast, near major rivers (Scheldt, Rhine–Meuse estuary), freshwater outlets (Noordzeekanaal) and coastal dumping grounds of harbor dredgings (Loswal Noord). In the coastal areas, north of the Nieuwe Waterweg, concentrations decreased by 26% between 1986 and 1996. In the southern coastal waters no significant change in sediment PAH concentrations was noted. The relatively high concentrations in the northern part of the offshore waters, the Oyster Grounds, which increased during 1986–1992, were attributed to potential influences from other sources, such as shipping and oil and gas production activities, according to Laane *et al.* (1999). Also relatively high PAH levels have been found in sediment and benthic fauna of the Dogger Bank, in the northern part of the North Sea (Den Besten *et al.* 2001).

As a result of biotransformation of PAHs in vertebrates and some invertebrates, food chain transfer and biomagnification of PAHs do not appear to exist in aquatic and terrestrial environments (Broman *et al.* 1990; Neff 1979; Clements *et al.* 1994). Although some primary consumers and detritivores may accumulate high levels of PAHs (Southworth *et al.* 1978; Dobroski and Epifanio 1980; Van Straalen *et al.* 1993; Van Hattum *et al.* 1998a; Thomann and Komlos 1999),

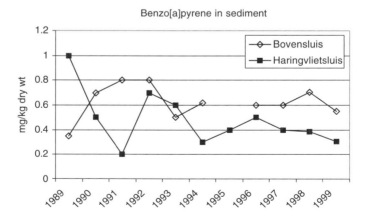

Figure 8.1   Concentrations of benzo[a]pyrene in sediments (mg/kg dry wt) in the Rhine-Meuse delta. Source: RIZA/RIKZ (2000). Reproduced with permission

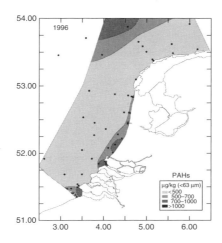

Figure 8.2 Spatial distribution of total PAHs (μg/kg dry wt) in sieved sediments ($< 63\ \mu$m) along the Dutch coast in 1996. Reprinted from: Laane *et al.* (1999) with permission from Elsevier Science

predators usually contain low levels (Clements *et al.* 1994; Hellou *et al.* 1991; Lemaire *et al.* 1993; Niimi and Dookran 1989).

During 1990–1996, extensive biomonitoring studies have been conducted on aquatic organisms in the Rhine–Meuse estuary, within the framework of sediment-remediation studies (see e.g. Den Besten *et al.* 1995). Measured PAH concentrations have been summarized in Table 8.1.

In Figure 8.3 an example is presented of benzo[a]pyrene and PCB 153 concentrations in aquatic organisms from various interconnected waters in the Rhine–Meuse estuary, based on Van Hattum *et al.* (1993, 1996, 1998b). The absence of biomagnification of PAHs in the food web is clearly illustrated, while for the PCBs a sharp increase with trophic level can be noted. The qualitative comparison of lipid-based concentrations shows that the highest PAH concentrations were encountered in aquatic plants (pondweed, *Potomogeton pectinatus*; waterweed, *Elodea* spp.), oligochaetes, isopods and freshwater clams (*Anodonta cygnea*). Lower concentrations were observed in other molluscs (*Dreissena polymorpha*, *Corbicula* spp.) and in chironomids. Concentrations below detection limits were observed in whole body homogenates of roach (*Rutilus rutilus*) and in the liver of 7 week-old cormorant chicks from the Biesbosch colony, feeding on roach and other cyprinids from this area. Due to variations in organism lipid content, a different ranking was seen when data are expressed on a wet weight basis. Wet weight concentrations of PAHs decreased in the order: isopods > oligochaetes > aquatic plants > *Anodonta* > other molluscs and chironomids > fish. For other PAHs, similar patterns of differences between species were observed. The process of decreasing concentrations with rising trophic level, which has been observed also for some trace metals, has been termed 'biominification' (Campbell *et al.* 1988).

TABLE 8.1 PAH concentrations in sediments (mg/kg dry wt; 1992–1997) and aquatic species (µg/kg wet wt; 1992–1996) in the Rhine-Meuse delta. Indicated are average values[1] or single observations. The Rhine-Meuse delta is indicated in the map in Figure 8.2 (South-West part of The Netherlands).

	n	dry wt (%)	Lipid (org-C) (%)	Naph	Acn	Acy	Flu	Phen	Anth	Flua	Pyr	BaA	Chry	BbF	BkF	BaP	DbA	BgP	IcdP	Sum PAHs
Sediments[2]																				
Hollandsch Diep	63	31.1	4.3	—	—	—	< 0.1	0.7	0.3	1.4	1.2	0.6	0.9	1.0	0.4	0.8	0.1	0.7	0.7	8.8
Biesbosch	13	42.7	7.9	0.7	0.3	< 0.1	0.2	0.9	0.4	1.7	1.3	0.8	1.1	1.2	0.5	0.9	0.1	0.8	0.7	11.7
Haringvliet	11	31.2	3.4	0.1	< 0.1	< 0.1	0.1	0.4	0.2	0.8	0.6	0.4	0.4	0.7	0.3	0.5	0.1	0.4	0.5	5.5
Aquatic macrophytes																				
Potamogeton pectinatus																				
Nieuwe Merwede	7	12.9	0.4	6	< 5	< 5	6	17	7	19	16	8	13	14	5	9	< 0.5	< 0.5	< 0.5	126
Dordtse Biesbosch	1		0.3	< 3	< 7	23	3	12	4	22	22	7	8	10	5	6	1	6	8	142
Amer	1	7.9	0.1	< 6	< 6	< 8	< 3	< 16	2	16	11	7	6	9	4	6	< 2	11	< 5	92
Brabantsche Biesbosch	1		0.3	< 3	< 7	28	3	18	5	31	34	13	14	19	9	13	3	13	20	230
Potamogeton nododus																				
Hollandsch Diep	1		0.3	< 3	< 7	8	< 2	7	2	16	19	6	7	4	5	6	2	8	11	105
Haringvliet	3	25.3	0.4	< 16	< 2	< 8	3	8	3	14	14	8	8	16	6	10	2	9	7	118
Elodea spp.																				
Amer	2	12.0	0.3	< 6	< 6	< 8	< 3	< 16	1	< 13	6	5	6	8	3	5	< 2	5	5	69
Dordtse Biesbosch	1	28.0	0.7	< 4	< 3	< 4	5	< 18	2	< 16	10	8	8	11	5	9	2	8	10	100
Zannichellia spp.																				
Amer	1	13.5	0.1	< 6	< 6	< 8	< 3	< 16	< 2	< 12	< 7	3	4	7	< 3	3	< 2	4	5	58
Invertebrates																				
Oligochaetes																				
Hollandsch Diep	1	11.4	1.0	< 5	< 5	< 6	< 4	34	22	109	102	36	49	35	15	25	4	19	16	475
Markermeer	2	12.1	1.1	< 8	< 8	< 10	< 5	< 40	3	< 36	15	3	3	6	2	2	< 1	2	< 6	85
Chironomids																				
Dordtse Biesbosch	1	12.1	1.1	< 17	< 14	< 15	< 11	< 80	< 8	< 70	< 43	< 6	< 9	15	4	2	< 2	6	< 10	169
Hollandsch Diep	1	8.5	0.9	< 5	< 4	< 4	< 3	< 25	5	40	50	8	11	10	3	6	1	11	4	168
Markermeer	2	6.4	0.6	< 35	< 28	< 14	< 23	< 170	< 17	< 150	< 90	< 12	< 20	2	1	1	1	2	< 21	163

Corbicula

Amer	3	17.9	3.7	<12	<16	<6	<33	3	102	220	110	100	79	28	31	3	23	16	753
Brabantsche Biesbosch	2	16.8	2.2	<3	15	3	18	4	92	190	69	72	38	15	15	3	17	13	568
Haringvliet	8	18.3	3.4	<25	<11	3	10	5	90	145	71	77	33	12	14	2	11	<10	492
Hollandsch Diep	7	20.5	2.9	<56	<68	<37	<280	<28	63	91	32	68	31	12	15	4	12	12	436
Nieuwe Merwede	9	13.0	1.5	<5	<5	<5	8	13	25	19	10	22	10	2	6	<0.5	<0.5	<0.5	125
Rhine Mainz-Duisburg	2	12.0	2.1	6	<5	12	45	5	51	49	18	51	10	6	6	0.3	0.3	3	266

Dreissena polymorpha

Brabantsche Biesbosch	2	11.8	1.8	<3	<8	2	14	3	57	128	44	55	32	15	18	3	10	12	411
Hollandsch Diep	1	22.8	1.8	<8	<17	<5	<40	<4	<35	22	14	33	11	6	7	<1	3	<5	157
IJsselmeer	1	5.0	0.5	<2	<2	<2	<12	<1	<10	<6	<1	1	1	0.3	0.2	0.2	0.1	<1	23
Nieuwe Merwede	1	14.4	1.9	<5	<5	<5	<5	4	<10	<2	11	19	7	2	6	<0.5	<0.5	<0.5	68

Fish

Roach

Nieuwe Merwede	2	24.2	2.7	<5	<5	<5	<5	<1	<10	<2	1.5	<1	1.3	<0.6	<0.6	<0.6	<0.6	<0.6	—

[1] Average values; in case only observations < detection limit (DL) were found this has been indicated; in other cases averages and sum PAHs were calculated with 0.5*DL as substitute values.
[2] For the sediments the value for organic carbon has been indicated in the fourth column. Source: sediment data from RIZA/RIKZ (2000); data on aquatic organisms from Van Hattum et al. 1993, 1996, 1998b.

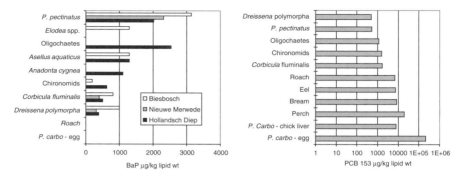

Figure 8.3   Benzo[a]pyrene (μg/kg lipid wt) and PCB 153 (μg/kg lipid wt) in organisms from adjacent waters in the Rhine–Meuse delta. PCB data are given for Hollandsch Diep and Biesbosch (cormorant data)

In line with the observations from the literature, the relative contribution of individual PAHs is species-dependent, as shown in Figure 8.4. The patterns in the aquatic macrophytes are more similar to the pattern in the sediment than the PAH distribution in the invertebrates where, especially in the chironomids, the contribution of the higher molecular weight PAHs is relatively low.

Apart from differences in uptake, storage and excretion mechanisms, differences in metabolization capacity are likely to be responsible for the decreased accumulation of higher PAHs in organisms at higher trophic levels. The differences between the oligochaetes and chironomids are in line with results from toxicokinetic experiments (summarized in Van Brummelen *et al.* 1998), demonstrating biotransformation in some insect species. As evidenced from the results for *Corbicula* and *Dreissena*, distinct differences are present even among closely related phylogenetic species.

8.3   BIOAVAILABILITY

8.3.1   LITERATURE ON THE BIOAVAILABILITY OF PAHs IN SEDIMENTS AND SOILS FOR UPTAKE IN FAUNA

All of the above-mentioned biological factors may have a large influence on the levels that are accumulated in organisms. At the same time, however, the main source of the accumulating PAHs in food chains (sediments and particulates) shows a considerable variability in bioavailability. Thomann and Komlos (1999) proposed a food web model for the accumulation of sediment-bound PAHs in invertebrates and fish, based on a detailed study by Burkhard and Sheedy (1995) on invertebrates (crayfish) and fish (sunfish, *Lepomis* sp.) in the Five Mile Creek (Alabama, USA), receiving coke plant effluents. In this model, lipid–organic carbon normalized BSAFs were predicted, matching the order of magnitude of measured ranges of 0.01–0.1 for crayfish and 0.0001–0.1 for sunfish. For the sunfish, BSAFs decreased markedly with hydrophobicity (K_{ow}) of the individual

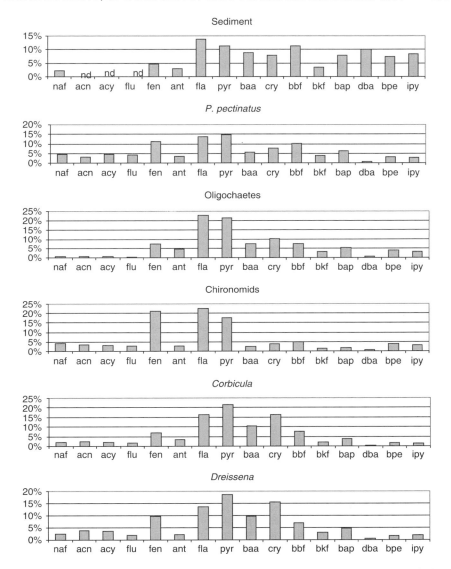

Figure 8.4   Relative concentration patterns (average percentage fraction of Σ PAHs) of individual PAHs in sediments, aquatic macrophytes and invertebrates

PAHs, with values of 0.0001–0.01 for PAHs with log $K_{ow} > 6$. Substituted PAHs tended to have higher BSAFs than unsubstituted PAHs. Decreased aqueous bioavailability, enhanced biotransformation and decreased gut assimilation efficiency for the more hydrophobic PAHs (log $K_{ow} > 5$) were assumed to be factors explaining this phenomenon. Although some components and assumptions of the model and the food web relations have been criticized (Watanabe and Bart 2001), the general conclusions still hold, and are in line with results from other studies.

Burckhard and Lukasewycz (2000) reported similarly low BSAFs (0.0001 – 0.007) for different parent PAHs in lake trout from Lake Superior. In recent studies reported by Baumard *et al.* (1998) on a food web in the Mediterranean (France, Spain), a linear decrease of BSAF (dry weight, uncorrected) with hydrophobicity (log K_{ow}) was found for mussels (*Mytilus galloprovincialis*) and especially in fish (*Mullus barbatus, Serranus scriba*). For benzo[a]pyrene, concentrations (in ng/g dry weight) were 0.07–612 in sediments ($n = 17$), 0.4–5.3 in mussels ($n = 12$), and 0.01–1.6 in *S. scriba* ($n = 6$) and *M. barbatus* ($n = 6$). Compared to the mussels, five- and four-ring PAHs were relatively low in both fish species. A similar observation was reported by Van der Oost *et al.* (1994), who observed mainly the lighter two- and three-ring PAHs in fish (eels) from inland waters in the vicinity of Amsterdam (The Netherlands), while four-ring PAHs dominated in the sediments.

In a study by Gewurtz *et al.* (2000) on sediments and invertebrates in Lake Erie, lipid–organic carbon normalized biota–sediment accumulation factors (BSAFs) for PAHs in mayfly larvae (0.1–10) and *Dreissena* (0.01–1) were higher than corresponding BSAFs for amphipods (0.001–1) and crayfish (0.001–0.5). The differential accumulation among the invertebrate species was attributed to differences in biotransformation. For exposure monitoring in organisms which are capable of metabolization of PAHs, the analysis of metabolites (hydroxylated PAHs) in excretion products, e.g. fish bile, is usually considered a better indicator of recent exposure to PAHs than the analysis of tissue residues (Ariese *et al.* 1994; Van der Oost *et al.* 1994; Stroomberg *et al.* 1999; Escartin and Porte 1999).

8.3.2   FIELD STUDIES ON THE BIOAVAILABILITY OF PAHs IN THE RHINE–MEUSE DELTA

The different methods that can be used to assess the bioavailability of PAHs have been described in Chapter 7. Whereas chemical methods are used to determine the distribution of PAHs among different sorption phases with different kinetic desorption properties, biological tests can be used to measure the actual accumulation in sediment dwelling organisms. The bioavailability of PAHs was investigated in two freshwater regions in The Netherlands: in the Rhine–Meuse delta and in Lake Ketelmeer. In one study, bioaccumulation bioassays were performed in which aquatic oligochaetes were exposed to sediment from different locations in the Rhine–Meuse delta (Den Besten 2002). A comparison was made between sublittoral waters, littoral zones and floodplains. Some characteristics of these sediments and the measured BSAFs for a number of PAHs are presented in Table 8.2. The BSAFs measured in the oligochaetes after a 28 day exposure period showed a considerable variation.

TABLE 8.2 Characteristics of the sediments sampled in the Rhine–Meuse delta [1]

Location/ sediment type	Sediment characteristics			BSAF [4]											
	Depth (m)[2]	OC[3] (%)	Silt (fraction < 63 mm)[3] (%)	Phenanthrene	Anthracene	Fluoranthene	Pyrene	Benzo[a]anthracene	Chrysene	Benzo[b]fluoranthene	Benzo[k]fluoranthene	Benzo[a]pyrene	Dibenzo[ah]anthracene	Benzo[ghi]perylene	Indeno[1,2,3,-cd]pyrene
River sediment (Meuse)	7.6	1.8	29	0.17	0.17	0.22	0.36	0.24	0.29	0.15	0.13	0.11	0.08	0.09	0.06
River sediment (Rhine–Meuse)	7.7	5.6	76	0.20	0.19	0.22	0.32	0.21	0.24	0.12	0.11	0.09	0.09	0.08	0.07
Littoral/sandy (delta)	2	1	12	0.57	0.42	0.86	1.83	1.16	1.03	0.62	0.55	0.47	0.41	0.32	0.14
Littoral/silty (delta)	1.5	1.4	21	0.20	0.30	1.20	1.54	0.94	1.03	0.46	0.42	0.39	0.25	0.23	0.11
Sublittoral/deep (delta)	8.4	4.4	74	0.08	0.11	0.27	0.39	0.25	0.27	0.16	0.14	0.11	0.12	0.10	0.05
Floodplain forest/silty (delta)	2	4.8	67	0.26	0.37	0.54	0.68	0.27	0.42	0.14	0.18	0.14	0.92	0.11	0.35
Meuse floodplain/dry sediment	–	1.8	5.2	1.93	1.91	2.37	2.57	2.28	2.64	2.35	2.33	2.74	0.26	1.81	1.16

[1] Data from Den Besten (2002) or unpublished data. Concentrations of individual PAHs in the sediment are in the range of 0.02–5 mg/kg dry sediment (not shown).
[2] Mean water depth at location where sediment was sampled.
[3] Mean percentage on dry weight basis.
[4] BSAF values represent mean values from three bioaccumulation bioassays per location.

For all organic contaminants the equilibrium-partitioning (EP) theory predicts a value of 1.7 (Van der Kooij *et al.* 1991). In the case of PCBs and some organochlorine pesticides, BSAFs close to 1.7 were found in the river sediments, whereas clearly elevated bioavailability was found in sediment from littoral zones and dry sediment from the floodplain of the River Meuse (Den Besten 2002). For most of the PAHs, the BSAFs measured in river sediment are well below the theoretical 1.7, while for sediment from littoral zones and dry sediment from the floodplain of the River Meuse BSAFs were much higher.

Bioaccumulation bioassays were also combined with a chemical assessment of the bioavailability of PAHs (and other organic contaminants). In 1998 and 1999 large sampling campaigns took place in Lake Ketelmeer, The Netherlands. A total of 19 sediment samples were collected from the lake. Part of the sediment samples were used in an exposure experiment in which oligochaetes were exposed to the sediment for 28 days. Another part of the samples was used for a chemical determination of the availability of PAHs using a 6 h Tenax extraction procedure (Cornelissen *et al.* 2001; Ten Hulscher and Wilkens 2001).

The results of this study are presented in Figure 8.5. Although the variability of compound masses present in the linear fraction in these sediments was not large, the BSAF values that were based on total sediment concentrations were much lower than the expected value of ~ 2 (based on the relative affinity of compounds for sediment oc and fat tissue). When using Tenax extraction to determine the concentration in the linear fraction, the BSAF values approach the theoretical value.

A similar sampling campaign was undertaken in the Sliedrechtse Biesbosch (a floodplain forest in The Netherlands) in 2000. Here also, accumulation studies with oligochaetes and availability measurements with Tenax were performed

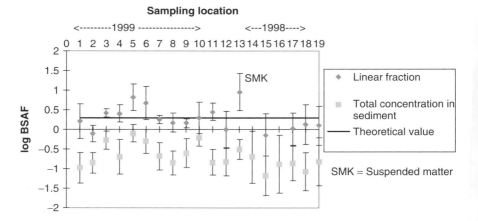

Figure 8.5   BSAF values for PAHs in 19 sediment samples. For each location the average and 95% CI of BSAF values are shown. Both BSAF values based on total concentrations in sediment and BSAF values based on linearly sorbed concentrations are shown

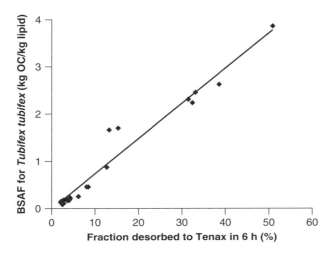

Figure 8.6 Relationship between the contaminant fraction that can be desorbed to Tenax, and the uptake by oligochaetes (one sediment sample, contaminants include PAHs, chlorobenzenes, PCBs, DDE, and DDD)

(Keijzers and Postma 2001). For the samples from this area, much larger differences in linear sorbed fractions were observed. At the same time, very good correlations can be shown between the Tenax-extractable fraction, and the observed BSAF values based on total concentrations in sediment (Figure 8.6; Ten Hulscher *et al.* 2003). Tenax extraction for 6 h is a convenient way to estimate the linear sorbed concentration in sediment. After 6 h, approximately 50% of linearly sorbed PAHs (and other relevant contaminants) are extracted (Cornelissen *et al.* 2001).

The high correlation presented in Figure 8.6 is a strong indication that the observation of low BSAF values in field sediments is caused by the fact that a large fraction of sorbed PAHs is not available for uptake in organisms, and that this can be reliably assessed using a 6 h extraction with Tenax.

Both examples illustrate the possibility of chemical methods to assess bioavailability. Although an exact measurement of bioavailability may not be expected, chemical methods such as the above illustrated extraction procedure can be used to evaluate contaminated sediments, and discriminate between situations with a high availability and those with a low availability. The explanation for the location-specific variation in the bioavailability of PAHs might, to some extent, be explained by differences in the grain size distribution of the sediments, resulting in a higher bioavailability in sandy sediments as compared with silty sediments. These findings are in line with results from others (Kukkonen and Landrum 1995; Harkey *et al.* 1994) and indicate that sorption of contaminants over different grain size classes is not solely dependent on the organic carbon content. Also, selective feeding of the oligochaetes on silt-rich sediment fractions could explain the higher BSAFs found for sandy sediments. These explanations may account for the differences in BSAFs between river sediments and the more sandy sediments from littoral zones. However, fluctuating hydrological conditions may also influence the bioavailability of contaminants.

8.3.3 BIOAVAILABILITY OF PAHs IN LANDFARMING EXPERIMENTS

The technique of landfarming is used to enhance natural biodegradation of organic contaminants, including PAHs. After the removal of sediment from a polluted site, the sediment is spread in a thin layer, dried and mixed a number of times in order to obtain a good soil structure. After this has been achieved the soil is left for a number of years in order to achieve further biodegradation of organic contaminants. The aim of these measurements was to study the effect of the landfarming on the PAH levels during the change of sediment to soil and the consequences for the bioavailability in bioassays with oligochaetes.

In the case of a pilot landfarming experiment with polluted sediment from the Rotterdam harbor 'Geulhaven' (GH), PAH concentrations were monitored for 82 months, and at various times samples were taken to perform accumulation bioassays with aquatic oligochaetes (soil samples were tested as described for sediments by Den Besten 2002). The results obtained for recently deposited

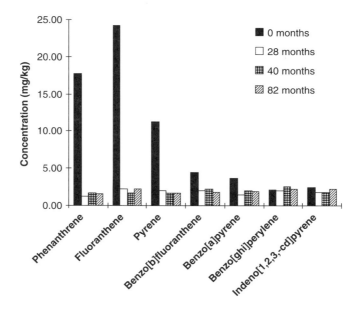

Figure 8.7 Concentrations of different PAHs during a landfarming experiment with sediment from Geulhaven (a harbor in the Rotterdam area)

sediment that stood clear of water are in line with observations made in landfarming experiments.

Bioavailability of PAHs was also measured in untreated sediment from the Rhine–Meuse delta (location Biesbosch) as a reference (only at one sampling time: together with GH 28 months samples). During the first 2 years the concentrations of most PAHs dropped markedly (Figure 8.7). Other PAHs, such as benzo[ghi]perylene and indenopyrene, did not decrease in concentration over a period of 7 years. Figure 8.8 shows the BSAFs measured at different periods of time after the start of landfarming. The BSAFs of PAH for the Geulhaven sediment show a trend in time comparable to that observed for the PAH levels in the sediment. At the start of the landfarming experiment they were elevated compared to untreated sediment (location BB) and close to the theoretical 1.7 (as a result of the removal and treatments at the start of the landfarming experiment), but they gradually dropped to values of 0.1 – 0.2 and to values lower than those found in sediment from location BB. However, the less

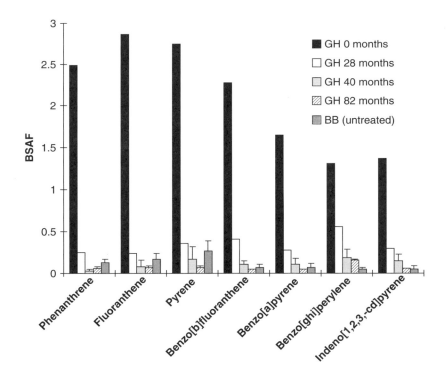

Figure 8.8 Biota–sediment accumulation factors (BSAFs) measured in bioaccumulation bioassays in which aquatic oligochaetes were exposed to sediment (soil) collected during a landfarming experiment with sediment from location Geulhaven (GH). Contaminated sediment from the Biesbosch area was used as reference. Bars represent mean values ($n = 2$ or 3). Data in part from Den Besten (1995) and partly unpublished results

degradable PAHs benzo[ghi]perylene and indeno[1,2,3-cd]pyrene (see relative change in concentration in Figure 8.7) showed a relatively high bioavailability after 40 and 82 months, which was higher than the values found in Biesbosch sediment. These changes in the bioavailability during landfarming were also found for PCBs and some organochlorine pesticides; they are most likely related to changes in the structure and sorption characteristics of the organic matter during the process of changing sediment to soil, and can be regarded as indicative of changes in the bioavailability of organic contaminants in water–floodplain interfaces.

8.4   EFFECTS OF PAHs IN SEDIMENT

In the foregoing sections, both biological and chemical aspects of bioavailability have been discussed. The last part of this chapter focuses on how the risks of PAHs can be assessed. PAHs that are released in the aquatic environment may cause risks both for ecosystems and for human health. Risks for human health may arise as a result of consumption of fishery products, e.g. mussels, shrimps or other crustaceans, and fish. For the freshwater environment especially the consumption of fish could be relevant. Tissue residues of PAHs or their metabolites may lead to additional exposure of human populations. However, there are only few data on the possible health effects through this exposure route and the risks may be low compared to exposure via air and consumption of other food types (e.g. vegetables). A significant factor in this is the biominification that can occur, e.g. in fish (see Section 8.2).

The same may be true for possible health effects on wildlife, except that there may be specialists that can be exposed to much higher concentrations of PAHs, e.g. birds that feed on freshwater mussels will have a much higher exposure to PAHs than birds that feed on fish (see also Chapters 12 and 13). In this light, it was not surprising that strong effects, such as decreased egg number and egg viability as well as behavioral problems of adults, were observed on the reproduction of the tufted duck, *Aythya fuligula*, which had been fed clams from the Haringvliet in long-term experiments (Scholten *et al*. 1989). Apart from heavy metals and PCBs, PAHs may also have played a role in these effects.

In addition to risks by foodchain poisoning, direct effects of PAHs on aquatic life may also be found. The mechanisms of PAH toxicity and the 'ecotoxicity' of PAHs towards freshwater species have been reviewed in a number of other papers (see e.g. Van Brummelen *et al*. 1998). In the latter review experimental toxicity data were described that suggest that a non-specific narcosis-like mode of action is the most important mechanism by which acute effects occur, except for additional effects through phototoxicity (see Chapter 14), e.g. in the case of anthracene and benzo[k]fluoranthene (Van Brummelen *et al*. 1998). However, it is likely that much more variation will be found in the type of mechanism by which effects develop during chronic exposure to relatively low concentrations of PAHs. Therefore, a better knowledge of the long-term effects that can develop as a result of the genotoxicity of PAHs, PAH-induced endocrine disruption and phototoxicity is crucial for a proper risk analysis. UV light-enhanced toxicity of contaminated sediments containing PAHs was demonstrated in bioassays with *Daphnia magna*, *Nitocra spinipes*, *Chironomus riparius* and *Hyalella azteca* (Wernersson *et al.* 2000). One promising approach to measure the effects of contaminants under realistic conditions, which may include also the phototoxicity of PAHs, is with *in situ* bioassays, in which test organisms are exposed in field cages (Burton 1999; Den Besten *et al.* 2002).

So far, assessments of the risks caused by sediment-bound PAHs have followed in most cases a 'black box' approach. Sediment quality is expressed by chemical parameters or, alternatively, as the degree of effects observed in bioassays in which test organisms are exposed to sediment samples or porewater isolated from the sediment. In the Rhine–Meuse delta the sediment quality Triad method was used to demonstrate the relation between effects on benthic macroinvertebrates, bioassays and sediment pollution. Sediment toxicity was evaluated by three different bioassays: a sediment-water bioassay with the midge *Chironomus riparius*, a sediment porewater bioassay with the water flea *Daphnia magna*, and a sediment porewater bioassay with fluorescence-producing bacteria, *Vibrio fischeri* (Microtox® assay). The results of the total research program have been described previously (Den Besten *et al.* 1995; Reinhold-Dudok van Heel and Den Besten 1999). Sediment samples from groyne sections of the Nieuwe Merwede and the deeper parts of the Hollandsch Diep and the Haringvliet were highly contaminated, contained low abundances of macroinvertebrates, a low species diversity, and often gave a strong response in bioassays. Silty sediments from creeks of the Dordtsche Biesbosch and the Brabantsche Biesbosch were also severely polluted, but contained higher abundances of macrofauna and less frequently gave a strong response in bioassays. In sandy sediments of the above-mentioned areas, more site-specific differences were observed with regard to macroinvertebrate abundances, sediment toxicity and contaminant levels.

Multivariate statistical analysis of the macrofauna data indicated that there are important differences in species composition between different rivers and creek systems that are part of the Rhine–Meuse delta (Peeters 2001). This difference could not be related to variation in sediment quality (chemical or on the basis of bioassay responses). However, within particular rivers, e.g. the Hollandsch Diep, a parameter such as the total density of chironomid larvae showed a significant negative correlation with PAH levels in the sediment, expressed as toxic units (see Figure 8.9).

The Triad studies in the delta of the rivers Rhine and Meuse showed a high frequency of bioassay responses (one of the three bioassays gave a strong effect in 78 out of 210 sampling points in sediments). The bioassay with *D. magna* appeared to be the most sensitive bioassay. In order to relate these effects to the levels of contaminants present in the sediment, a toxic unit analysis was performed on the results obtained for Hollandsch Diep and Dordtsche

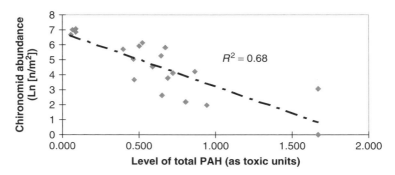

Figure 8.9 Relation between total abundance of chironomids and PAH levels in sediment at different locations in the Hollandsch Diep. Regression coefficient was statistically significant at $p < 0.0.5$. Data from Den Besten (1997)

Biesbosch. This was done by comparing the levels of heavy metals, pesticides and PAHs with data in literature for the 'no-observed-effect concentrations' (NOECs) of specific contaminants in each of the bioassays used (Den Besten *et al.* 1995). This toxic unit analysis indicated that the effects on *D. magna* could be attributed mainly to the heavy metal pollution and to a lesser extent to PAHs. The effects on *C. riparius* could be related to heavy metals, pesticides and PAHs, while effects on *V. fischeri* could be explained mostly by the presence of PAHs in the sediment (Table 8.3).

The toxic units analysis is based on literature data of the toxicity of each of the PAH compounds for a specific test organism. Similar approaches have been proposed to assess the total risk caused by PAHs, by calculating toxic equivalent quantities (TEQs) by relating the toxicity of PAHs to that of benzo[a]pyrene

TABLE 8.3 Toxic unit analysis for effects observed in bioassays with sediment samples from two areas in the Rhine–Meuse delta[1]

Bioassay	Contaminant group	Sum of toxic units	
		Hollandsch Diep	Dordtsche Biesbosch
Daphnia magna	Metals	2.38	3.90
	OCBs	0.02	0.13
	PAHs	0.55	0.62
Chironomus riparius	Metals	0.79	1.38
	OCBs	0.13	2.88
	PAHs	0.62	0.67
Vibrio fischeri	Metals	0.18	0.32
	OCBs	< 0.01	< 0.01
	PAHs	1.55	1.89

[1] Presented are median TU values. Data are from Den Besten (1997). Toxic units were calculated as described by Den Besten *et al.* (1995).

using biomarker responses measured in cells isolated from mussels (Faucet *et al.* 2001).

8.5 CONCLUSION

From the bioaccumulation data presented and the available literature, it can be concluded that PAH concentrations in general seem to decrease with trophic level, due to species-specific differences in toxicokinetics and an increased biotransformation, especially in vertebrates. The uptake of PAHs in aquatic organisms is influenced greatly by differences in the bioavailability of PAHs in sediment. Finally, the data presented in this chapter suggest strongly that PAHs can contribute significantly to effects on aquatic freshwater species.

ACKNOWLEDGMENTS

Paul van Noort provided the detailed interpretation of the Sliedrechtse Biesbosch data (Figure 8.6). Paul van Noort and Chiel Cuypers are thanked for their valuable comments on the manuscript.

REFERENCES

Ariese F, Verkaik M, Hoornweg GP, van de Nesse RJ, Jukema-Leenstra SR, Hofstraat JW, Gooijer C and Velthorst NH (1994) Trace analysis of 3-hydroxy benzo[a]pyrene in urine for the biomonitoring of human exposure to polycyclic aromatic hydrocarbons. *Journal of Analytical Toxicology*, **18**, 195–204.

Baumard P, Budzinski H, Garrigues P, Sorbe JC, Burgeot T and Bellocq J (1998) Concentrations of PAHs (polycyclic aromatic hydrocarbons) in various marine organisms in relation to those in sediments and to trophic level. *Marine Pollution Bulletin*, **36**, 951–960.

Broman D, Naf C, Lundbergh I and Zebuhr Y (1990) An *in situ* study on the distribution, biotransformation and flux of polycyclic aromatic hydrocarbons (PAHs) in an aquatic food-chain (*Seston, Mytilus edulis* L, *Somateria mollissima* L) from the Baltic — an ecotoxicological perspective. *Environmental Toxicology and Chemistry*, **9**, 429–442.

Burkhard LP and Lukasewycz MT (2000) Some bioaccumulation factors and biota–sediment accumulation factors for polycyclic aromatic hydrocarbons in lake–trout. *Environmental Toxicology and Chemistry*, **19**, 1427–1429.

Burkhard LP and Sheedy BR (1995) Evaluation of screening procedures for bioconcentratable organic chemicals in effluents and sediments. *Environmental Toxicology and Chemistry*, **14**, 689–629.

Burton GA Jr (1999) Realistic assessments of ecotoxicity using traditional and novel approaches. *Aquatic Ecosystem Health and Management*, **2**, 1–8.

Campbell PGC, Lesis AG, Chapman PM, Fletcher WK, Imber BE, Luoma SN, Stokes PM and Winfrey M (1988) *Biologically Available Metals in Sediments*. Publication No. 27694. National Research Council of Canada, Ottawa.

Clements WH, Oris JT and Wissing TE (1994) Accumulation and food-chain transfer of fluoranthene and benzo[a]pyrene in *Chironomus riparius* and *Lepomis macrochirus*. *Archives of Environmental Contamination and Toxicology*, **26**, 261–266.

Cornelissen G, Rigterink H, ten Hulscher DEM, Vrind BA and van Noort PCM (2001) A simple Tenax extraction method to determine the availability of sediment-sorbed organic compounds. *Environmental Toxicology and Chemistry*, **20**, 706–711.

Den Besten PJ (1997) *Biotic Effects Caused by Sediment Pollution in the Delta of the Rivers Rhine and Meuse (Netherlands). Part II: Hollandsch Diep and Dordtsche Biesbosch*. Institute for Inland Water Management and Waste Water Treatment (RIZA), Lelystad, The Netherlands (in Dutch). Report No. 97.098, pp. 144.

Den Besten PJ (2002) Contamination in intertidal areas: risks of increased bioavailability. *Aquatic Ecosystem Health and Management* (in press).

Den Besten PJ (1995) *Bioavailability of Contaminants in Sediment from the Dordtsche Biesbosch and Geulhaven. Results of Bioaccumulation Assays with Oligochaetes*. Report 95.176X. Institute for Inland Water Management and Waste Water Treatment (RIZA), Lelystad, The Netherlands (in Dutch).

Den Besten PJ, Schmidt CA, Ohm M, Ruys MM, van Berghem JW and van de Guchte C (1995) Sediment quality assessment in the deltas of the rivers Rhine and Meuse based on field observations, bioassays and food chain implications. *Journal of Aquatic Ecosystem Health*, **4**, 257–270.

Den Besten PJ, de Valk S, van Weerlee E, Nolting RF, Postma JF and Everaarts JM (2001) Bioaccumulation and biomarkers in the sea star *Asterias rubens* (Echinodermata: Asteroidea): a North Sea field study. *Marine Environmental Research*, **51**, 365–387.

Den Besten PJ, Naber A, Grootelaar EMM and van de Guchte C (2002) *In situ* bioassays with *Chironomus riparius*: laboratory–field comparisons of sediment toxicity and effects during wintering. *Aquatic Ecosystem Health and Management* (in press).

Dobroski CJ and Epifanio CE (1980) Accumulation of benzo[a]pyrene in a larval bivalve via trophic transfer. *Canadian Journal of Fisheries and Aquatic Science*, **37**, 2318–2322.

Escartin E and Porte C (1999) Biomonitoring of PAH pollution in high-altitude mountain lakes through the analysis of fish bile. *Environmental Science and Technology*, **33**, 406–409.

Faucet J, Heude C, Mathieu M, Maurice M and Burgeot T (2001) Measure of toxic equivalent quantities (TEQs) of PAHs in the mussel. Abstract, SETAC, Madrid, May 2001.

Gandrass J and Salomons W (eds) (2001) *Dredged Material in the Port of Rotterdam — Interface between Rhine Catchment Area and North Sea*. GKSS Research Centre, Geesthacht (Germany).

Gewurtz SB, Lazar R and Haffner GD (2000) Comparison of polycyclic aromatic hydrocarbon and polychlorinated biphenyl dynamics in benthic invertebrates of Lake Erie, USA. *Environmental Toxicology and Chemistry*, **19**, 2943–2950.

Harkey G, van Hoof P and Landrum P (1995) Bioavailability of polycyclic aromatic hydrocarbons from a historically contaminated sediment core. *Environmental Toxicology and Chemistry*, **14**, 1551–1560.

Hellou J, Upshall C, Ni IH, Payne JF and Huang YS (1991) Polycyclic aromatic hydrocarbons in harp seals (*Phoca groenlandica*) from the north-west Atlantic. *Archives of Environmental Contamination and Toxicology*, **21**, 135–140.

Keijzers CM and Postma JF (2001) Ecotoxicological and chemical risk assessment for a water–bank–soil gradient. *Sliedrechtse Biesbosch — 2000*. AquaSense Report No. 1639. AquaSense, The Netherlands (in Dutch).

Kukkonen J and Landrum PF (1995) Measuring assimilation efficiencies for sediment-bound PAH and PCB congeners by benthic organisms. *Aquatic Toxicology*, **32**, 75–92.

Laane RWPM, Sonneveldt HLA, Van der Weyden AJ, Loch JPG and Groeneveld G (1999) Trends in the spatial and temporal distribution of metals (Cd, Cu, Zn and Pb) and organic compounds (PCBs and PAHs) in Dutch coastal zone sediments from 1981 to 1996: a model case study for Cd and PCBs. *Journal of Sea Research*, **41**, 1-17.

Lemaire P, den Besten PJ, O'Hara SCM and Livingstone DR (1993) Comparative metabolism of benzo[a]pyrene by microsomes of hepatopancreas of the shore crab *Carcinus maenas* L. and digestive gland of the common mussel *Mytilus edulis* L. *Polycyclic Aromatic Compounds*, **3**(suppl), 1133-1140.

Meador JP, Stein JE, Reichert WL and Varanasi U (1995) Bioaccumulation of polycyclic aromatic hydrocarbons by marine organisms. *Critical Reviews in Environmental Contamination and Toxicology*, **43**, 79-165.

Neff JM (1979) *Polycyclic Aromatic Hydrocarbons in the Aquatic Environment*. Applied Science Publishers, London.

Neilson A and Hutzinger O (eds) (1998) PAHs and related compounds. *The Handbook of Environmental Chemistry*, vol.3, Part J. Springer Verlag, Berlin.

Niimi AJ and Dookhran GP (1989) Dietary absorption efficiencies and elimination rates of polycyclic aromatic hydrocarbons (PAHs) in rainbow trout (*Salmo gairdneri*). *Environmental Toxicology and Chemistry*, **8**, 719-722.

Peeters ETHM, Dewitte AMCM, Koelmans AA, van de Velde JA and den Besten PJ (2001) Evaluation of bioassays vs. contaminant concentrations in explaining the macroinvertebrate community structure in the Rhine-Meuse delta, The Netherlands. *Environmental Toxicology and Chemistry*, **20**, 2883-2891.

Reinhold-Dudok van Heel HC and den Besten PJ (1999) The relation between macroinvertebrate assemblages in the Rhine-Meuse delta (The Netherlands) and sediment quality. *Aquatic Ecosystem Health and Management*, **2**, 19-38.

RIZA/RIKZ (2000) Jaarboek monitoring rijkswateren 1999. Rijks Instituut voor Integraal Zoetwaterbeheer en Afcalwaterbehandeling/Rijksinstituut voor Kust en Zee. Lelystad/Den Haag (in Dutch: CD-ROM database also available at www.watermarkt.nl).

Scholten MCT, Foekema E, Chr. de Kock W and Marquenie JM (1989) *Reproduction Failure in Tufted Ducks Feeding on Mussels from Polluted Lakes*. MT-TO Laboratory for Applied Marine Research, Den Helder, The Netherlands.

Southworth GR, Beauchamp JJ and Schmieder PK (1978) Bioaccumulation potential of polycyclic aromatic hydrocarbons in *Daphnia pulex*. *Water Research*, **12**, 973-977.

Stroomberg GJ, de Knecht JA, Ariese F, Van Gestel CAM and Velthorst NH (1999) Pyrene metabolites in the hepatopancreas and gut of the isopod *Porcellio scaber*, a new biomarker for polycyclic aromatic hydrocarbon exposure in terrestrial ecosystems. *Environmental Toxicology and Chemistry*, **18**, 2217-2224.

Ten Hulscher TEM and Wilkens M (2001) *Monitoring of the Remediation of Ketelmeer-Oost — T_0 Situation, Additional Assessments. Environmental Chemistry of Organic Micropollutants*. RIZA Report 2000.142X. RIZA, Lelystad, The Netherlands (in Dutch).

Ten Hulscher TEM, Postma J, Den Besten PJ, Stroomberg GJ, Belfroid A, Wegener JW, Faber JH, Van der Pol JJC, Hendeiks AJ and Van Noort PCM (2003) Tenax extraction mimics bioavailability of organic contaminants for aquatic and terrestrial worms. *Environmental Toxicology and Chemistry* (in press).

Thomann RV and Komlos J (1999) Model of biota-sediment accumulation factor for polycyclic aromatic hydrocarbons. *Environmental Toxicology and Chemistry*, **18**, 1060-1068.

Van Brummelen TA, van Hattum B, Crommentuijn T and Kalf DF (1998) Bioavailability and ecotoxicity of PAHs. In Neilson A, Hutzinger O, (eds), *PAHs and Related Compounds. The Handbook of Environmental Chemistry*, vol. 3, Part J. Springer Verlag, Berlin, pp. 203-263.

Van der Kooij LA, van de Meent D, van Leeuwen CJ and Bruggeman WA (1991) Deriving quality criteria for water and sediment from the results of aquatic toxicity tests and product standards: application of the equilibrium partitioning method. *Water. Research*, **25**, 697-705.

Van der Oost R, van Schooten FJ, Ariese F, Heida H, Satumalay K and Vermeulen NPE (1994) Bioaccumulation, biotransformation and DNA-binding of PAHs in feral eel (*Anguilla anguilla*) exposed to polluted sediments — a field survey. *Environmental Toxicology and Chemistry*, **13**, 859–870.

Van Hattum B, Curto Pons MJ and Cid Montañés JF (1998a) Polycyclic aromatic hydrocarbons in freshwater isopods and field-partitioning between abiotic phases. *Archives of Environmental Contamination and Toxicology*, **35**, 257–267.

Van Hattum B, Burgers I, Swart K, van der Horst A, Wegener JWM and den Besten PJ (1998b) *Biomonitoring of Micropollutants in Foodchains in the Amer and the Haringvliet*. Report E-98/08. Institute for Environmental Studies, Vrije Universiteit Amsterdam, The Netherlands (in Dutch).

Van Hattum B, Burgers I, Swart K, van der Horst A, Wegener JW, Leonards P, Rijkeboer M and den Besten PJ (1996) *Biomonitoring of Micropollutants in Foodchains of Hollandsch Diep, Dordtsche Biesbosch and Brabantsche Biesbosch*. Report E-96/12. Institute for Environmental Studies, Vrije Universiteit Amsterdam, The Netherlands (in Dutch).

Van Hattum B, Leonards P, Burgers I and van der Horst B (1993) *Biomonitoring of Micropollutants in Foodchains of the Nieuwe Merwede and the Dordtse Biesbosch*. Report E-92/19. Institute for Environmental Studies, Vrije Universiteit Amsterdam, The Netherlands (in Dutch).

Van Straalen NM, Verweij RA and van Brummelen TC (1993) PAH concentrations in forest floor invertebrates in the vicinity of a blast furnace plant. In Garrigues P, Lamotte M (eds), *Polycyclic Aromatic Compounds. Synthesis, Properties, Analytical Measurements, Occurrence and Biological Effects*. Proceedings of the 13th International Symposium on Polynuclear Aromatic Hydrocarbons. Gordon and Breach, Yverdon, Switzerland, pp. 1001–1006.

Watanabe KH, Bart HL (2001) Comments on model of biota–sediment accumulation factor for polycyclic aromatic hydrocarbons. *Environmental Toxicology and Chemistry*, **20**, 1867–1868.

Wernersson A-S, Dave G and Nilsson E (2000) Assessing pollution and UV-enhanced toxicity in Torsviken, Sweden, a shallow bay exposed to contaminated harbor sediment and harzardous waste leachate. *Aquatic Ecosystem Health and Management*, **3**, 301–316.

Bioaccumulation of PAHs in Marine Invertebrates

JAMES P. MEADOR
Northwest Fisheries Science Center, National Oceanic and Atmospheric Administration, Seattle, WA, USA

9.1 INTRODUCTION

Polycyclic aromatic hydrocarbons (PAHs) are ubiquitous, worldwide contaminants that are most heavily concentrated in urbanized coastal areas (see Chapters 1 and 2). Even though PAHs occur naturally, the highest concentrations are mainly due to human activities and the primary sources are combustion products and petroleum. In this review, several important processes that affect bioaccumulation of PAHs in marine invertebrates will be addressed. Some of the more important points that will be examined concern the role of organic carbon and lipid in the control of PAH partitioning, the different modes of exposure, the rates of uptake and elimination, seasonal variation, the role of biotransformation, the types of PAHs, the efficiency of absorption, and experimental artifacts.

Two factors, lipid and organic carbon, play an important role in the partitioning behavior of PAHs in sediment, water, and tissue. The more hydrophobic a compound is, the more likely it is to associate with non-polar phases (e.g. lipid and organic carbon). The octanol–water partition coefficient (K_{ow}) is the single best predictor of PAH partitioning and can be used to determine its behavior and bioavailability in the environment. Accumulation of PAHs from the environment occurs in all marine organisms; however, a wide range in tissue concentrations results due to variations in environmental concentrations, time of exposure, and species ability to metabolize these compounds. In invertebrates, the highest concentrations can be found in the internal organs, such as the hepatopancreas, and tissue concentrations appear to follow seasonal cycles, which may be related to variations in lipid content, spawning cycles, or environmental flux (Jovanovich and Marion 1987; Maruya *et al*. 1997; Miles and Roster 1999).

Bioavailability and organism physiology are the two important variables that have a major effect on contaminant body residues. Unlike metals and ionizable

PAHs: An Ecotoxicological Perspective. Edited by Peter E.T. Douben.

organics, the proportion of PAHs available for uptake is mainly affected by a few environmental variables, such as organic carbon, contact time (aging), desorption rate, PAH source (petrogenic vs. pyrogenic), and sediment surface area. Organismal factors, including lipid levels and the rates of uptake and elimination, are the main determinants of total tissue accumulations.

For a given species, the amount of contaminant found in tissue is often associated with the amount of anthropogenic activity occurring in the species' habitat (Pittinger *et al.* 1985; O'Connor 2002). PAHs accumulate in sediments due to their association with organic carbon and in biota that have high rates of uptake or are unable to efficiently metabolize the parent compounds. The ability to predict tissue concentrations in feral organisms is important in the assessment of possible toxic effects and the protection of sensitive species. Generation of critical body residues for various modes of action may be a useful first approximation of potential deleterious effects. For species that have a weak or non-existent ability to metabolize PAHs, assessing tissue concentrations would be relatively simple for different biological responses. For those species that readily biotransform PAHs, determination of critical body residues for PAHs provides a unique challenge. Assessment of metabolite concentrations or DNA adducts may be a reasonable approach for correlating exposure to PAHs and adverse biological effects, such as impaired growth or reproduction.

9.2 UPTAKE AND ELIMINATION OF PAHs

9.2.1 UPTAKE OF PAHs

Species exhibit different rates of uptake, which is a major determinant of the final body residue. It is generally believed that the process of uptake of these neutral hydrophobic compounds is passive and controlled by diffusion pressure (fugacity) because of the differential between the environmental matrix and tissue concentrations. Uptake from water is generally accomplished by ventilation over the gill structure, although diffusion through the integument may also contribute to tissue concentrations (Landrum and Stubblefield 1991). As the organism directs water over the gill surface to extract oxygen, contaminants are taken up because of the high surface area and lipid-rich membranes.

Ingestion of prey organisms, detritus, and sediment is also important for PAH accumulation; however, the factors that determine the bioavailable fraction that will be accumulated from the ingested materials appear to be more complex and are less well known. It has been demonstrated that the rates of uptake vary little over a wide range of PAHs and are therefore not strongly linked to chemical hydrophobicity (Bender *et al.* 1988; Landrum 1988).

The predominate exposure matrix can be an important determinant for bioaccumulation, especially under non-equilibrium conditions. Benthic species are likely to exhibit higher tissue residues than pelagic species because they often ingest sediment, detritus, or other benthic species and may be exposed to high concentrations found in porewater. Benthic species can be infaunal (live under the sediment surface) or epibenthic (live on top of the sediment surface). Some infauna are exposed to porewater and some build tubes and pump overlying water through their burrows. Studies that compare different species under identical conditions can be very informative and lead to hypotheses regarding the degree to which bioaccumulation varies and the mechanisms that control tissue concentrations.

For the more water-soluble PAHs (log $K_{ow} \leq 5.5$), the main route of uptake is believed to be through ventilated water, whereas the hydrophobic compounds (log $K_{ow} \geq 5.5$) are taken in mainly through ingestion of food or sediment (Landrum 1989; Landrum and Robbins 1990; Meador *et al.* 1995a). Of course there are many variables, such as chemical hydrophobicity, absorption efficiency, feeding rate, and ventilatory volume, which may affect the final tissue concentration. The route of uptake may be an important issue for short-term (acute) events; however, under long-term exposure and equilibrium conditions between water, prey, and sediment, the route of uptake may be immaterial because tissue residues are expected to be a function of thermodynamics, not kinetics (Bierman 1990; Di Toro *et al.* 1991). The key assumption of equilibrium between these different compartments may rarely occur for PAHs in the environment.

Even though the route (and presumably the rate) of uptake is predicted to have no influence on tissue concentrations under equilibrium conditions (Bierman 1990; Di Toro *et al.* 1991), the rate of uptake can be an important factor under pre-equilibrium conditions. For example, one study with a deposit-feeding polychaete (*Abarenicola pacifica*) found a correlation between the 48 h BSAF and the ingestion rate (Penry and Weston 1998). Because equilibrium conditions may be rare in the marine environment, the kinetics and route of uptake in addition to the factors that affect them should always be considered when assessing bioaccumulation and making predictions.

9.2.1.1 Feeding mode

There are a variety of feeding modes, including sediment ingestion (selective and non-selective), detritus feeding, predation, suspension feeding, and filter feeding. Each of these modes may have a dramatic impact on the degree that the organism is exposed to contaminants and final bioaccumulation values, especially if disequilibrium prevails among compartments (water, tissue, and sediment). There are a number of studies that demonstrate the importance of the mode of feeding by invertebrates in relation to the degree of bioaccumulation (Roesijadi *et al.* 1978; Foster *et al.* 1987; Meador *et al.* 1995a; Hickey *et al.* 1995; Kaag *et al.* 1997). For example, Foster *et al.* (1987) demonstrated large differences in tissue concentrations between two clams, one a deposit feeder and the other a filter feeder. This observation is similar to the results from Hickey *et al.* (1995), who examined two bivalves, a deposit feeder (*Macomona liliana*) and a filter feeder (*Austrovenus stutchburyi*). Concentrations of PAHs along with BSAF values were generally higher in the deposit feeder, even though the filter feeder exhibited lipid values that were approximately twice that found in the deposit feeder.

A study by Varanasi *et al.* (1985) observed large differences in BAF values between infaunal amphipods (*Eohaustorius washingtonianus* and *Rhepoxynius abronius*) and a deposit-feeding clam (*Macoma nasuta*) for benzo[a]pyrene (B[a]P). The amphipods exhibited higher BAFs than the clam, even though their metabolic capacity for this compound was much greater than the capacity of the clam. Several factors may explain this apparent paradox, including a higher rate of uptake leading to higher body residues, or greater exposure to porewater. Also, the clam may have closed its valves, which would reduce its uptake of B[a]P from ingested sediment or overlying water.

Another study with *Rhepoxynius abronius*, an infaunal amphipod that does not ingest sediment, and an infaunal deposit-feeding polychaete (*Armandia brevis*) found similar accumulations of low molecular weight PAHs (LPAHs) by the two species (Meador *et al.* 1995a) but substantially more accumulation of high molecular weight PAHs (HPAHs) by the polychaete. Because these were laboratory bioassays, there was probably no prey available for the amphipod. The results of this study suggest that deposit- and non-deposit-feeding infaunal invertebrates will acquire most of their body burden of LPAHs through porewater, regardless of feeding strategy; however, ingestion of sediment or food may be the dominant route of uptake when hydrophobic compounds exceed a log K_{ow} of approximately 5.5. This conclusion is supported by other similar studies (Landrum 1989; Landrum and Robbins 1990).

In the case where an organism is in contact with sediment, but does not ingest sediment, equilibrium or steady-state tissue concentration may be very different, reflecting only its interaction with overlying water. Consider a clam living in sediment that experiences little exposure because it filters overlying water that contains much lower concentrations than expected when in equilibrium with the sediment. Comparative studies, such as the ones mentioned here, are useful for exploring the factors that control bioaccumulation; however, several parameters, including differences in uptake, metabolism, lipid content, behavior, organism health, and myriad environmental conditions need to be considered before conclusions can be generated.

9.2.1.2 Absorption efficiency

There are very few studies that have examined absorption efficiency (AE) of PAHs in marine invertebrates. One study with the marine polychaete (*Abarenicola pacifica*) reported mean AE values of 33% for phenanthrene and 21% for benzo[a]pyrene, although the results were highly variable (Penry and Weston 1998). Some of this variability may be methodological; however, the authors concluded that differences between individuals were caused by physiological disparity. In support of this, Weston and Mayer (1998) demonstrated a three-fold difference (9 – 28%) in the proportion of B[a]P solubilized by digestive fluid from a closely related species (*Arenicola brasiliensis*). Another study found relatively high AE values (69%) for B[a]P in the deposit-feeding polychaete *Nereis succinea* using radiolabeled organic matter and B[a]P (Ahrens *et al.* 2001). The results of this study also indicated an increasing AE with increasing organism size. In support of this is the work by Kukkonen and Landrum (1995), who reported AE values for B[a]P of 56% for a freshwater amphipod (*Diporeia* spp.) and 13% for a freshwater oligochaete (*Lumbriculus variegatus*). A recent review of AE lists additional studies with other invertebrate species and organic contaminants (Wang and Fisher 1999), including values for freshwater species showing similar AE values for B[a]P.

One interesting question concerns the absorption efficiency of contaminants as a function of chemical hydrophobicity. Understanding this relationship may help to interpret bioaccumulation patterns observed for different compounds. A study that examined amounts of various PAHs solubilized from contaminated sediment using the digestive fluids of a deposit-feeding polychaete (*Arenicola marina*) revealed conflicting results (Mayer *et al.* 1996). For one site, an inverse relationship between the percentage solubilized and the K_{ow} was observed. For another site, the amount released from sediment by digestive fluid was relatively constant for the 10 PAHs examined, except for pyrene, which was approximately three-fold higher. In all cases, the amount of PAH solubilized was less than 10% after 4 h. Many factors could explain the difference between sites, including the source of PAHs (pyrogenic vs. petrogenic).

The proportional amount of a compound that is taken up by invertebrates from ingested items or ventilated water is difficult to determine. It is even more difficult to determine absorption efficiency for compounds that are metabolized, such as PAHs. Because the role of metabolism is often not addressed, the apparent absorption efficiency may be severely underestimated, due to effective elimination of these compounds. In order to avoid confounding the estimate of AE, the parent compound plus metabolites must be determined. Additional research that includes uptake and elimination kinetics is needed to better assess absorption efficiency of PAHs for the different routes of uptake, especially the dietary route. These data will help greatly in predicting bioaccumulation from different environmental matrices.

9.2.2   ELIMINATION OF PAHs

The elimination of PAHs can occur by biotransformation of parent compounds, diffusion when the concentration gradient favors outward flux, and excretion of parent or metabolite compounds. These processes are collectively termed 'elimination' (Meador *et al.* 1995b). Metabolism is the main form of elimination for PAHs in many invertebrate species. For those species that do not metabolize PAHs, outward diffusion is the main route. Invertebrates exhibit a wide range in ability to biotransform PAHs, even among closely related species. Without information about a species' ability to biotransform different PAH compounds, an adequate comparison of bioaccumulation among species and with theoretical values cannot be made. To make such comparisons, one would have to use species that do not biotransform PAHs or generate metabolism-adjusted bioaccumulation factors for species that are able to metabolize parent compounds.

9.2.2.1 Metabolism in determining tissue residues

One of the most important considerations is the amount of accumulated PAH that is metabolized. For those species that have strong metabolism of PAHs, they may accumulate large amounts of these compounds, show effects, but not contain any measurable concentration of the parent compounds. In those species that are able to metabolize these compounds, mutagenic metabolites are often formed, which can be more toxic than the parent compounds (for more details, see Chapter 5).

Among invertebrate taxa, the ability to metabolize PAHs is highly variable. Even within some taxa (e.g. crustaceans and polychaetes) there is substantial variability in the ability to convert parent PAHs to metabolites. For example, the American lobster (*Homarus americanus*) exhibits little or no ability to metabolize PAHs (Foureman *et al.* 1978; Bend *et al.* 1981), whereas the amphipod *Rhepoxynius abronius* has a fairly well-developed ability to biotransform these compounds (Reichert *et al.* 1985). These differences are likely due to the variable amount or activity of the cytochrome P450-dependent mixed-function oxidase system found in tissues (Bend *et al.* 1977).

Polychaetes can also exhibit wide differences in metabolic conversion of PAHs. For example, one study of three polychaete species found substantial variability in benzo[a]pyrene (B[a]P) metabolism, with one species exhibiting almost no metabolites (< 10%) after 6 days of exposure and another species exhibiting B[a]P concentrations that were more than 90% metabolites (Kane Driscoll and McElroy 1996). This study also found that two species within the same family (Spionidae) exhibited widely different proportions of parent compound and metabolites after short-term exposure. The content of B[a]P in the tissue of *Scolecolepides viridis* was only 50% after 6 days' exposure, whereas the parent compound comprised 91% of the total in *Spio setosa* for the same exposure period. Kane Driscoll and McElroy (1996) also present a summary table of results, based on their current work and past studies, showing large differences in metabolic ability for PAHs in 11 species of polychaetes.

A recent publication examining the metabolic conversion of PAHs has substantially increased the available information on biotransformation of PAHs in marine invertebrates and highlighted some important points (McElroy *et al.* 2000). This study examined the conversion of B[a]P in 10 species from three major taxa (annelids, molluscs, and crustaceans) after 7 days of exposure to sediment-associated PAHs. Previously, it was believed that molluscs in general possessed a weak to non-existent ability to metabolize PAHs. The results of this study demonstrated that two gastropod species (*Hydrobia toteni* and *Ilyanassa obsoleta*) possess relatively strong metabolic systems, due to the high percentage (50–70%) of B[a]P metabolites found. The results for two bivalves showed weaker metabolism; however, metabolites comprise 20–30% of the total B[a]P found in tissues. The work by McElroy *et al.* (2000) also confirms the observation that annelid and crustacean species display highly variable biotransformation abilities for PAHs. Four species of annelids contained 7–96% B[a]P metabolites and the two crustaceans studied had metabolized 20% and 60% of the B[a]P accumulated after 7 days.

In some invertebrate species, the enzymes responsible for metabolism of PAHs may be increased (induced) by exposure to these compounds. Lee *et al.* (1981) found increases in mixed function oxygenase and cytochrome P450 in polychaetes (*Nereis virens*) and crabs (*Callinectes sapidus*, *Sesarma cinerum*, and *Uca pugilator*) exposed to PAHs; however, the results were not consistent. Another study concluded that the metabolism of B[a]P was not inducible in two species of polychaetes by exposure to 3-methylcholanthrene, a strong inducer of cytochrome P450 (Kane Driscoll and McElroy 1997). For those species that have an inducible cytochrome P450 system, the tissue concentrations of PAHs could vary substantially according to previous exposure history. This is an important feature when attempting to characterize bioaccumulation in organisms from the field compared to those observed in the laboratory.

Accumulation factors (BCFs and BAFs) generally increase with chemical hydrophobicity in those species with a weak or non-existent ability to biotransform PAHs. Often a reverse pattern is observed in some species of declining accumulation factors with increasing hydrophobicity, which may be due to metabolism, insufficient time for accumulation, reduced uptake due to declining bioavailability, or depletion of porewater concentrations by kinetic limitations. Metabolism may explain this pattern, because it is suspected that HPAHs are more rapidly metabolized than LPAHs due to differences in enzyme affinity (Schnell *et al.* 1980). Additionally, closely related PAHs may be metabolized at different rates, producing unexpected bioaccumulation patterns, such as that noted by Baumard *et al.* (1999a) for benzo[a]pyrene and benzo[e]pyrene.

9.2.2.2   Steady-state residues

The elimination rate constant (k_2) is the key parameter for determining the time to steady-state tissue concentrations. This can be expressed with the equation $TSS_{50} = 0.693/k_2$, which is the time to achieve a tissue concentration that is 50% of the steady-state concentration. This value is also used to determine the tissue half-life. The time it takes an organism to reach 95% of steady state can be determined with $2.99/k_2$, or 4.3-fold the time to 50% steady state.

As determined by fugacity calculations (thermodynamics), an aquatic organism should come into equilibrium with its exposure concentration, given sufficient time (Mackay and Paterson 1981). Of course, this assumes that the exposure concentration is constant, the environmental matrices are in equilibrium with each other, and diffusive loss is the only form of elimination. At steady state or equilibrium, the rates of uptake and elimination are equal. Equilibrium tissue concentrations occur at their thermodynamic maximum when elimination is solely by passive diffusion. Steady-state tissue concentrations are constant over time, but lower than the thermodynamic maximum because of metabolism. As mentioned earlier, there are myriad factors that can affect the rates of uptake and elimination having a direct bearing on steady-state tissue concentrations.

Time to steady state or equilibrium tissue concentrations is highly variable between species. Even within a species, the rate of elimination may be variable leading to differences in time to steady state. This subject was addressed by Meador *et al.* (1995b) and a table of half-lives for several invertebrate species was included to highlight the extensive variation found for species and compounds. Because the rate of elimination is conditional (i.e. a function of several factors, such as exposure history, temperature, and lipid content), variation in time to steady state among individuals within a species is also expected.

At least one study has found that very small benthic invertebrates achieve steady state within hours to a few days (Lotufo 1998), which is considerably faster than most other marine invertebrates, even those with relatively strong cytochrome P450 systems (Meador *et al.* 1995b). The study by Lotufo (1998) demonstrated very high rates of elimination leading to tissue half-lives of approximately 5 h, which may be due to this small species' large surface area/volume ratio. Based on these few studies, the rate of elimination may also be allometric.

9.3 FACTORS THAT AFFECT UPTAKE AND ELIMINATION OF PAHs

Several factors can influence the amount of PAHs accumulated by organisms, including organism behavior, laboratory artifacts, organism size, and seasonal changes in physiology, behavior, and environmental inputs. Two well-known factors, temperature and lipid content, that affect the rates of uptake and elimination are not addressed separately, but are components of some of the factors discussed below. Making predictions about bioaccumulation generally requires some knowledge about each of these factors, which are not always available. Even if most factors that affect bioaccumulation are quantified, other less well-known factors, such as the interaction between bioturbation and the amount of available compound, can make bioaccumulation predictions less accurate. The following is a summary of the major factors that can affect the uptake and elimination of PAHs and the resulting tissue residues.

9.3.1 SEASONALITY

As mentioned above, lipid content is one important factor that can determine the amount of hydrophobic compound that is accumulated. Lipid content is not static in organisms. It can vary seasonally because of food availability/quality or reproductive status and lipid content can decline with poor health as reserves are utilized. The influence of lipid content of bioaccumulation and trophic transfer of organic contaminants has been reviewed recently by Landrum and Fisher (1998).

Other factors, such as population size and feeding behavior, which can change with food composition, may also influence seasonal exposure to contaminants through changes in the rate of feeding and choice of food. Braumard *et al.* (1999a) found elevated BSAFs for mussels from the Baltic in March compared to those in August and October. The authors noted that mussels tend to have a higher rate of filtration at the end of winter, which could explain the higher tissue concentrations. They also noted that these results could not be due to lipid content in this species because there was no seasonal variation observed for this parameter. Braumard *et al.* (1999a) also hinted that digestive enzymes may follow seasonal cycles, an area certainly worthy of additional investigation.

Environmental flux may also explain seasonal variation in bioaccumulation, such as that proposed for variable PAH concentrations found in mussels nearby a defunct fuel depot (Miles and Roster 1999). The authors concluded that rainfall enhanced PAH bioavailability to the mussels by increasing the flow of contaminated groundwater. A similar mechanism for seasonally variable PAH contamination was also proposed by Braumard *et al.* (1999a) to explain differences in PAH concentrations in mussels during summer and fall. In contrast to Miles and Roster (1999), Maruya *et al.* (1997) found that BSAFs decreased during the rainy season, which was attributed to increased runoff of PAHs associated with soot particles, which have lower bioavailability than petrogenic PAHs. Seasonal variations in fossil fuel hydrocarbons have been noted for sediment (Stainken *et al.* 1983) and water concentrations (Witt 1995), which may also explain some of the seasonal variation observed in the amounts bioaccumulated.

9.3.2   BIOTURBATION

The presence of organisms in the sediment has been shown to increase PAH concentrations in overlying water, which may be an important factor in assessing bioaccumulation for some species. For example, it has been shown that the presence of a tubiculous polychaete (*Nereis virens*) can enhance the flux of sediment sorbed benz[a]anthracene to the water column (McElroy *et al.* 1990). This increased flux to the water column could elevate tissue concentrations in those animals that take in PAHs through ventilation of water over gill membranes. Ciarelli *et al.* (1999) also found that bioturbation by one species enhanced bioaccumulation in another. In this study the authors showed a linear relationship between fluoranthene in *Mytilus edulis* and the density of an amphipod (*Corophium volutator*).

9.3.3   PHYSICAL/CHEMICAL FACTORS

Several properties of sediment can influence the amount of PAH that is bioaccumulated. Such factors as contaminant contact time, organic carbon content of sediment and porewater, contaminant source, redox state, grain size, and others may impact bioavailability.

The amount of time a contaminant has been in contact with sediment particles as a factor in determining the bioavailable fraction has been examined in a few studies (Landrum 1989; Kukkonen and Landrum 1998). This is an important parameter when comparing bioaccumulation factors from laboratory studies with spiked sediments to those generated from field-collected sediments (see also Section 7.3.2). Ideally, laboratory studies with spiked sediments should allow sufficient contact time to increase the likelihood that sediment and water will achieve equilibrium. In the field, recent contaminant inputs to sediments may lead to higher bioavailability to organisms compared to other field sediments that contain aged contaminants. It has been suggested by some authors that it is the rate and extent of desorption of a contaminant from sediment to water that correlates with bioavailability and the amount accumulated (Cornelissen *et al.* 1998; Lamoureux and Brownawell 1999; Kraaij *et al.* 2001), which may vary with sediment aging.

There is also some indication that BSAF values vary with total organic carbon (TOC) content of sediment (McElroy and Means 1988; Ferraro *et al.* 1990). These studies have observed an inverse correlation between BSAFs and TOC in sediment. Mechanistic support for this observation comes from the work of Weston and Mayer (1998), who demonstrated that the proportion of sediment-associated B[a]P solubilized by the digestive fluids of a polychaete (*Arenicola brasiliensis*) was inversely correlated to the organic carbon content of the test sediment. This implies that as TOC increases, less of the compound is soluble and available for uptake across the gut wall into the organism. Presumably, this would cause lower BSAFs for those sediments with higher TOC content. Contrary to this, Penry and Weston (1998) reported BSAFs that were two to three times higher in *Abarenicola pacifica* feeding on high-TOC sediments, compared to worms feeding on low-TOC sediments. These authors also noted that the ingestion rate was inversely correlated to sediment TOC content, but that BSAFs increased with increasing ingestion rate.

Interestingly, the BSAF can increase with increasing TOC, even if the rate of uptake is lower. For example, if a worm accumulates 100 ng/g in a 1% TOC sediment for a given period of time and only 75 ng/g in a 2% TOC sediment for the same period, the BSAF will increase from 0.1 to 0.15, assuming that lipid content, absorption efficiency, and sediment concentration remain unchanged. In general, it can be difficult to make comparisons of bioaccumulation between studies with widely different TOC values, exposure histories, and species with different modes and rates of feeding.

The source of PAHs has been implicated as a factor in bioavailability. Several studies that have shown reductions in water or tissue concentrations of PAHs as they may relate to PAH type and source (Farrington *et al.* 1983; McGroddy and Farrington 1995; Meador *et al.* 1995a; Maruya *et al.* 1997; Naes *et al.* 1999). Sediment-associated PAHs that come from petroleum sources are generally bioavailable to benthic organisms; however, fresh petroleum would likely result in different bioaccumulation patterns compared to weathered petroleum, because of proportional change among compounds. Pyrogenic PAHs that are derived from combustions sources (e.g. soot, coal, or an aluminum smelter) may produce very different bioaccumulation patterns than those observed for petrogenic sources. One study found a gradient in BSAF values in three different benthic invertebrates, noting decreasing values as samples were taken closer to an aluminum smelter (Naes *et al.* 1999). Depending on the species, BSAFs decreased from five- to 10-fold from one end of a fjord to the other, where the smelter was located, and may be related to the types and sources of PAHs found along the gradient.

   One recent study has examined the ratios of various PAHs in mussel tissue in an attempt to determine sources, which is also called 'fingerprinting' (Hellou *et al.* 2002). These authors found that the PAH profile in tissue differed from that in sediment, which may indicate variations in sources (e.g. combustion sources in sediment and higher levels of petroleum sources in the water column). Fingerprint analysis in tissue could be a valuable tool for assessing the sources for bioaccumulation; however, several factors, including metabolism, bioavailability, weathering, and differences in absorption efficiency, need to be considered.

9.3.4   ORGANISM BEHAVIOR

In some cases it has been observed that species may change their behavior when tissue residues approach toxic levels or contaminants are detected in the environment. As tissue residues approach critical levels, the organism may alter its rate of ingestion or ventilation, causing tissue concentrations to increase or decrease dramatically. A comparison of bioaccumulation factors between individuals of a species that were exposed to non-toxic concentrations may be very different compared to those exposed to toxic levels. This phenomenon has been observed by Landrum *et al.* (1994) and Meador and Rice (2001). Another mechanism whereby an organism can alter its expected tissue concentration is by avoidance of contaminants it can detect. Even on small spatial scales, exposure avoidance can be a critical determinant for assessing tissue residues. One example is a bivalve that can close its shell when contaminants are detected (Doherty *et al.* 1987).

9.3.5   ORGANISM SIZE

The size of the organism can have an important effect on bioaccumulation, especially on a temporal scale. The rates of feeding and respiration are allometric and consequently are influential on the rates of contaminant uptake. One study with a freshwater amphipod demonstrated that uptake clearance of benzo[a]pyrene was highly correlated to the surface/volume ratio (Landrum and Stubblefield 1991). Smaller organisms may also have a higher feeding rate, which may lead to higher bioaccumulation factors (Peter Landrum, personal communication). Additionally, the surface area/volume ratio may also be important for the rate of elimination for contaminants, as suggested by the work of Lotufo (1998).

9.3.6   SPATIAL CORRELATIONS

One major consideration when assessing bioaccumulation is the correlation between tissue and sediment concentrations. By convention, it is usually the top 2 cm of sediment that is homogenized and analyzed for contaminants; however, benthic invertebrates feed in a heterogeneous fashion. Some species (e.g. polychaetes) may feed below this 2 cm horizon, while other species (e.g. *Macoma* spp.) glean detritus at the surface. Contaminant concentrations below the 2 cm level may be higher or lower compared to that found within the 2 cm zone, as a function of past inputs. Additionally, the very surface (top 0.5 cm) may contain very high concentrations due to recent inputs.

Many invertebrate species are known to select fine, organic-rich particles when feeding on sediment (Fauchald and Jumars 1979), which are likely to contain high concentrations of contaminants, such as PAHs. This phenomenon may be very important in sandy sediment, which probably contains low percentages of fine material enriched with hydrophobic organic compounds. It is possible that analytical determinations of a sandy sediment would indicate that it is comparatively clean; however, organisms inhabiting the site may exhibit relatively high concentrations of contaminants because of their selective feeding on fine particles.

9.3.7   LABORATORY ARTIFACTS

Laboratory bioassays that attempt to determine the bioaccumulation potential for a species exhibit a few unique problems. In the field such factors as pH, salinity, oxygen content, and temperature can be variable, whereas in the laboratory these are controlled to rigid specifications. Each of these factors can influence the amount of contaminant that is bioaccumulated by an organism, either by changing the amount of the bioavailable fraction or by altering the rates of uptake and elimination. Other important factors to consider when comparing laboratory and field studies include organism behavior, sediment properties, excess hydrogen sulfide or ammonia, and contaminant heterogeneity. Many of these factors are discussed above. Consequently, comparisons between laboratory and field studies for a given species may be inaccurate because of this variability.

The turnover rate of water supplied to the experimental chamber is also an important consideration. Some sediment bioaccumulation tests are performed with no water changes (static) and others have a continuous renewal of water (flow-through). The importance of this variable depends in part on the species and chemical being tested. Some benthic invertebrates interact extensively with the water column and may derive a good deal of their acquired tissue residue from filtering overlying water. A steady supply of clean water to the chambers will likely lead to an underestimation of the bioaccumulation potential; however, under static conditions equilibrium may be established, which may lead to an overestimation because equilibrium conditions may rarely occur in the environment. If a species does not interact substantially with overlying water, the type of water supply may have no impact on bioaccumulation.

For an infaunal deposit-feeding species, the accumulation of contaminants may be the same regardless of the exposure regime. For example, Landrum (1989) found no difference in bioaccumulation for a freshwater amphipod exposed to sediment-associated PAHs under static and flow-through conditions. What may be critical is the degree to which a chemical partitions between water and sediment. For very hydrophobic compounds, most of the accumulation will come from ingestion of sediment or prey, so flow-through conditions may not affect the results. The less hydrophobic compounds that are predominately accumulated from water will probably be influenced more by flow conditions.

Many species that are used for bioaccumulation bioassays require additional food in the experimental chamber. The amount of contaminant that is accumulated from sediment can be severely underestimated if the added food does not come into equilibrium with the contaminants in water and sediment. While not designed to study bioaccumulation, one study examined the toxicity response as a function of food ration in the common bioassay species, *Neanthes arenaceodentata* (Bridges *et al.* 1997). For the contaminated sediment treatments, the higher the food ration, the lower the percentage mortality, suggesting that the higher amounts of food were somewhat protective. The result of this is reduced bioaccumulation and concomitant biological effects. For long-term bioaccumulation tests with deposit-feeding invertebrates, additional sediment should be added to experimental chambers as a food source, not the commonly used flake food for fish (Boese and Lee 1992).

9.4 OCCURRENCE AND BIOACCUMULATION POTENTIAL OF PAHs

9.4.1 OBSERVED DATA

Observed concentrations of PAHs in marine invertebrates from different sites around the world are shown in Table 9.1. This table is by no means exhaustive, but is presented to show that many organisms in diverse locations can accumulate PAHs, often to high concentrations. The table is divided into two groups: in one, mussels and oysters, and in the other, benthic invertebrates. There is such a large body of literature on epibenthic marine bivalves that a separate

TABLE 9.1 Occurrence of PAHs in marine bivalves and invertebrates from field collections

Species	Feeding mode	Area	Total PAH (ng/g)	Sites/ PAHs	d-w	Reference	Type
Mussels and oysters							
Mytilus edulis	FF	Norway	500–12845	11/32	d	1	ws
Mytilus galloprovincialis	FF	Mediterranean	25–390	23/14	d	2	ws
Mussels and oysters	FF	USA (all coasts)	77–1100*	214/24	d	3	ws
Mussels and oysters	FF	USA (all coasts)	192–503#	97–191/44	d	4	ws,my
Mytilus edulis,	FF	Gulf of Naples, Italy	205	6/16	w	5	local
Mytilus galloprovincialis	FF	Mediterranean, Spain	190–5490	6/ns	w	6	ws
Mytilus edulis	FF	Northern Baltic Sea	440	3/19	d	7	local
Mytilidae	FF	Finland (Archipelago Sea)	nd–150	7/7	w	8	ws
Mytilus galloprovincialis	FF	Gulf of Mexico, USA	36–7530	4/17	w	9	seep
Crassostrea virginica	FF	Greece	77–110	57/17	w	10	ws
	FF	Florida, USA	361–11026	14/> 25	d	11	local
M. edulis, M. galloprovincialis and C. gigas	FF	France	nd–300000¥	110/ns	d	12	ws, my
Mytilus edulis	FF	Netherlands	45–100	2/6	w	13	local, my
Mytilus edulis	FF	Scotland	54–2803	27/10	w	14	ws
Mytilus edulis	FF	Puget Sound, WA	40–63600	9/24	d	15	local
Mytilus spp.	FF	San Francisco Bay, CA	180–4100	6/34	d	16	local
Benthic invertebrates							
Macoma balthica	DF/FF	Scheldt, Netherlands	947 (449)	2/12	d	17	local
Crangon crangon	scav	Scheldt, Netherlands	410 (285)	2/12	d	17	local
Nereis diversicolor	omn	Scheldt, Netherlands	785 (409)	2/12	d	17	local
Homarus americanus	scav	Nova Scotia, Canada	235–73000	1/10	w	18	local, my
Littorina littorea	herb	Southern Norway	595–1430	4/27	d	1	ws
Patella vulgata	herb	Southern Norway	674–15462	2/31	d	1	local
Asterias rubens	pred	Southern Norway	325–458	2/19	d	1	local
Macropipus tuberculatus	omn	Spain	60–930	6/ns	w	6	ws

Values are ranges or mean concentrations, some with standard deviations in parenthesis. Values represent all sites and PAHs measured. Sites/PAHs are the number of sites sampled and PAHs analyzed. nd, not detected; ns, not stated; d-w, dry weight or wet weight; *15th–85th percentile; #, annual mean values; ¥, mean for most stations = 3–5 ppm. 'Type' refers to type of sampling; ws, widespread; my, multiyear; seep, natural petroleum seepage areas. Feeding types are: DF, deposit feeder; FF, filter feeder; Omn, omnivore; herb, herbivore; Pred, predator. All studies examined the soft tissue of whole organisms except Ref 16 (hepatopancreas). Many studies were surveys and included samples from relatively uncontaminated to heavily polluted locations. ns, not stated.

1, Knutzen and Sortland (1982); 2, Baumard et al. (1998); 3, O'Connor (2002); 4, NOAA (1998); 5, Cocchieri et al. (1990); 6, Porte and Albaiges (1993); 7, Broman et al. (1990); 8, Rainio et al. (1986); 9, Wade et al. (1989); 10, Iosifidou et al. (1982); 11, Fisher et al. (2000); 12, Claisse (1989); 13, Stronkhorst (1992); 14, Mackie et al. (1980); 15, Krishnakumar et al. (1994); 16, Miles and Roster (1999); 17, Stronkhorst et al. (1994); 18, Uthe and Musial (1986).

section was warranted. Mussels and oysters have been the target for several large multi-year monitoring programs around the world (e.g. Mussel Watch in the USA) and analyzed for a range of compounds (O'Connor 2002; Claisse 1989). Bivalves can accumulate high levels of PAHs and other contaminants in polluted environments, primarily due to their high rate of filtration and low metabolic capacity for these compounds (Livingston 1994).

Bioaccumulation factors (BAFs and BSAFs) in marine invertebrates are shown in Table 9.2. This table shows results from field monitoring studies and laboratory exposure to field-contaminated and spiked sediments. Table 9.2 also contains total PAH concentrations in some field-collected invertebrates in the same fashion as Table 9.1. Several locations around the world are represented and a wide range of marine invertebrates with diverse modes of feeding are included. Additionally, the time for exposure is also listed, because this factor can have a very large effect on the bioaccumulation factor. Bioconcentration factors in field-collected animals are uncommon and are not listed here, mainly due to the difficulty in accurately determining temporally variable water concentrations.

The lipid content in many invertebrates approximates 5% dry weight (Boese and Lee 1992). Using this value for lipids and assuming a TOC in sediment 1–3% of dry weight, most of the BAF values reported in Table 9.2 would likely be 1.7–5 times higher than their respective BSAF values.

It is difficult to draw any strong conclusions from such a table because frequently many PAH compounds were examined and mean values hide important patterns that may be observed with single compounds. Also, the source for PAH exposure varies widely with some sites containing mainly petroleum-based PAHs while others are contaminated with combustion PAHs.

Based on the range of BSAFs, it appears that some values fall into the range expected for equilibrium conditions. Invertebrates from all types of studies (field and laboratory) exhibited values close to the theoretical maximum, but mostly for the LPAHs, which can achieve steady-state or equilibrium partitioning between tissue and the exposure matrix relatively rapidly (Meador *et al.* 1995b). When the mean BSAF values are examined, it appears that most values are relatively consistent in the 0.2–0.4 range. This is very close to values reported for invertebrates in the review paper of Tracey and Hansen (1996) for PAHs.

In general, BSAFs in benthic invertebrates for PAHs appear to occur at values approximately one order of magnitude below those expected at equilibrium and those reported for other non-metabolized neutral organic compounds, such as PCBs (Bierman 1990; Tracey and Hansen 1996). Without detailed analysis, it is not possible to determine which factors (e.g. reduced exposure time, metabolism, or reduced bioavailability) are important for this observation.

The extensive US Mussel Watch dataset (NOAA 1998) was examined to determine average BSAF values for bivalves. Only those sites that had chemistry for all components of the BSAF equation were used. Sixty-five sites, covering the entire USA, were selected at random from the dataset and many sites had data for multiple years (1986–1994). For all species, the mean and standard error of the mean

TABLE 9.2 Bioaccumulation factors for PAHs accumulated by marine invertebrates

Species	Feeding type	Area/type	Number of PAHs	Total PAH (ng/g) dry	BAF (dry wt)	BSAF (range)	BSAF (mean)	Time (days)	Reference
Mytilus edulis	FF	Western Baltic Sea/F	Several	88–3880	0.02–53	0.01–0.59	0.17		1
Macomona liliana	DF	Northern New Zealand/F	9 PAHs	18–203	0.09–1.7	0.04–0.13	0.10		2
Austrovenus stutchburyi	FF	Northern New Zealand/F	9 PAHs	9–47	0.01–0.58	0.002–0.05	0.04		2
Macoma inquinata	DF/FF	Northwest USA/LS	3 PAHs			0.6–2.4	1.3	60	3
Abarenicola pacifica	DF	Northwest USA/LS	3 PAHs			1.5–3.7	2.4	60	3
Macoma balthica	DF/FF	Chesapeake Bay, USA/LS	chyr, naph			0.17–0.78	0.17	12	4
Rhepoxynius abronius	omn	New York, USA/LF	24 PAHs			0.001–0.5	0.052	10	5
Armandia brevis	DF	New York, USA/LF	24 PAHs			0.002–0.9	0.18	10	5
Yoldia limatula	DF	New York USA/LF	several			1–3		35	6
Palaemonetes pugio	omn	Louisiana USA/LS	B[a]P, phen			nd–1.6	0.23	14	7
Rangia cuneata	FF	Louisiana USA/LS	B[a]P, phen			nd–1.5	0.42	14	7
Ampelisca abdita	sus	Rhode Island USA/LS	BA, B[a]P			0.13–0.15	0.14	10	8
Macoma nasuta	DF/FF	Northwest USA/LS	B[a]P		0.25			7	9
E. washingtonianus	detr	Northwest USA/LS	B[a]P		3.0			7	9
R. abronius	omn	Northwest USA/LS	B[a]P		1.5			7	9
Macoma nasuta	DF/FF	Northwest USA/LF	16 PAHs		0.06–0.19			28	9
E. washingtonianus	detr	Northwest USA/LF	16 PAHs		0.1–0.45			7	9
R. abronius	omn	Northwest USA/LF	16 PAHs		0.09–0.5			7	9
Nereis virens	omn	California and New Jersey/LF	Fluoranth			0.8–3.3		15	10
Macoma nasuta	DF/FF	California and New Jersey/LF	Fluoranth			0.6–3.8		15	10
Abarenicola pacifica	DF	Northwest USA/LS	B[a]P			1.1–2.3		68	11
Potamocorbula amurensis	FF	San Francisco, USA/F	18 PAHS	130–860		0.6–5.4	0.3		12

Species	Feeding type	Area/type	PAHs	Conc.		BSAF range	Mean BSAF	Time	Ref
Tapes japonica	FF	San Francisco, USA/F	18 PAHS	95–450		0.007–2.7	0.15		12
Polychaetes (several spp.)	DF/omn	San Francisco, USA/F	18 PAHS	310–1790		0.04–2.0	0.2		12
Macoma nasuta	DF/FF	Los Angeles, USA/LF	5 PAHs			0.05–1.0	0.4	28	13
Leptocheirus plumulosus	detr/sus	Chesapeake Bay, USA/LS	Fluoranth				0.32	26	14
Corophium volutator	detr	Netherlands/LF	8 PAHs			0.5–1.7		25	15
Stichopus tremulus	DF	Norway/F	12 PAHs	237–797		0.004–0.67	0.18		16
Chlamys septemradiata	FF	Norway/F	12 PAHs	75–304		0.007–0.43	0.11		16
Ascidia sp.	FF	Norway/F	12 PAHs	193–671		0.004–0.47	0.09		16
Arenicola marina	DF	Netherlands/LF	7 PAHs	370–3100	0.76			60–90	17
Macoma balthica	DF/FF	Chesapeake Bay, USA/F	14 PAHs			0.5–8.1 Ψ	2.5		18
Nereis succinea	omn	Chesapeake Bay, USA/F	14 PAHs			0.41–6.0 Ψ	2.4		18
Streblospio benedicti	DF	South Carolina, USA/F	3 PAHs	860–2000	0.2–1.4	0.08–0.4	0.22		19
Schizopera knabeni	DF	Louisiana, USA/LS	Fluoranth			0.51–0.80	0.62	1	20
Coullana sp.	omn	Louisiana, USA/LS	Fluoranth			0.22–0.67	0.43	1	20
Mytilus sp.	FF	Baltic and Mediterranean	14 PAHs	25–2420		0.001–7.4	0.52		21

Feeding types are: DF, deposit feeder; FF, filter feeder; omn, omnivore; sus, suspension feeder; detr, detritivore. Area/type list, the area found or where research was conducted and type of exposure (F, samples from field; L, exposures in laboratory to field contaminated sediment; LS, exposures in laboratory to spiked sediments). Values are ranges of total PAHs found in tissues. Time is the exposure period; no value is given for field collections. Fluoranth, fluoranthene; phen, phenanthrene; B[a]P, benzo[a]pyrene; BA, benz[a]anthracene. All analyses conducted with whole organisms or the soft tissue of clams. Mean BSAFs includes all PAHs and locations studied. Ψ, excluding site CB, which produced BSAF values of 106 (M. balthica) and 98 (N. succinea). An approximate conversion of wet weight to dry weight concentration is ([PAH] dry weight = [PAH] wet weight ∗ 5).

1, Baumard et al. (1999a); 2, Hickey et al. (1995); 3, Augenfeld et al. (1982); 4, Foster et al. 1987; 5, Meador et al. (1995a); 6, Lamoureux and Brownawell (1999); 7, Mitra et al. (2000); 8, Fay et al. (2000); 9, Varanasi et al. (1985); 10, Brannon et al. (1993); 11, Weston (1990); 12, Maruya et al. (1997); 13, Ferraro et al. (1990); 14, Kane Driscoll et al. (1998); 15, Kraaij et al. (2001); 16, Naes et al. (1999); 17, Kaag et al. (1997); 18, Foster and Wright (1988); 19, Ferguson and Chandler (1998); 20, Lotufo (1998); 21, Baumard et al. (1999b).

(SEM) BSAF for the LPAHs was 0.7 (0.2) ($n = 117$) and the mean and SEM for the HPAHs was 0.25 (0.09) ($n = 121$). The majority of sites were sampled for *Crassostrea virginica* (oyster) or a mussel (Mytilidae). Two bivalves from this dataset were compared and one was found to exhibit slightly higher BSAFs. The mean (SEM) LPAH BSAF for *C. virginica* was 0.8 (0.2) ($n = 49$) and was 0.2 (0.09) ($n = 51$) for the HPAHs. Conversely, the mean (SEM) LPAH BSAF for *Mytilus edulis* was 0.35 (0.05) and 0.16 (0.04) for the HPAHs (both $n = 55$). Based on a t-test, the mean LPAH BSAF for the oyster was significantly ($\alpha = 0.05$) higher than that for the mussel. There are many factors that could explain these differences, such as differences in geography, metabolism, temporal exposure, and sources (e.g. greater exposure to petroleum); however, a definitive answer is not possible.

One recent review of BSAF values found no differences between laboratory- and field-exposed benthic invertebrates for PAHs (ERDC-WES 2002). The mean and standard deviation (SD) PAH BSAF values for laboratory-exposed and field-exposed benthic invertebrates were 0.34 (0.95) $n = 167$ and 0.36 (0.98) $n = 183$, respectively. The median values for each group were identical (0.08). Laboratory studies of bioaccumulation are generally conducted for relatively short periods of time (10–28 days; Lee *et al.* 1993); hence the field-exposed organisms were expected to be closer to steady state. Because these are mean values and the variance is high, it is difficult to make any meaningful conclusions regarding the differences between individual species exposed in the laboratory vs. those collected from the field. Consequently, differential metabolic ability by the species examined may contribute to the high variability observed.

Variability in assessing bioaccumulation can arise in field or laboratory studies. For example, the BSAF values for a given species (e.g. *Macoma* spp. for PAHs) can be highly variable across studies, which may be due to artifacts such as lack of gut purging, field vs. laboratory sources of contamination, static vs. flow-through testing conditions, or variable exposure time. Additionally, bioaccumulation is likely to vary by season (as described above) due to changes in temperature, salinity, organismal lipid content, runoff conditions, storm events, and other naturally occurring factors.

9.4.2 TROPHIC TRANSFER

Because many species can metabolize PAHs, it is generally assumed that trophic transfer and biomagnification (an increase in tissue concentration over two or more trophic levels) is not important for food webs (Suedel *et al.* 1994). Biomagnification of PAHs is not expected for food webs involving fish (see Chapter 11); however, species from the lower trophic levels that are not able to effectively metabolize these compounds may exhibit food web transfer. Predatory molluscs and polychaetes that prey on other polychaetes and molluscs would likely have higher PAH tissue residues than other similar species that only ingest sediment.

Food web transfer of PAH metabolites is another area that has received little attention. Even though parent PAHs may not be biomagnified, prey species may contain high levels of metabolites that could be accumulated by predators. This was examined by McElroy and Sisson (1989), who fed polychaetes (*Nereis virens*) containing benzo[a]pyrene and accompanying metabolites to winter flounder (*Pseudopleuronectes americanus*) and found that fish had accumulated the metabolites. Additional research assessing the bioaccumulation of PAH metabolites from ingested prey will add greatly to our understanding of this very important process of trophic transfer and the potential for toxic effects.

9.4.3 ALKYLATED PAHs

Unfortunately, the vast majority of aquatic studies concerning PAHs only focus on a select set of compounds, usually 8–24 parent compounds. There are hundreds of aromatic compounds, many of which occur in environmental samples. Besides the usual parent PAH compounds, there are dozens of alkylated naphthalenes, phenanthrenes, fluorenes, chrysenes, dibenzothiopenes and others that increase in hydrophobicity and bioaccumulation potential with increasing alkylation. Consequently, consideration of only the limited set of parent compounds can severely underestimate PAH bioaccumulation and toxicity. There is very little information on the metabolism of alkylated PAHs in invertebrates and the ability of these compounds to be accumulated by higher trophic levels.

A few studies have highlighted the degree that alkylated PAHs are accumulated. For example, the parent PAH compounds comprised 5–70% of the total PAH concentration in the stomach contents of juvenile chinook salmon residing in two different urban estuaries (Varanasi *et al.* 1993). Because of the very high total PAH concentrations (up to 150 ppm wet wt) in fish stomach contents, which generally consists of small benthic invertebrates, an analysis of only parent PAHs would severely underestimate the potential exposure in some samples. This observation is also supported by NOAA's Mussel Watch program (NOAA 1998), showing very high percentages of alkylated PAHs in bivalves at some sites around the USA and other studies with mussels, including Miles and Roster (1999) and Hellou *et al.* (2002).

9.5 CONCLUSIONS AND RECOMMENDATIONS

For an accurate assessment of the potential impacts from PAHs, scientists and regulators need reliable information regarding bioaccumulation. Understanding the factors responsible for the amount bioaccumulated, such as toxicokinetics, organism behavior and physiology, and environmental factors, will help greatly when attempting to recognize patterns, generate predictions, and determine toxicity thresholds. Knowledge of the mechanisms that control accumulation will help scientists to develop predictive models of contaminant accumulation for both acute events, such as oil spills, and long-term chronic exposure found in many urban coastal ecosystems.

Currently, there are several important areas of research that will enhance our understanding of the nature of PAH bioaccumulation, which will be invaluable for predicting tissue concentrations and adverse effects. These include several environmental factors, such as the role of sediment particle size on bioavailability, the behavior of petroleum- vs. combustion-type PAHs, and the degree of equilibrium in the field. In addition to these are several organism-related parameters, including absorption efficiency from different sources, variability in digestive physiology among species, trophic transfer of parent compounds and metabolites, intra- and interspecific toxicokinetic differences, understanding and controlling artifacts, spatial and temporal variables in uptake, the role of alkylated PAHs, and the effects of organism behavior and health on uptake and elimination rates. Research on these various factors will help greatly to enhance our understanding of the mechanisms of PAH accumulation in marine organisms and promote more reliable information for toxicity assessment.

ACKNOWLEDGMENTS

I would like to thank Anne McElroy for her help in defining the scope of this chapter, John Stein for his thorough review, and the editor, Peter Douben, for his patience.

REFERENCES

Ahrens MJ, Hertz J, Lamoureux EM, Lopez GR, McElroy AE and Brownawell B (2001) The effect of body size on digestive chemistry and absorption efficiencies of food and sediment-bound organic contaminants in *Nereis succinea* (Polychaeta). *Journal of Experimental Marine Biology and Ecology*, **263**, 185–209.

Augenfeld JM, Anderson JW, Riley RG and Thomas BL (1982) The fate of polyaromatic hydrocarbons in an intertidal sediment exposure system: bioavailability to *Macoma inquinata* (Mollusca: Pelecypoda) and *Abarenicola pacifica* (Annelida: Polychaeta). *Marine Environmental Research*, **7**, 31–50.

Baumard P, Budzinski H, Garrigues P, Dizer H and Hansen PD (1999a) Polycyclic aromatic hydrocarbons in recent sediments and mussels (*Mytilus edulis*) from the western Baltic Sea: occurrence, bioavailability and seasonal variation. *Marine Environmental Research*, **47**, 17–47.

Baumard P, Budzinski H, Garrigues P, Narbonne JF, Burgeot T, Michel X and Bellocq J (1999b) Polycyclic aromatic hydrocarbon (PAH) burden of mussels (*Mytilus* spp.) in different marine environments in relation with sediment PAH contamination, and bioavailability. *Marine Environmental Research*, **47**, 415–439.

Baumard P, Budzinski H and Garrigues P (1998). Polycyclic aromatic hydrocarbons in sediments and mussels of the western mediterranean sea. *Environmental Toxicology and Chemistry* **17**, 765–776.

Bend JR, James MO, Little PJ and Foureman GL (1981) *In vitro* and *in vivo* metabolism of benzo[a]pyrene by selected marine crustacean species. In Dawe CJ (ed.), *Phyletic Approaches to Cancer*. Japan Scientific Society Press, Tokyo, pp. 179-194.

Bend JR, James MO and Dansette PM (1977) *In vitro* metabolism of xenobiotics in some marine animals. In Kraybill HF, Dawe CJ, Harshberger JC and Tardiff RG (eds), *Aquatic Pollutants and Biologic Effects with Emphasis on Neoplasia*. New York Academy of Sciences, New York, pp. 505-521.

Bender ME, Hargis WJ, Huggett RJ and Roberts MH (1988) Effects of polynuclear aromatic hydrocarbons on fishes and shellfish: an overview of research in Virginia. *Marine Environmental Research*, **24**, 237-241.

Bierman VJ Jr (1990) Equilibrium partitioning and biomagnification of organic chemicals in benthic animals. *Environmental Science and Technology*, **23**, 1407-1412.

Boese BL and Lee H II (1992) Synthesis of methods to predict bioaccumulation of sediment pollutants. ERL-N Contribution No. N232. US Environmental Protection Agency, Washington DC.

Brannon JM, Price CB, Reilly FJ Jr, Pennington JC and McFarland VA (1993) Effects of sediment organic carbon on distribution of radiolabeled fluoranthene and PCBs among sediment, interstitial water, and biota. *Bulletin of Environmental Contamination and Toxicology*, **51**, 873-880.

Bridges TS, Farrar JD and Duke BM (1997) The influence of food ration on sediment toxicity in *Neanthes arenaceodentata* (Annelida: Polychaeta). *Environmental Toxicology and Chemistry*, **16**, 1659-1665.

Broman D, Näf C, Lundbergh I and Zebühr Y (1990) An *in situ* study on the distribution, biotransformation and flux of polycyclic aromatic hydrocarbons (PAHs) in an aquatic food chain (seston: *Mytilus edulis* L. — *Somateria mollissima* L.) from the Baltic: an ecotoxicological perspective. *Environmental Toxicology and Chemistry*, **9**, 429-442.

Ciarelli S, van Straalen NM, Klap VA and van Wezel AP (1999) Effects of sediment bioturbation by the estuarine amphipod *Corophium volutator* on fluoranthene resuspension and transfer into the mussel (*Mytilus edulis*). *Environmental Toxicology and Chemistry*, **18**, 318-328.

Claisse D (1989) Chemical contamination of French Coasts: the results of a ten years mussel watch. *Marine Pollution Bulletin*, **20**, 523-528.

Cocchieri RA, Arnese A and Minicucci AM (1990) Polycyclic aromatic hydrocarbons in marine organisms from Italian central Mediterranean coasts. *Marine Pollution Bulletin*, **21**, 15-18.

Cornelissen G, Rigterink H, Ferdinandy M and van Noort P (1998) Rapidly desorbing fractions of PAHs in contaminated sediments as predictor of the extent of bioremediation. *Environmental Science and Technology*, **32**, 966-970.

Di Toro DM, Zarba CS, Hansen DJ, Berry WJ, Swartz RC, Cowan CE, Pavlou SP, Allen HE, Thomas NA and Paquin PR (1991) Technical basis for establishing sediment quality criteria for non-ionic organic chemicals using equilibrium partitioning. *Environmental Toxicology and Chemistry*, **10**, 1541-1583.

Doherty FG, Cherry DS and Cairns J Jr (1987) Valve closure responses of the Asiatic clam *Corbicula fluminea* exposed to cadmium and zinc. *Hydrobiologia*, **153**, 159-167.

ERDC-WES (2002) ERDC-WES BSAF database, V McFarland, personal communication. Engineering Research and Development Center, Waterways Experimental Station, Vicksburg, MS.

Farrington JW, Goldberg ED, Risebrough RW, Martin JH and Bowen VT (1983) US 'mussel watch' 1976-1978: an overview of the trace-metal, DDE, PCB, hydrocarbon, and artificial radionuclide data. *Environmental Science and Technology*, **17**, 490-496.

Fauchald K and Jumars PA (1979) The diet of worms: a study of polychaete feeding guilds. *Oceanography and Marine Biology Annual Reviews*, **17**, 193-284.

Fay AA, Brownawell BJ, Elskus AA and McElvoy AE (2000). Critical body residues in the marine amphipod *Ampelisca abdita*: Sediment exposures with nonionic organic contaminants. *Environmental Toxicology and Chemistry*, **19**, 1028–1035.

Ferguson PL and Chandler GT (1998) A laboratory and field comparison of sediment polycyclic aromatic hydrocarbon bioaccumulation by the cosmopolitan estuarine polychaete *Streblospio benedicti* (Webster). *Marine Environmental Research*, **45**, 387–401.

Ferraro SP, Lee H II, Ozretich RJ and Specht DT (1990) Predicting bioaccumulation potential: a test of a fugacity-based model. *Archives of Environmental Contamination and Toxicology*, **19**, 386–394.

Fisher WS, Oliver LM, Winstead JT and Long ER (2000) A survey of oysters *Crassostrea virginica* from Tampa Bay, Florida: associations of internal defense measurements with contaminant burdens. *Aquatic Toxicology*, **51**, 115–138.

Foster GD, Baksi SM and Means JC (1987) Bioaccumulation of trace organic contaminants from sediment by Baltic clams (*Macoma balthica*) and soft-shelled clams (*Mya arenaria*). *Environmental Toxicology and Chemistry*, **6**, 969–976.

Foster GD and Wright DA (1988) Unsubstituted polynuclear aromatic hydrocarbons in sediments, clams, and clam worms from Chesapeake Bay. *Marine Pollution Bulletin*, **19**, 459–465.

Foureman GL, Ben-Zvi S, Dostal L, Fouts JR and Bend JR (1978) Distribution of ^{14}C-benzo[a]pyrene in the lobster, *Homarus americanus*, at various times after a single injection into the pericardial sinus. *Bulletin Mt Desert Island Biological Laboratory*, **18**, 93–96.

Hellou J, Steller S, Zitko V, Leonard J, King T, Milligan TG and Yeats P (2002). Distribution of PACs in surficial sediments and bioavailability to mussels, *Mytilus edulis*, of Halifax Harbour. *Marine Environmental Research* **53**, 357–379.

Hickey CW, Roper DS, Holland PT and Trower TM (1995) Accumulation of organic contaminants in two sediment-dwelling shellfish with contrasting feeding modes: deposit- (*Macomona liliana*) and filter-feeding (*Austrovenus stutchburyi*). *Archives of Environmental Contamination and Toxicology*, **29**, 221–231.

Iosifidou, HG, Kilikidis SD and Kamarianos AP (1982) Analysis of polycyclic aromatic hydrocarbons in mussels (*Mytilus galloprovincialis*) from the Thermaikos Gulf, Greece. *Bulletin of Environmental Contamination and Toxicology*, **28**, 535–541.

Jovanovich MC and Marion KR (1987) Seasonal variation in uptake and depuration of anthracene by the brackish water clam, *Rangia cuneata*. *Marine Biology*, **95**, 395–403.

Kaag NHBM, Foekema EM, Scholten MC Th and van Straalen NM (1997) Comparison of contaminant accumulation in three species of marine invertebrates with different feeding habits. *Environmental Toxicology and Chemistry*, **16**, 837–842.

Kane Driscoll S and McElroy AE (1996) Bioaccumulation of benzo[a]pyrene in three species of polychaete worms. *Environmental Toxicology and Chemistry*, **15**, 1401–1410.

Kane Driscoll S and McElroy AE (1997) Elimination of sediment associated benzo[a]pyrene and its metabolites by polychaete worms exposed to 3-methylcholanthrene. *Aquatic Toxicology*, **39**, 77–91.

Kane Driscoll SB, Schaffner LC and Dickhut RM (1998) Toxicokinetics of fluoranthene to the amphipod *Leptocheirus plumulosus* in water-only and sediment exposures. *Marine Environmental Research* **45**, 269–284.

Knutzen J and Sortland B (1982) Polycyclic aromatic hydrocarbons (PAH) in some algae and invertebrates from moderately polluted parts of the coast of Norway. *Water Research*, **16**, 421–428.

Kraaij R, Ciarelli S, Tolls J, Kater BJ and Belfroid A (2001) Bioavailability of lab-contaminated and native polycyclic aromatic hydrocarbons to the amphipod

Corophium volutator relates to chemical desorption. *Environmental Toxicology and Chemistry*, **20**, 1716–1724.

Krishnakumar PK, Casillas E and Varanasi U (1994) Effect of environmental contaminants on the health of *Mytilus edulis* from Puget Sound, Washington, USA. I. Cytochemical measures of lysosomal responses in the digestive cells using automatic image analysis. *Marine Ecology Progress Series*, **106**, 249–261.

Kukkonen J and Landrum PF (1995) Measuring assimilation efficiency for sediment-bound PAH and PCB congeners by benthic organisms. *Aquatic Toxicology*, **32**, 75–92.

Kukkonen J and Landrum PF (1998) Effect of particle–xenobiotic contact time on bioavailability of sediment-associated benzo[a]pyrene to benthic amphipod, *Diporeia* spp. *Aquatic Toxicology*, **42**, 229–242.

Lamoureux EM and Brownawell BJ (1999) Chemical and biological availability of sediment-sorbed hydrophobic organic contaminants. *Environmental Toxicology and Chemistry*, **18**, 1733–1741.

Landrum PF (1988) Toxicokinetics of organic xenobiotics in the amphipod, *Pontoporeia hoyi*: role of physiological and environmental variables. *Aquatic Toxicology*, **12**, 245–271.

Landrum PF (1989) Bioavailability and toxicokinetics of polycyclic aromatic hydrocarbons sorbed to sediments for the amphipod, *Pontoporeia hoyi*. *Environmental Science and Technology*, **23**, 588–595.

Landrum PF and Fisher SW (1998) Influence of lipids on the bioaccumulation and trophic transfer of organic contaminants in aquatic organisms. In Arts M and Wainman B (eds), *Lipids in Freshwater Ecosystems*. Springer-Verlag, New York, pp. 203–234.

Landrum PF and Robbins JA (1990) Bioavailability of sediment associated contaminants: a review and simulation model. In Baudo R, Giesy JP and Muntau H (eds), *Sediments: Chemistry and Toxicity of In-place Pollutants*. Lewis, Chelsea, MI, pp. 237–263.

Landrum PF, Dupuis WS and Kukkonen J (1994) Toxicokinetics and toxicity of sediment-associated pyrene and phenanthrene in *Diporeia* spp.: examination of equilibrium-partitioning theory and residue-based effects for assessing hazard. *Environmental Toxicology and Chemistry*, **13**, 1769–1780.

Landrum PF and Stubblefield CR (1991) Role of respiration in the accumulation of organic xenobiotics by the amphipod, *Pontoporeia hoyi*. *Environmental Toxicology and Chemistry*, **10**, 1019–1028.

Lee H II, Boese BL, Pelletier J, Winsor M, Specht DT and Randall RC (1993) *Guidance Manual: Bedded Sediment Bioaccumulation Tests*. US EPA Report No. 600/R-93/183, US EPA, ERL-Narragansett, Rhode Island, 231 pp.

Lee RF, Singer SC and Page DS (1981) Responses of cytochrome P-450 systems in marine crab and polychaetes to organic pollutants. *Aquatic Toxicology*, **1**, 355–365.

Livingstone DR (1994) Recent developments in marine invertebrate organic xenobiotic metabolism. *Toxicology and Ecotoxicology News*, **1**, 88–95.

Lotufo GR (1998) Bioaccumulation of sediment-associated fluoranthene in benthic copepods: uptake, elimination and biotransformation. *Aquatic Toxicology*, **44**, 1–15.

Mackay D (1982) Correlation of bioconcentration factors. *Environmental Science and Technology*, **16**, 274–278.

Mackay D and Paterson S (1981) Calculating fugacity. *Environmental Science and Technology*, **15**, 1006–1014.

Mackie PR, Hardy R, Whittle KJ, Bruce C and McGill AS (1980) The tissue hydrocarbon burden of mussels from various sites around the Scottish coast. In Bjørseth A, Dennis AJ (eds), *Polynuclear Aromatic Hydrocarbons: Chemistry and Biological Effects*. Battelle, Columbus, OH. pp. 379–393.

Maruya KA, RW Risebrough and AJ Horne (1997) The bioaccumulation of polynuclear aromatic hydrocarbons by benthic invertebrates in an intertidal marsh. *Environmental Toxicology and Chemistry*, **16**, 1087–1097.

Mayer LM, Chen Z, Findlay RH, Fang J, Sampson S, Self RFL, Jumars PA, Quetel C and Donard OFX (1996) Bioavailability of sedimentary contaminants subject to deposit-feeder digestion. *Environmental Science and Technology*, **30**, 2641–2645.

McElroy AE and Means JC (1988) Factors affecting the bioavailability of hexachloro-biphenyls to benthic organisms. In Adams WJ, Chapman GA and Landis WG (eds), *Aquatic Toxicology and Hazard Assessment*, Vol. 10. ASTM STP 971. American Society for Testing and Materials, Philadelphia, PA, pp. 149–158.

McElroy AE, Farrington JW and Teal JM (1990) Influence of mode of exposure and the presence of a tubiculous polychaete on the fate of benz[a]anthracene in the benthos. *Environmental Science and Technology*, **24**, 1648–1655.

McElroy AE, Leitch K and Fay A (2000) A survey of *in vivo* benzo[a]pyrene metabolism in small benthic marine invertebrates. *Marine Environmental Research*, **50**, 33–38.

McElroy AE and Sisson JD (1989) Trophic transfer of benzo[a]pyrene metabolites between benthic marine organisms. *Marine Environmental Research*, **28**, 265–269.

McGroddy S and Farrington JW (1995) Sediment–porewater partitioning of polycyclic aromatic hydrocarbons (PAHs) in three cores from Boston Harbor, Massachusetts. *Environmental Science and Technology*, **29**, 1542–1550.

Meador JP and Rice CA (2001) Impaired growth in the polychaete *Armandia brevis* exposed to tributyltin in sediment. *Marine Environmental Research*, **51**, 113–129.

Meador JP, Casillas E, Sloan CA and Varanasi U (1995a) Comparative bioaccumulation of polycyclic aromatic hydrocarbons from sediment by two infaunal invertebrates. *Marine Ecology Progress Series*, **123**, 107–124.

Meador JP, Stein JE, Reichert WL and Varanasi U (1995b) A review of bioaccumulation of polycyclic aromatic hydrocarbons by marine organisms. *Reviews in Environmental Contamination and Toxicology*, **143**, 79–165.

Miles AK and Roster N (1999) Enhancement of polycyclic aromatic hydrocarbons in estuarine invertebrates by surface runoff at a decommissioned military fuel depot. *Marine Environmental Research*, **47**, 49–60.

Mitra S, Klerks PL, Bianchi TS, Means J and Carman KR (2000) Effects of estuarine organic matter biogeochemistry on the bioaccumulation of PAHs by two epibenthic species. *Estuaries*, **23**, 864–876.

Naes K, Hylland K, Oug E, Forlin L and Ericson G (1999) Accumulation and effects of aluminum smelter-generated polycyclic aromatic hydrocarbons on soft-bottom invertebrates and fish. *Environmental Toxicology and Chemistry*, **18**, 2205–2216.

National Oceanic and Atmospheric Administration (NOAA) (1998) Chemical contaminants in oysters and mussels by O'Connor TP. In *NOAA's State of the Coast Report*, Silver Spring, MD. Data online, http://ccmaserver.nos.noaa.gov/NSandT/NSandTdata.html

O'Connor TP (2002) National distribution of chemical concentrations in mussels and oysters in the USA. *Marine Environmental Research*, **53**, 117–143.

Penry DL and Weston DP (1998) Digestive determinants of benzo[a]pyrene and phenan-threne bioaccumulation by a deposit-feeding polychaete. *Environmental Toxicology and Chemistry*, **17**, 2254–2265.

Pittinger CA, Buikema AL, Horner SG and Young RW (1985) Variation in tissue burdens of polycyclic aromatic hydrocarbons in indigenous and relocated oysters. *Environmental Toxicology and Chemistry*, **4**, 379–387.

Porte C and Albaiges J (1993) Bioaccumulation patterns of hydrocarbons and polychlo-rinated biphenyls in bivalves, crustaceans, and fishes. *Archives of Environmental Contamination and Toxicology*, **26**, 273–281.

Rainio K, Linko RR and Ruotsila L (1986) Polycyclic aromatic hydrocarbons in mussel and fish from the Finnish archipelago sea. *Bulletin of Environmental Contamination and Toxicology*, **37**, 337–343.

Reichert WL, Eberhart B-T and Varanasi U (1985) Exposure of two species of deposit-feeding amphipods to sediment-associated [^{3}H]benzo[a]pyrene: uptake, metabolism and covalent binding to tissue macromolecules. *Aquatic Toxicology*, **6**, 45–56.

Roesijadi G, Anderson JW and Blaylock JW (1978) Uptake of hydrocarbons from marine sediments contaminated with Prudhoe Bay crude oil: influence of feeding type of test species and availability of polycyclic aromatic hydrocarbons. *Journal of the Fisheries Research Board of Canada*, **35**, 608–614.

Schnell JV, Gruger EH and Malins DC (1980) Monooxygenase activities of coho salmon (*Oncorhynchus kisutch*) liver microsomes using three polycyclic aromatic hydrocarbons substrates. *Xenobiotica*, **10**, 229–234.

Stainken D, Multer HG and Mirecki J (1983) Seasonal patterns of sedimentary hydrocarbons in the Raritan Bay–lower New York Bay. *Environmental Toxicology and Chemistry*, **2**, 35–42.

Stronkhorst J (1992) Trends in pollutants in blue mussel *Mytilus edulis* and flounder *Platichthys flesus* from two Dutch estuaries, 1985–1990. *Marine Pollution Bulletin*, **24**, 250–258.

Stronkhorst J, Vos PC and Misdorp R (1994) trace metals, PCBs, and PAHs in benthic (epipelic) diatoms from intertidal sediments; a pilot study. *Bulletin Environmental contamination and toxicology*. **52**, 818–824.

Suedel BC, Boraczek JA, Peddicord RK, Clifford PA and Dillon TM (1994) Trophic transfer and biomagnification potential of contaminants in aquatic ecosystems. *Reviews on Environmental Contamination and Toxicology*, **136**, 22–89.

Tracey GA and Hansen DJ (1996) Use of biota–sediment accumulation factors to assess similarity of non-ionic organic chemical exposure to benthically-coupled organisms of differing trophic mode. *Archives of Environmental Contamination and Toxicology*, **30**, 467–475.

Uthe JF and Musial CJ (1986) Polycyclic aromatic hydrocarbon contamination of American lobster, *Homarus americanus*, in the proximity of a coal-coking plant. *Bulletin of Environmental Contamination and Toxicology*, **37**, 730–738.

Varanasi U, Casillas E, Arkoosh MR, Hom T, Misitano DA, Brown DW, Chan S-L, Collier TK, McCain BB and Stein JE (1993) *Contaminant Exposure and Associated Biological Effects in Juvenile Chinook Salmon* (Oncorhynchus tshawytscha) *from Urban and Non-urban Estuaries of Puget Sound*. NOAA Technical Memorandum NMFS-NWFSC-8, SilverSpring, MD.

Varanasi U, Reichert WL, Stein JE, Brown DW and Sanborn HR (1985) Bioavailability and biotransformation of aromatic hydrocarbons in benthic organisms exposed to sediment from an urban estuary. *Environmental Science and Technology*, **19**, 836–841.

Wade TL, Kennicutt MC and Brooks JM (1989) Gulf of Mexico hydrocarbon seep communities: Part III. Aromatic hydrocarbon concentrations in organisms, sediments and water. *Marine Environmental Research*, **27**, 19–30.

Wang W-X and Fisher NS (1999) Assimilation efficiencies of chemical contaminants in aquatic invertebrates: a synthesis. *Environmental Toxicology and Chemistry*, **18**, 2034–2045.

Weston DP (1990) Hydrocarbon bioaccumulation from contaminated sediment by the deposit feeding polychaete *Abarenicola pacifica*. *Marine Biology*, **107**, 159–169.

Weston DP and Mayer LM (1998) *In vitro* digestive fluid extraction as a measure of the bioavailability of sediment-associated polycyclic aromatic hydrocarbons: sources of variation and implications for partitioning models. *Environmental Toxicology and Chemistry*, **17**, 820–829.

Witt G (1995) Polycyclic aromatic hydrocarbons in water and sediment of the Baltic Sea. *Marine Pollution Bulletin*, **31**, 237–248.

10

Uptake and Accumulation of PAHs by Terrestrial Invertebrates

RUDOLF K. ACHAZI[1] AND CORNELIS A. M. VAN GESTEL[2]
[1]*Freie Universität Berlin, Berlin, Germany*
[2]*Vrije Universiteit, Amsterdam, The Netherlands*

10.1 INTRODUCTION

The background concentration of individual polycyclic aromatic hydrocarbons (PAHs) is estimated to be in the range of $1-10$ µg/kg dry soil, while total PAH levels are in the range $0.5-1.5$ mg/kg dry soil. Lowest concentrations measured today are frequently 10 times higher. PAHs accumulate in topsoils in the order: arable soils < mineral soil below forest < permanent grassland < urban soils < specifically contaminated sites (Knoche *et al.* 1995; Wilke 2000). In the sparsely populated biosphere reserve, Spreewald, median concentrations of total PAH in topsoil increase as follows: 269 µg/kg in arable land < 441 µg/kg in grassland < 688 µg/kg in forest < 1378 µg/kg in the litter layer. There seems to be a specific correlation between total organic content (TOC) of soil and the concentration of phenanthrene, chrysene, benzo[b]fluoranthene, indeno[1,2,3-cd]pyrene, and dibenzo[ah]anthracene in grassland and forest soils, but not in arable soil, probably due to tilling during cultivation and the removal of soil material at harvesting (Terytze *et al.* 1998).

In temperate zones, concentrations of individual PAHs in soils are strongly correlated with each other. The most abundant PAHs seem to be benzo[jkf]fluoranthene, chrysene, fluoranthene, and pyrene. Due to the similarity of this pattern, benzo[a]pyrene concentration is often used to estimate total PAH concentrations. According to Wilke (2000), there is a close relationship between the benzo[a]pyrene concentration and the sum of the 16 EPA PAHs in the A, B, and Oa soil horizons. Close to point sources, PAH concentrations may also correlate with the concentrations of heavy metals such as Pb on roadsides (Thomas *et al.* 1984) or Cd, Cu, Hg and Zn in the vicinity of industrial areas (Schulte 1996).

PAHs: An Ecotoxicological Perspective. Edited by Peter E.T. Douben.
© 2003 John Wiley & Sons Ltd

PAHs in soil and litter may be transferred to soil invertebrates, directly by soil ingestion and oral or dermal uptake from porewater, or indirectly by feeding on living and decaying plant material. The bioavailability of soil PAHs depends on their hydrophobicity and binding to the soil matrix (see Chapter 5 for more detail), consisting of the mineral constituents and organic matter.

10.2   UPTAKE OF PAHs BY TERRESTRIAL INVERTEBRATES FROM SOIL

10.2.1   GENERAL PRINCIPLES AND OBSERVATIONS

The uptake routes for the different species of soil invertebrates depend on their affiliation to one of the three levels of the ecological hierarchy: the micro-, meso- and macrofauna and their feeding habits. Free-living nematodes, as representatives of the microfauna, inhabit the soil porewater and feed on bacteria and algae, detritus, fungi, and sometimes also on higher plants and small animals. Enchytraeids and springtails (Collembola) represent major groups of the mesofauna level. Enchytraeids (small annelids) prefer the same microhabitat as nematodes and also live on microorganisms, fungi, and decaying plant material. In contrast, springtails, as wingless insects (micro-arthropods), inhabit the airspace of soils and leaf litter. They feed on algae, fungi, yeast and decaying organic material. Earthworms (annelids) and isopods (crustaceans), on the other hand, belong to the macrofauna. The lumbricids are classified according to their main habitat as epigeic, endogeic and anecic, living either in the litter and fragmentation layer of soil, in the organic topsoil, or in the mineral soil, respectively. All three types live in close contact with the litter and fragmentation layer and collect food from these layers and plant material from the soil surface. Isopods live predominantly in the litter and fragmentation layer and as juveniles on top of the organic soil. They feed on decaying plant material.

The uptake of hydrophobic organic contaminants in all soil porewater-dwelling invertebrates is via passive diffusion through the integument and from contaminated food. In members of the macrofauna group, additional uptake is possible by resorption of these compounds from the soil material passing the gut. In airspace-dwelling species, such as springtails and isopods, uptake takes place mainly with contaminated food and through the atmosphere via the gaseous phase. In springtails additional uptake from porewater by the collophor, a tube-like organ on the ventral side, might also to be expected (Hopkin 1997).

Chapter 7 considered partitioning and bioavailability in general. The equilibrium partitioning (EqP) theory assumes that the concentration of a chemical in an organism depends solely on its concentration in the water phase. For terrestrial animals, such as earthworms, only the dissolved fraction of the chemical in the soil porewater is available for uptake by passive diffusion. In *Eisenia andrei* the uptake rate clearly decreases with log K_{ow} if based on total PAH concentrations in soil, but is independent of log K_{ow} if based on soil solution concentrations. This is also true for other species, e.g. *Lumbricus rubellus*, *Eisenia fetida*, and *Lumbricus terrestris* (Van Brummelen *et al.* 1996b; Ma *et al.* 1998; Jager *et al.* 2000; Krauss *et al.* 2000). Also, the correction with lipid content is discussed in Chapter 7. For *Lumbricus rubellus* there is a linear relationship for the uptake of PAHs with intercept 0 between lipid-normalized concentrations in the earthworm and organic matter-normalized concentrations of the PAHs in soil. Therefore the biota–soil accumulation factor (BSAF) is independent of the log K_{ow}. This is in agreement with the EqP theory, at least for low-molecular weight PAHs. Bioaccumulation factors established from the concentrations in worm lipid and in interstitial soil water increase with the log K_{ow} and range from 80 for anthracene to 18,000 for dibenzo[a]anthracene. Bioconcentration factors (BCFs) measured in a short-term water-only exposure experiment for phenanthrene, fluoranthene, pyrene and benzo[a]pyrene, were in good agreement with BCFs generated by kinetic modeling. Experimental and fitted BCFs provided evidence that the uptake for *Lumbricus rubellus* of PAHs with log $K_{ow} < 5$ is mediated through direct contact with the soluble phase. For PAHs of higher log K_{ow}, the dietary uptake may provide an additional route of exposure (Ma *et al.* 1998), but will not exceed 10% of the total uptake in earthworms (Belfroid *et al.* 1994). However, accumulation depends on the access to food, as the BAF for phenanthrene and fluoranthene in *Lumbricus rubellus* is higher in starving than in fed animals (Ma *et al.* 1995).

For isopods the situation is different. If isopods and earthworms are exposed to the same soil material contaminated with PAHs, accumulation is different. *Porcellio scaber* displayed elevated levels of phenanthrene, fluoranthene, pyrene, benz[a]anthracene, chrysene, benzo[b]fluoranthene, benzo[k]fluoranthene, and benzo[a]pyrene, whereas *Oniscus asellus* had elevated levels of fluoranthene and pyrene, and *Philoscia muscorum* elevated levels of phenanthrene, fluoranthene, pyrene, and chrysene. In *Lumbricus rubellus* low levels of phenanthrene and high levels of the high-molecular weight PAHs benz[a]anthracene, chrysene, benzo[b]fluoranthene, benzo[k]fluoranthene, and benzo[a]pyrene were present, besides fluoranthene and pyrene. The EqP model is only true for the earthworm–soil system, while BASFs for *Porcellio scaber* and *Philoscia muscorum* are clearly negatively correlated with increasing PAH ring number and increasing octanol–water partition coefficient. This deviation from the EqP theory observed in isopods may be the combined result of: (a) a restricted contact to the soil solution due to their adaptation to water loss; (b) a short contact time of food and intestinal tissues resulting in non-equilibrium conditions; and (c) the metabolism of the PAHs (Van Brummelen *et al.* 1996b). The importance of the latter factor has been confirmed by Van Brummelen and Van Straalen (1996), who found benzo[a]pyrene elimination half-lives in *Porcellio scaber* of approximately 1 day. There seems to be no significant correlation between the BSAFs in isopods and earthworms and soil properties such as organic matter content or grain-size fractions. This means that BSAFs, corrected for lipid and organic matter content, are site-independent. However, BSAFs were clearly affected by the stratum and were four to five times higher in the mineral zone than in the litter zone, the fragmentation zone, and the humus layer (Van Brummelen *et al.* 1996b).

The toxicity of PAHs to the springtail *Folsomia fimetaria* increased with lipophilicity (log K_{ow}) when based on the estimated porewater concentrations, suggesting a narcotic mode of action for most substances. Anthracene, however, showed a deviating position in the regression plot, suggesting a more specific mode of action. The absence of toxicity for the non-toxic compounds benz(a)anthracene, perylene, indeno[1,2,3-cd]pyrene, chrysene, dibenzo[b]fluoranthene, dibenzo[k]fluoranthene, dibenz[ah]anthracene could be explained by a limited water solubility. This indicates that these compounds also act by narcosis and that their toxicity is governed by concentrations in the porewater (Sverdrup *et al.* 2002c).

10.2.2   THE EFFECT OF AGING ON UPTAKE AND TOXICITY OF PAHs

In earthworms, bioaccumulation of PAHs is affected by the aging and pre-treatment of soils and soil materials. Aging of spiked soil with PAHs reduces bioavailability, including that to earthworms (Alexander 1995; Hatzinger and Alexander 1995); more details are covered in Section 7.3.2. Bioavailability of phenanthrene for uptake by earthworms was further decreased when soils were subjected to wetting and drying during the aging period (White *et al.* 1997, 1998); e.g. the uptake of anthracene by *Eisenia fetida* was reduced from 32.2% in unaged spiked soil to 13.7% after 203 days (Tang *et al.* 1998; Tang and Alexander 1999). Uptake rates, BSAFs ($C_{worm/lipidcontent}$:$C_{soil/TOC}$), and leachability declined from spiked soil > source soil > treated soil > aged soil. The binding of PAHs to soil affects not only the bioavailability but also the toxicity (Stroo *et al.* 2000). Upon aging of pyrene- and phenantrene-amended natural soil (2.8% organic matter) for periods up to 120 days at 20°C, both chemicals degraded with half-lives of > 200 and 60 days, respectively. Aging did not, however, lead to a reduction in toxicity to *Folsomia fimetaria* when EC_{50} values were expressed on the basis of measured total soil concentrations (Sverdrup *et al.* 2002b). According to the authors, a decreased bioavailability of the parent PAH by aging may have been counteracted by the formation of toxic metabolites. Another possibility is that the bioavailability of the non-degraded fraction was still high, perhaps as a consequence of the fairly low organic matter content of the soil. No attempts were made, however, to confirm one of these options by measuring metabolites or bioavailable fractions in the test soils.

The bioavailability of unaged and aged anthracene, fluoranthene, and pyrene to *Eisenia fetida* in soils of different locations was correlated with their extractability by organic solvents such as *n*-butanol, ethyl acetate, methanol, and propanol. The extent of exposure of earthworms to PAHs is correlated with the extractability with these organic solvents but not with their extractability by vigorous methods, e.g. Soxhlet extraction. The decline in extractability by these mild extractants as a consequence of aging was not the result of significant covalent complexing with soil organic matter, biodegradation, or volatilization, because > 88% was recovered by Soxhlet extraction (Kelsey and Alexander 1997; Tang and Alexander 1999). This conclusion does not correspond to results in field-contaminated soils, where regression analyses showed close and significant relationships between log C_{worm} and log C_{soil} for *Lumbricus terrestris*. According to Krauss *et al.* (2000), the concentration of a PAH in earthworms can be predicted from its total concentration in soil.

The uptake of PAHs by members of the mesofauna has hardly been investigated. Preliminary data are available for *Enchytraeus crypticus*. This micro-annelid can be cultured on agar-agar indefinitely and kept in artificial fresh water for up to 9 days at 12°C. The uptake of ^{14}C-benzo[a]pyrene in this species was studied in agar-agar and in a water test system. The uptake at 0.1 and 1 mg benzo[a]pyrene/L agar-agar increased linearly with time and reached an equilibrium after 18 days. In steady-state conditions the BCFs were 230–275. Elimination rates were high after transfer of the worms to benzo[a]pyrene-free agar-agar and the concentration in the animals decreased to 20% in 2 weeks. In the aqueous test system, equilibrium was obtained after 4–5 days and the BCF at 4 and 14 μg ^{14}C-benzo[a]pyrene/L water was 478 and 385, respectively (Achazi and Beylich, 1997).

10.3 UPTAKE OF PAHs WITH FOOD

The uptake of hydrophobic organic compounds from food has been investigated intensively for isopods and to a lesser extent for earthworms, enchytraeids and springtails. In the isopod *Porcellio scaber*, uptake of benzo[a]pyrene increased linearly with increasing concentration in the food and reached an equilibrium within 4 weeks. Consumption of food was not affected by benzo[a]pyrene, as the animals did not display avoidance behavior. The efficiency of food assimilation of 25.6% corresponded to that of poplar leaves. The observed concentrations were about 30–40 times lower than those in the food, resulting in biomagnification factors (BMFs) of 0.033–0.044 (Van Brummelen *et al.* 1991).

When free access to contaminated oats (50 mg ^{14}C-benzo[a]pyrene/kg) was provided as food, the benzo[a]pyrene uptake in *Enchytraeus crypticus* increased linearly up to 30 days and declined slightly thereafter. At day 30, BMF was 0.44–1 (Achazi and Beylich 1997). The uptake of benzo[a]pyrene in the springtail *Folsomia candida* was analyzed by feeding ^{14}C-benzo[a]pyrene-spiked yeast (*Saccharomyces cerevisiae*) for 4 days. The final concentration in the animals was 0.88–0.55 mg benzo[a]pyrene/kg fresh weight (fwt), giving a BMF of 0.11 (Bauer and Pohl 1998).

10.4   THE EFFECT OF EARTHWORM ACTIVITY ON PAH CONCENTRATION IN SOIL

Earthworms function as ecosystem engineers, mixing soil, improving soil general structure by providing a looser configuration, and increasing aeration and drainage (Edwards 1998). This activity also affects the concentration of hydrophobic organic components in soil. In phenanthrene-spiked artificial soil (10 mg/kg), a steady decrease was visible to very low levels after 56 days. The decrease was much faster in the presence of *Eisenia fetida*. In fluoranthene-spiked soil (10 mg/kg) there was no decrease at all in the absence of earthworms. In the presence of *Eisenia fetida* the decline started after a lag-phase of 15 days. The effect of organic matter content (10–40%) was not pronounced. Reproduction took place without any relation to the change in PAH concentrations (Eijsackers *et al.* 2001).

In spiked native soil, a continuous decrease over 56 days for added phenanthrene from 95 mg/kg to 4 mg/kg was measured. The decrease was promoted by the introduction of *Lumbricus rubellus* in the presence of food and increased further if food was withdrawn. The fluoranthene concentration did not decrease in the absence of earthworms, in contrast to phenanthrene. The addition of worms increased the disappearance rate significantly. This disappearance was not dependent on the ring number, as both two- and five-ring PAHs disappeared (Ma *et al.* 1995). Such a relationship with ring number was not, however, found when composting field soil, contaminated with PAH concentrations up to 4445 mg/kg, in bench-scale units, in the presence of 30% corn cobs and cow manure, or activated or autoclaved sludge, in different concentrations for 12 weeks. All amendment conditions resulted in decreased concentrations of PAHs with two to four rings, whereas no reduction of five- and six-ring PAHs occurred (Potter *et al.* 1999).

The enhanced disappearance of PAHs in the presence of earthworms is not the result of the metabolizing capacities of the earthworms, as the biotransformation system of terrestrial annelids is weak and hardly inducible by PAHs (Achazi *et al.* 1998; Berghout *et al.* 1991). Instead, the improved biodegradation of PAHs could be due to improved aerobic soil conditions, intimate mixing of microorganisms with soil in the gut, dispersing these microorganisms with the worm castings and improving the working conditions for microorganisms. Leaching and volatilization are of minor importance (Ma *et al.* 1995).

10.5   THE TOXICITY OF PAHs TO SOIL INVERTEBRATES

Data on the toxicity of PAHs to terrestrial invertebrates are compiled in Table 10.1. The effect concentrations are based on nominal values given in the references. Earthworms were usually exposed to PAHs in artificial soil. Data for springtails and enchytraeids were obtained with spiked field soil or

TABLE 10.1 Data on toxicity of polycyclic aromatic hydrocarbons to terrestrial invertebrates

Compound	Test species	Endpoint	Experiment duration	Matrix	Effect concentration	Reference
Acenaphthene	E. fetida	Mortality	48 h	Filter paper	LC_{50} 49 µg/cm²	1
Acenaphthene	F. fimetaria	Mortality	21 d	Soil, 2.8% OM	LC_{50} 107 mg/kg	18
Acenaphthene	F. fimetaria	Reproduction	21 d	Soil, 2.8% OM	EC_{10} 31 mg/kg	18
Acenaphthylene	F. fimetaria	Mortality	21 d	Soil, 2.8% OM	LC_{50} 145 mg/kg	18
Acenaphthylene	F. fimetaria	Reproduction	21 d	Soil, 2.8% OM	EC_{10} 23 mg/kg	18
Acridine	E. crypticus	Mortality	21 d	Soil, 2.8% OM	LC_{50} 3800 mg/kg	6
Acridine	E. crypticus	Reproduction	21 d	Soil, 2.8% OM	EC_{50} 1500 mg/kg	6
Acridine	E. veneta	Mortality	28 d	Soil, 2.8% OM	LC_{50} 863 mg/kg	19
Acridine	E. veneta	Growth	28 d	Soil, 2.8% OM	EC_{50} 125 mg/kg	19
Acridine	F. fimetaria	Mortality	21 d	Soil, 2.8% OM	LC_{50} 910 mg/kg	16
Acridine	F. fimetaria	Reproduction	21 d	Soil, 2.8% OM	EC_{50} 460 mg/kg	16
Anthracene	E. andrei	Mortality	2 w	Soil, 10% OM	LC_{50} > 1000 mg/kg	5
Anthracene	F. candida	Mortality	24 h	Soil, 10% OM	LC_{50} 1000 mg/kg	5
Anthracene	F. candida	Reproduction	24 h	Soil, 10% OM	EC_{50} 1000 mg/kg	5
Anthracene	F. fimetaria	Mortality	21 d	Soil, 2.8% OM	LC_{50} 67 mg/kg	18
Anthracene	F. fimetaria	Reproduction	21 d	Soil, 2.8% OM	EC_{10} 5 mg/kg	18
Benz[a]anthracene	O. asellus	Growth	47 w	Food, > 90% OM	EC_{10} 20 mg/kg	9
Benz[a]anthracene	O. asellus	Growth	47 w	Food, > 90% OM	LOEC 90 mg/kg	9
Benz[a]anthracene	F. fimetaria	Mortality	21 d	Soil, 2.8% OM	LC_{50} > 980 mg/kg	18
Benzo[a]pyrene	E. crypticus	Cocoon fertility	4 d	Agar-agar	LOEC 1 mg/L	8
Benzo[a]pyrene	E. crypticus	Cocoon fertility	7 d	Food (oats)	LOEC 75 mg/kg	7
Benzo[a]pyrene	E. crypticus	Reproduction	30 d	Soil, 3.9% OM	LOEC 100 mg/kg	7
Benzo[a]pyrene	E. fetida	Mortality	4 w	Soil, 3.9% OM	LOEC 10 mg/kg	4
Benzo[a]pyrene	E. fetida	Cocoon prod.	4 w	Soil, 3.9% OM	LOEC 1 mg/kg	4
Benzo[a]pyrene	E. fetida	NRR*	4 w	Soil, 2.0% OM	LOEC 26.4 mg/kg	12
Benzo[a]pyrene	O. asellus	Growth	9 w	Food, > 90% OM	LOEC 100 mg/kg	10
Benzo[a]pyrene	P. scaber	assimilation	4 w	Food, > 90% OM	LOEC 125 mg/kg	11
Benzo[a]pyrene	P. scaber	Growth	4 w	Food, > 90% OM	LOEC 125 mg/kg	11
Benzo[a]pyrene	P. scaber	Growth	9 w	Food, > 90% OM	LOEC 100 mg/kg	10
Benzo[a]pyrene	F. candida	Reproduction	28 d	Soil, 5.5% OM	LOEC 10 mg/kg	15
Benzo[a]pyrene	F. fimetaria	Mortality	21 d	Soil, 2.8% OM	LC_{50} > 840 mg/kg	18

(continued overleaf)

TABLE 10.1 *(continued)*

Compound	Test species	Endpoint	Experiment duration	Matrix	Effect concentration	Reference
Benzo[b]fluoranthene	F. fimetaria	Mortality	21 d	Soil, 2.8% OM	LC_{50} > 360 mg/kg	18
Benzo[k]fluoranthene	F. fimetaria	Mortality	21 d	Soil, 2.8% OM	LC_{50} > 560 mg/kg	18
Carbazole	E. crypticus	Mortality	21 d	Soil, 2.8% OM	LC_{50} > 2100 mg/kg	6
Carbazole	E. crypticus	Reproduction	21 d	Soil, 2.8% OM	EC_{50} 52 mg/kg	6
Carbazole	E. veneta	Mortality	28 d	Soil, 2.8% OM	LC_{50} 106 mg/kg	19
Carbazole	E. veneta	Growth	28 d	Soil, 2.8% OM	EC_{50} 54 mg/kg	19
Carbazole	F. fimetaria	Mortality	21 d	Soil, 2.8% OM	LC_{50} 2500 mg/kg	16
Carbazole	F. fimetaria	Reproduction	21 d	Soil, 2.8% OM	EC_{50} 35 mg/kg	16
Chrysene	E. fetida	Mortality	14 d	Soil, 10% OM	LC_{50} > 1000 mg/kg	13
Chrysene	E. fetida	Reproduction	21 d	Soil, 10% OM	NOEC > 1000 mg/kg	13
Chrysene	F. fimetaria	Mortality	21 d	Soil, 2.8% OM	LC_{50} > 1030 mg/kg	18
Dibenz[ah]anthracene	F. fimetaria	Mortality	21 d	Soil, 2.8% OM	LC_{50} 67 mg/kg	18
Dibenzofuran	E. crypticus	Mortality	21 d	Soil, 2.8% OM	LC_{50} 390 mg/kg	6
Dibenzofuran	E. crypticus	Reproduction	21 d	Soil, 2.8% OM	EC_{50} 130 mg/kg	6
Dibenzofuran	E. veneta	Mortality	28 d	Soil, 2.8% OM	LC_{50} 78 mg/kg	19
Dibenzofuran	E. veneta	Growth	28 d	Soil, 2.8% OM	EC_{50} 61 mg/kg	19
Dibenzofuran	F. fimetaria	Mortality	21 d	Soil, 2.8% OM	LC_{50} 50 mg/kg	16
Dibenzofuran	F. fimetaria	Reproduction	21 d	Soil, 2.8% OM	EC_{50} 23 mg/kg	16
Dibenzothiophene	E. crypticus	Mortality	21 d	Soil, 2.8% OM	LC_{50} > 2700 mg/kg	6
Dibenzothiophene	E. crypticus	Reproduction	21 d	Soil, 2.8% OM	EC_{50} 110/120 mg/kg	6
Dibenzothiophene	E. veneta	Mortality	28 d	Soil, 2.8% OM	LC_{50} 133 mg/kg	19
Dibenzothiophene	E. veneta	Growth	28 d	Soil, 2.8% OM	EC_{50} 44 mg/kg	19
Dibenzothiophene	F. fimetaria	Mortality	21 d	Soil, 2.8% OM	LC_{50} 21 mg/kg	16
Dibenzothiophene	F. fimetaria	Reproduction	21 d	Soil, 2.8% OM	EC_{50} 11 mg/kg	16
Fluoranthene	E. crypticus	Cocoon fertility	27 d	Agar-agar	LOEC 10 mg/kg	7
Fluoranthene	E. crypticus	Mortality	21 d	Soil, 2.8% OM	LC_{50} > 2400 mg/kg	6
Fluoranthene	E. crypticus	Reproduction	14 d	Agar-agar	LOEC 10 mg/L	7
Fluoranthene	E. crypticus	Reproduction	21 d	Soil, 2.8% OM	EC_{50} 61 mg/kg	6
Fluoranthene	E. crypticus	Reproduction	30 d	Soil, 3.9% OM	LOEC 1200 mg/kg	7
Fluoranthene	E. fetida	Mortality	48 h	Filter paper	LC_{50} 2160 μg/cm^2	1
Fluoranthene	E. veneta	Mortality	28 d	Soil, 2.8% OM	LC_{50} 416 mg/kg	19
Fluoranthene	E. veneta	Growth	28 d	Soil, 2.8% OM	EC_{50} 166 mg/kg	19

Compound	Species	Endpoint	Duration	Medium	Value	Reference
Fluoranthene	*F. fimetaria*	Mortality	21 d	Soil, 2.8% OM	LC$_{50}$ 81 mg/kg	16,18
Fluoranthene	*F. fimetaria*	Reproduction	21 d	Soil, 2.8% OM	EC$_{50}$ 51 mg/kg	16
Fluoranthene	*F. fimetaria*	Reproduction	21 d	Soil, 2.8% OM	EC$_{10}$ 37 mg/kg	18
Fluorene	*O. asellus*	Growth	47 w	Food, > 90% OM	EC$_{10}$ 370 mg/kg	9
Fluorene	*O. asellus*	Growth	47 w	Food, > 90% OM	LOEC 208 mg/kg	9
Fluorene	*E. crypticus*	Mortality	21 d	Soil, 2.8% OM	LC$_{50}$ 1800 mg/kg	6
Fluorene	*E. crypticus*	Reproduction	21 d	Soil, 2.8% OM	EC$_{50}$ 51 mg/kg	6
Fluorene	*E. eugeniae*	Mortality	2 w	Soil, 10% OM	LC$_{50}$ 197 mg/kg	2
Fluorene	*E. eugeniae*	Mortality	48 h	Filter paper	LC$_{50}$ 47 µg/cm^2	2
Fluorene	*E. fetida*	Mortality	2 w	Soil, 10% OM	LC$_{50}$ 173 mg/kg	1,2
Fluorene	*E. fetida*	Mortality	48 h	Filter paper	LC$_{50}$ 171 µg/cm^2	1,2
Fluorene	*E. fetida*	Mortality	8 w	Horse manure	LOEC 1500 mg/kg	3
Fluorene	*E. fetida*	Reproduction	8 w	(> 90% OM)	LOEC 750 mg/kg	3
Fluorene	*E. veneta*	Mortality	28 d	Soil, 2.8% OM	LC$_{50}$ 69 mg/kg	19
Fluorene	*E. veneta*	Growth	28 d	Soil, 2.8% OM	EC$_{50}$ 50 mg/kg	19
Fluorene	*F. fimetaria*	Mortality	21 d	Soil, 2.8% OM	LC$_{50}$ 39 mg/kg	16,18
Fluorene	*F. fimetaria*	Reproduction	21 d	Soil, 2.8% OM	EC$_{50}$ 14 mg/kg	16
Fluorene	*F. fimetaria*	Reproduction	21 d	Soil, 2.8% OM	EC$_{10}$ 7.7 mg/kg	18
Fluorene	*P. excavatus*	Mortality	2 w	Soil, 10% OM	LC$_{50}$ 170 mg/kg	2
Fluorene	*P. excavatus*	Mortality	48 h	Filter paper	LC$_{50}$ 78 µg/cm^2	2
Indeno[1,2,3-cd]pyrene	*F. fimetaria*	Mortality	21 d	Soil, 2.8% OM	LC$_{50}$ > 910 mg/kg	18
Naphthalene	*E. fetida*	Mortality	48 h	Filter paper	EC$_{50}$ 4750 µg/cm^2	1
Naphthalene	*F. fimetaria*	Mortality	21 d	Soil, 2.8% OM	LC$_{50}$ 167 mg/kg	18
Naphthalene	*F. fimetaria*	Reproduction	21 d	Soil, 2.8% OM	EC$_{10}$ 20 mg/kg	18
Perylene	*F. fimetaria*	Mortality	21 d	Soil, 2.8% OM	LC$_{50}$ > 560 mg/kg	18
Phenanthrene	*E. crypticus*	Mortality	21 d	Soil, 2.8% OM	LC$_{50}$ > 2000 mg/kg	6

(continued overleaf)

TABLE 10.1 (continued)

Compound	Test species	Endpoint	Experiment duration	Matrix	Effect concentration	Reference
Phenanthrene	E. crypticus	Reproduction	21 d	Soil, 2.8% OM	EC_{50} 87 mg/kg	6
Phenanthrene	E. fetida	Mortality	21 d	Soil 10% OM	LOEC 1000 mg/kg	13
Phenanthrene	E. fetida	Reproduction	21 d	Soil, 10% OM	EC_{50} 241 mg/kg	13
Phenanthrene	E. fetida	Reproduction	21 d	Soil, 10% OM	LOEC 330 mg/kg	13
Phenanthrene	E. veneta	Mortality	28 d	Soil, 2.8% OM	LC_{50} 134 mg/kg	19
Phenanthrene	E. veneta	Growth	28 d	Soil, 2.8% OM	EC_{50} 94 mg/kg	19
Phenanthrene	F. candida	Mortality	7 d	Soil, 10% OM	LC_{50} 321 mg/kg	13
Phenanthrene	F. candida	Mortality	14 d	Soil, 10% OM	LC_{50} 189 mg/kg	13
Phenanthrene	F. candida	Mortality	21 d	Soil, 10% OM	LC_{50} 158 mg/kg	13
Phenanthrene	F. candida	Mortality	28 d	Soil, 10% OM	LC_{50} 145 mg/kg	13
Phenanthrene	F. candida	Mortality	33 d	Soil, 10% OM	LC_{50} 380 mg/kg	14
Phenanthrene	F. candida	Reproduction	28 d	Soil, 10% OM	EC_{50} 124 mg/kg	13
Phenanthrene	F. candida	Reproduction	28 d	Soil, 10% OM	NOEC 75 mg/kg	13
Phenanthrene	F. candida	Reproduction	33 d	Soil, 10% OM	EC_{50} 175 mg/kg	14
Phenanthrene	F. candida	Reproduction	34 d	Soil, 10% OM	LOEC 220 mg/kg	14
Phenanthrene	F. fimetaria	Mortality	21 d	Soil, 2.8% OM	LC_{50} 41 mg/kg	16,17,18
Phenanthrene	F. fimetaria	Reproduction	21 d	Soil, 2.8% OM	EC_{50} 30 mg/kg	16,17
Phenanthrene	F. fimetaria	Reproduction	21 d	Soil, 2.8% OM	EC_{10} 23 mg/kg	18
Pyrene	E. crypticus	Mortality	21 d	Soil, 2.8% OM	LC_{50} > 2300 mg/kg	6
Pyrene	E. crypticus	Reproduction	21 d	Soil, 2.8% OM	EC_{50} 42 mg/kg	6
Pyrene	E. veneta	Mortality	28 d	Soil, 2.8% OM	LC_{50} 155 mg/kg	19
Pyrene	E. veneta	Growth	28 d	Soil, 2.8% OM	EC_{50} 71 mg/kg	19
Pyrene	F. fimetaria	Mortality	21 d	Soil, 2.8% OM	LC_{50} 53 mg/kg	16,17,18
Pyrene	F. fimetaria	Reproduction	21 d	Soil, 2.8% OM	EC_{50} 16 mg/kg	16,17
Pyrene	F. fimetaria	Reproduction	21 d	Soil, 2.8% OM	EC_{10} 10 mg/kg	18

* NRR, neutral red retention time; see text for further explanation. OM, Organic matter; d, days; h, hours; w, weeks.
Sources: 1, Neuhauser et al. (1985); 2, Neuhauser et al. (1986); 3, Neuhauser and Callahan (1990); 4, Schaub and Achazi (1996); 5, Römbke et al. (1994); 6, Sverdrup et al. 2002a; 7, Achazi et al. (1995a); 8, Achazi et al. (1995b); 9, Van Brummelen et al. (1996a); 10, Van Brummelen and Struijtzand (1993); 11, Van Straalen and Verweij (1991); 12, Eason et al. (1999); 13, Bowmer et al. (1993); 14, Crouau et al. (1999); 15, Bauer and Pohl (1998); 16, Sverdrup et al. (2001); 17, Sverdrup et al. (2002c); 18, Sverdrup et al. (2002c); 19, Sverdrup et al. (2002d).

agar-agar as substrate. Feeding experiments were performed with isopods and enchytraeids. Data of filter-paper tests are included, but no data obtained with field-contaminated soils containing other pollutants besides PAHs.

Adult *Eisenia fetida* seem to be very insensitive to PAHs, as the LOECs for effects on survival and reproduction in soil are, respectively, 1500 and 750 mg/kg soil dwt (dry weight) for fluorene (Neuhauser and Callahan 1990) and 1000 and 330 mg/kg soil dwt for phenanthrene (Bowmer *et al*. 1993). Benzo[a]pyrene affects the survival at much lower concentrations (LOEC 10 mg/kg soil dwt) and cocoon production even at 1 mg/kg dwt (Schaub and Achazi 1996). *Eisenia veneta* seems to be slightly more sensitive to PAHs compared with *Eisenia fetida*, e.g. the LC_{50} for fluorene is 69 mg/kg compared to 173 mg/kg (Sverdrup *et al*. 2002d).

In isopods, benzo[a]pyrene was not toxic to female and male *Porcellio scaber* in terms of acute mortality up to a concentration of 125 mg/kg fwt in the diet; however, the growth efficiency of adult males decreased from 11% in controls to 1.7% at 125 mg/kg fwt food, corresponding to increased food assimilation. In females, growth efficiency was slightly lower than in the controls at higher concentrations. The metabolic rate was low and not affected by benzo[a]pyrene (Van Brummelen *et al*. 1991; Van Straalen and Verweij 1991). By increasing the exposure time to 9 weeks, the fresh weight of *Oniscus asellus* and *Porcellio scaber* was reduced at 100 mg benzo[a]pyrene/kg dwt in the diet by 20% and 27% and dry weight by 15% and 30%. The mortality of *Oniscus asellus* was increased at 316 mg/kg dwt food after 5 weeks, but no effect on *Porcellio scaber* was observed (Van Brummelen and Stuijfzand 1993). Long-term dietary exposure of *Oniscus asellus* and *Porcellio scaber* to fluorene, phenanthrene, fluoranthene, benzo[a]anthracene, and benzo[a]pyrene had no effect on survival. In female *Porcellio scaber*, a slight weight gain in solvent control animals (−10%) was recorded, a stronger effect at the highest fluoranthene concentration (−18%) and a small effect on protein content at the highest benzo[a]pyrene concentration. A small effect on the dry weight of *Oniscus asellus* after an exposure up to 47 weeks was observed, beginning at concentrations in the diet of 28.5 mg benzo[a]anthracene/kg dwt and 200 mg fluorene/kg dwt. Reproduction was stimulated by phenanthrene, fluoranthene, benzo[a]anthracene, and benzo[a]pyrene. A larger proportion of females became gravid, which resulted in a higher number of juveniles/females. Exposure did not affect the brood size, weight of the mother, or survival rate of starving juveniles (Van Brummelen *et al*. 1996a).

PAH-contaminated sites are usually dominated by homocyclic PAHs, whereas heterocyclic PAHs, also called PACs (polycyclic aromatic compounds), typically constitute 1–10% of the total concentration. N-, S-, and O-substituted PACs constitute a very heterogeneous group of chemicals with respect to physicochemical properties. They are commonly found in high concentrations at sites contaminated by creosote/tar. Sverdrup *et al*. (2001, 2002a) investigated the effect of four PAHs (pyrene, fluoranthene, phenanthrene, fluorene) and four PACs (carbazole, dibenzothiophene, dibenzofuran, acridine) on survival and reproduction of the micro-annelid *Enchytraeus crypticus* and the micro-arthropod *Folsomia fimetaria* in sandy loam spiked with these compounds in increasing concentrations. The recovery of the PACs from the spiked samples by vigorous extraction decreased gradually during the exposure time of 21 days and ranged from 15% to 75% for the various substances.

In the case of *Enchytraeus crypticus,* adult survival was not affected by concentrations up to 1800 mg/kg dry soil of these compounds, with the exception of dibenzofuran (EC_{50} 390 mg/kg dwt). *Folsomia fimetaria* was much more sensitive, with LC_{50}s in the range 21–80 mg/kg, with the exception of acridine (910 mg/kg dwt) and carbazole (2500 mg/kg dwt). For both species reproduction is the more sensitive endpoint. The EC_{50}s for *Enchytraeus crypticus* were in the range 51–130 mg/kg dwt, and for *Folsomia fimetaria* in the range 10–50 mg/kg dwt, except for acridine (1500 mg/kg dwt resp. 460 mg/kg dwt). For *Folsomia fimetaria,* the EC_{10}/NOEC ratios had an average of 0.92 and may be used interchangeably, but not in *Enchytraeus crypticus,* where the NOECs were about two-fold higher than the calculated EC_{10} (Sverdrup *et al.* 2001, 2002a).

When toxic effects were expressed on the basis of estimated porewater concentrations, a different rank of toxicity was established. Now pyrene was the most toxic for *Folsomia fimetaria,* followed by dibenzothiophene, fluorene, fluoranthene, phenanthrene and carbazole (41–99 μg/L), and dibenzofuran and acridine (170 and 300 μg/L, respectively.). *Enchytraeus crypticus* was less sensitive, but as for *Folsomia fimetaria* the toxicity increased with increasing lipophilicity, indicating that the compounds had a non-specific mode of toxic action and that the organisms were mainly exposed through pore water. The toxicities of the compounds to *Folsomia fimetaria* were close to those expected for aquatic invertebrates such as *Daphnia magna* for non-polar narcotics (Sverdrup *et al.* 2001, 2002a).

The springtail *Folsomia candida,* used as the standard organism in the ISO test (ISO 1999) on reproduction toxicity for springtails, seems to be less sensitive than *Folsomia fimetaria,* with an LC_{50} of 380 mg phenanthrene/kg and an EC_{50} of 175 mg phenanthrene/kg dry soil. In field-polluted soil containing pentachlorophenol and 2,4,6-trichlorophenol, the EC_{50} equalled a nominal concentration of 70 mg/kg phenanthrene (Crouau *et al.* 1999).

As already observed for *Oniscus asellus,* reproduction of *Folsomia candida* is positively affected by hydrophobic organic components. In field soil spiked with 10 mg benzo[a]pyrene/kg dwt and 2 mg PCB52/kg dwt, the reproduction rate of *Folsomia candida* increased to 178% and 184%, respectively (Kratz and Klepka 1998). This positive effect may be the result of the estrogenic activity of these compounds or their metabolites, which was also reported for *o, p'*-DDT at concentrations of 0.1–1000 mg/kg soil dwt (Tschirpke 1998).

The different sensitivities of micro-arthropods and micro-annelids is also valid for soils from different sites contaminated with PAHs. Depending on the origin and the soil type used for dilution, the LC_{50} values for the effect on *Folsomia candida* and *Enchytraeus crypticus* reproduction are in the ranges 50–260 and 130–550 mg/kg EPA-PAH, respectively, while the EC_{50}s are in the ranges 10–210 and 65–380 mg/kg EPA-PAH. These varying sensitivities in different soil materials were also found for soils contaminated with TNT and other explosives (Achazi *et al.* 2000; Römbke *et al.* 2000; Schäfer and Achazi 1999).

The sensitivity and reproducibility of standardized biotests with *Folsomia candida, Enchytraeus crypticus* and *Eisenia fetida* were compared in an international ring test with up to nine participating groups. Field soils contaminated with mineral oils, TNT and PAHs were used. The soil material with the highest concentration of PAHs (Σ PAH: 2268 mg/kg dwt) reduced the survival rate of all three species and decreased the reproduction rate to less than 20%. After dilution to 1134 mg PAH/kg soil dwt, survival of all three species was only slightly impaired. The reproduction rate of *Enchytraeus crypticus* increased to 142% and decreased in *Folsomia candida* and *Eisenia* fetida to 82% and 54%, respectively The following order of sensitivity can be established on the basis of the results of this ring test with the different soils: $Enchytraeus < Folsomia < Eisenia$ for the mortality test and $Enchytraeus < Folsomia \cong Eisenia$ for the reproduction test (Hund-Rinke *et al.* 2000; Warnecke *et al.* 2002). This observation corresponds to the results of an investigation concerned with the relative sensitivity of *Eisenia veneta*, *Folsomia fimetaria* and *Enchytraeus crypticus*, which states that the enchytraeid was the least sensitive of the three species (Sverdrup *et al.* 2002d).

Erstfeld and Snow-Ashbrook (1999) and Snow-Ashbrook and Erstfeld (1998) made an inventory of nematode, mite and springtail abundance and diversity on a heavily contaminated site containing PAH levels of 5284–80,465 mg/kg dry soil. Organic matter content of these soils ranged between 5.7% and 67%. PAH contamination seemed to stimulate rather than to inhibit most organisms, except for mite (Acarina) abundance. The apparent positive effect could at least partly be attributed to the large difference in soil organic matter content of the sites, with highest abundances being found in the most organic soils. The stimulatory effect may, according to the authors, also be explained from the increase in PAH-degrading microorganisms, which are an important food source for microbivorous nematodes and arthropods. Similar stimulatory effects were also seen in bioassays exposing *Eisenia andrei* to these field soils and using growth as the endpoint (Erstfeld and Snow-Ashbrook 1999). Eason *et al.* (1999), on the other hand, found a strong reduction of earthworm (*Eisenia fetida*) survival upon exposure to field-contaminated soil containing over 14,000 mg PAH/kg dry soil (approx. 2% OM). Charrois *et al.* (2001) found 100% mortality of *Eisenia fetida* within 6 days of exposure to soils containing 1320–1500 mg PAH/kg (1.7–4.9% OM). Apparently, the low organic matter content of the soils in the latter two studies may be responsible for the high toxicity. In addition, the form in which the PAHs were present, probably mainly soot and coal particles in the first two studies, may also have reduced bioavailability and toxicity.

Van der Kraan and Van Wingerden (1996) exposed eggs of the large marsh grasshopper *Stethophyma grossum* to solutions of phenanthrene (200 mg/L) or a mixture of anthracene, phenanthrene, fluorene, benz[a]anthracene and chrysene (1.4, 4.5, 1.9, 0.3 and 0.3 mg/L). Exposure took place at stages before and after diapause. No effects were seen on the hatching of grasshopper eggs, but both treatments resulted in nymph mortality and delayed moulting of first instar nymphs. The severity of the effects was dependent on duration and development stage during exposure. The treatment with the high phenanthrene concentration completely blocked female fecundity.

Besides direct toxic effects, such as reduced survival, growth or reproduction, exposure to PAHs may also affect other physiological traits, making organisms more susceptible to additional stresses. This was demonstrated by Sjursen *et al.* (2001), subjecting *Folsomia fimetaria* to drought stress following a 3 week exposure period to different PAHs and PACs in a natural soil (Sverdrup *et al.* 2001). Sublethal concentrations of dibenzothiophene, fluorene, fluoranthene and pyrene caused a dose-related decrease in drought tolerance of exposed adult springtails, whereas no such relationship was seen for acridine, dibenzofuran and carbazole. In the case of dibenzothiophene, drought effects were observed at concentrations below the threshold (NOEC) for reproduction effects, suggesting a synergistic interaction (Sjursen *et al.* 2001).

REFERENCES

Achazi RK, Düker C, Henneken M and Rothe B (1995a) Einfluss von anthropogenen Schadstoffen auf Zeitschrift für Umweltchemie und Ökotoxikologie terrestrische Invertebraten. 2. Einfluss von Benzo[a]pyrene, Fluoranthen und Cadmium auf Lebenszyklusparameter von *Enchytraeus crypticus* in Labor–Testsystemen. *Verhandlung der Gesellschaft für Ökologie*. **24**, 535–540.

Achazi RK, Chroszcz G, Dücker C, Hennecken M, Rothe B, Schaub K and Steudel I (1995b) The effect of fluoranthene, benzo[a]pyrene and cadmium upon survival rate and life cycle parameters of two terrestrial annelids in laboratory test systems. *Newsletter on Enchytraeidae*, **4**, 7–14.

Achazi RK, Chroszcz G, Dücker C, Hennecken M, Khan MA, Pilz C, Throl C and Steudel I (1996) Untersuchungen zur Ökotoxizität von Rieselfeldböden mit dem Enchytraeen-Testsystem. In *Rieselfelder in Berlin und Brandenburg*. Kratz W (ed), Landschaftsentwicklung und Umweltforschung, Technische Universität Berlin, **101**, 167–179.

Achazi RK and Beylich A (1997) Einfluss von anthropogenen Schadstoffen (PAK und PCB) auf terrestrische Invertebraten urbaner Ökosysteme: Abschlussbericht. BMBF-Vorhaben 07 OTX 08D/2, Berlin, 1997, 174 pp.

Achazi RK, Flenner C, Livingston DR, Peters LD, Schaub K and Scheiwe E (1998) Cytochrome P450 and dependent activities in unexposed and PAH-exposed terrestrial annelids. *Comparative Biochemistry and Physiology*, **121C**, 339–350.

Achazi RK, Römbke J, Riepert F (2000) Collembolen als Testorganismen. In Heiden S, Erb R, Dott W and Eisenträger A (eds), *Toxikologische Beurteilung von Böden — Leitstungsfähigkeit biologischer Testverfahren*. Spektrum Akademischer Verlag, Heidelberg, Berlin, pp. 83–103.

Alexander M (1995) How toxic are toxic chemicals in soil? *Environmental Science and Technology*, **29**, 2713–2717.

Bauer H and Pohl D (1998) *Bodenökologische Untersuchungen zur Wirkung und Verteilung von organischen Stoffgruppen (PAK, PCB) in ballungsraumtypischen Ökosystemen.* GSF-Forschungszentrum für Umwelt und Gesundheit, Forschungsbericht des Projektträgers 1/98, München 165 pp.

Belfroid A, Van den Berg M, Seinen W, Hermens J and Van Gestel K (1995) Uptake, bioavailability and elimination of hydrophobic compounds in earthworms (*Eisenia andrei*) in field contaminated soil. *Environmental Toxicology and Chemistry* **14**, 605–612.

Belfroid A, Meiling J, Sijm D, Hermens J, Seinen W and Van Gestel K (1994) Uptake of aromatic hydrocarbons from food by earthworms (*Eisenia andrei*). *Archives of Environmental Contamination and Toxicology*, **27**, 260–265.

Berghout AGRV, Wenzel E, Büld J and Netter KJ (1991) Isolation, partial purification, and characterization of the cytochrome P450-dependent monooxygenase system from the midgut of the earthworm *Lumbricus terrestris*. *Comparative Biochemistry and Physiology*, **100C**, 389–396.

Bowmer DT, Roza P, Henzen L and Degeling C (1993) *The development of chronic toxicological tests for PAH-contaminated soils using the earthworm Eisenia fetida and the springtail Folsomia candida.* Report IMW-R 92/387. TNO Institute of Environmental Sciences, Delft, The Netherlands.

Charrois JWA, McGill WB and Froese KL (2001) Acute ecotoxicity of creosote-contaminated soils to Eisenia fetida: A survival-based approach. *Environmental Toxicology and Chemistry*, **20**, 2594–2603.

Crouau Y, Chenon P and Gisclard C (1999) The use of *Folsomia candida* (Collembola, Isotomidae) for the bioassay of xenobiotic substances and soil pollutants. *Applied Soil Ecology*, **12**, 103–111.

ISO (1999) Soil quality — Inhibition of reproduction of Collembola (*Folsomia candida*) by soil pollutants. ISO 11267. International Standardization Organization, Geneva, Switzerland.

Eason CT, Svendsen C, O'Halloran K and Weeks JM (1999) An assessment of the lysosomal neutral red retention test and immune function assay in earthworms (*Eisenia andrei*) following exposure to chlorpyrifos, benzo[a]pyrene (B[a]P), and contaminated soil. *Pedobiologia*, **43**, 641–645.

Edwards CA (1998) Breakdown of animal, vegetable and industrial organic wastes by earthworms. In Edwards CA and Neuhauser EF (eds), *Earthworms in Waste and Environmental Management*. Academic Press, The Hague, The Netherlands, pp. 21–31.

Eijsackers H, Van Gestel CAM, De Jonge S, Muijs B and Slijkerman D (2001) Polycyclic aromatic hydrocarbon-polluted dredged peat sediments and earthworms: a mutual interference. *Ecotoxicology*, **10**, 35–50.

Erstfeld KM and Snow-Ashbrook J (1999) Effects of chronic low-level PAH contamination on soil invertebrate communities. *Chemosphere*, **39**, 2117–2139.

Hatzinger PB and Alexander M (1995) Effect of aging of chemicals in soil on their biodegradability and extractability. *Environmental Science and Technology*, **29**, 537–545.

Hopkin SP (1997) *Biology of Springtails (Insecta: Collembola)*. Oxford University Press, Oxford.

Hund-Rinke K, Römbke J, Riepert F and Achazi RK (2000) Beurteilung der Lebensraumfunktion von Böden mit Hilfe von Regenwurmtests. In Heiden S, Erb R, Dott W, Eisenträger A (eds), *Toxikologische Beurteilung von Böden — Leitstungsfähigkeit biologischer Testverfahren*. Spektrum Akademischer Verlag, Heidelberg, Berlin, pp. 59–81.

Jager T, Anton Sanchez FA, Muijs B, Van der Velde EG and Posthuma L (2000) Toxicokinetics of polycyclic aromatic hydrocarbons in *Eisenia andrei* (Oligochaeta) using spiked soil. *Environmental Toxicology and Chemistry*, **19**, 953–961.

Knoche H, Klein M, Kördel W, Wahle U, Hund K, Müller J and Klein W (1995) Literaturstudie zur Ableitung von Bodengrenzwerten für polyzyklische aromatische Kohlenwasserstoffe (PAK). *Umweltbundesamt*, **71**, Berlin, 73 pp.

Kelsey JW and Alexander M (1997) Declining bioavailability and inappropriate estimation of risk of persistent compounds. *Environmental Toxicology and Chemistry*, **16**, 582–585.

Krauss M, Wilcke W and Zech W (2000) Availability of polycyclic aromatic hydrocarbons (PAHs) and polychlorinated biphenyls (PCBs) to earthworms in urban soils. *Environmental Science and Technology*, **34**, 4335–4340.

Kratz W and Klepka S (1998) Die Wirkung von Benzo[a]pyrene und PCB52 in Rieselfeldböden auf die Populationsentwicklung von *Folsomia candida* (Willem 1902) (Collembola, Insecta). *Zeitschrift für Umweltchemie und Ökotoxikologie*, **10**, 375.

Ma WC, Immerzeel J and Boldt J (1995) Earthworm and food interactions on bioaccumulation and disappearance in soil of polycyclic aromatic hydrocarbons: studies on phenanthrene and fluoranthene. *Ecotoxicology and Environmental Safety*, **32**, 226–232.

Ma W, Van Kleunen A, Immerzeel J and De Maagd PG (1998) Bioaccumulation of polycyclic aromatic hydrocarbons by earthworms: assessment of the equilibrium partitioning theory in *in situ* studies and water experiments. *Environmental Toxicology and Chemistry*, **17**, 1730–1737.

Neuhauser EF, Loehr RC, Malecki MR, Milligan DL and Durkin PR (1985) The toxicity of selected organic chemicals to the earthworm *Eisenia fetida*. *Journal of Environmental Quality*, **14**, 383–388.

Neuhauser EF, Durkin PR, Malecki MR and Anatra M (1986) Comparative toxicity of ten organic chemicals to four earthworm species. *Comparative Biochemistry and Physiology*, **1**, 197–200.

Neuhauser EF and Callahan CA (1990) Growth and reproduction of the earthworm *Eisenia fetida* exposed to sublethal concentrations of organic chemicals. *Soil Biology and Biochemistry*, **22**, 175–179.

Potter CL, Glaser JA, Chang LW, Meier JR, Dosani MA and Herrmann RF (1999) Degradation of polynuclear aromatic hydrocarbons under bench-scale compost conditions. *Environmental Science and Technology*, **33**, 1717–1725.

Römbke J, Bauer C and Marschner A (1994) Verhalten und Wirkung von sechs Umweltchemikalien in terrestrischen Labortests. *Ecoinforma*, **6**, 269–282.

Römbke J, Riepert F and Achazi R (2000) Enchytraeen als Testorganismen. In Heiden S, Erb R, Dott W, Eisenträger A (eds), *Toxikologische Beurteilung von Böden — Leistungsfähigkeit biologischer Testverfahren*. Spektrum Akademischer Verlag, Heidelberg, Berlin, pp. 105–129.

Schäfer R and Achazi RK (1999) The toxicity of soil samples containing TNT and other ammunition-derived compounds in the enchytraeid and Collembola biotest. *Environmental Science and Pollution Research*, **6**, 213–219.

Schaub K und Achazi RK (1996) Die akute und subakute Toxizität von Rieselfeldböden und polycyclischen aromatischen Kohlenwasserstoffen (PAK) im Regenwurmtest. In *Rieselfelder in Berlin und Brandenburg*. Kratz W (ed.), *Landschaftsentwicklung und Umweltforschung, Technische Universität Berlin*, **101**, 181–187.

Schulte G (1996) *Stadtböden Schadstoffbelastung und Schadstoffmobilität*. Umweltforschungszentrum Halle/Leipzig, UFZ-Bericht 11, 96 pp.

Sjursen H, Sverdrup LE and Krogh PH (2001) Effects of polycyclic aromatic compounds on the drought tolerance of *Folsomia fimetaria* L. (Collembola, Isotomidae). *Environmental Toxicology and Chemistry*, **20**, 2899–2902.

Snow-Ashbrook J and Erstfeld KM (1998) Soil nematode communities as indicators of the effects of environmental contamination with polycyclic aromatic hydrocarbons. *Ecotoxicology*, **7**, 363–370.

Stroo HF, Jensen R, Loehr RC, Nakles DV, Fairbrother A and Liban CB (2000) Environmentally acceptable endpoints for PAHs at a manufactured gas plant site. *Environmental Science and Technology*, **34**, 3831–3836.

Sverdrup LE, Kelley AE, Krogh PH, Jensen J, Scott-Fordsmand JJ, Nielsen T and Stenersen J (2001) Effects of eight polycyclic aromatic compounds on the survival and reproduction of the springtail *Folsomia fimetaria* L. (Collembola, Isotomidae). *Environmental Toxicology and Chemistry*, **20**, 1332–1338.

Sverdrup LE, Stenersen J, Kelley AE, Jensen J and Krogh PH (2002a) Effects of eight polycyclic aromatic compounds on the survival and reproduction of *Enchytraeus crypticus* (Oligochaeta, Clitellata). *Environmental Toxicology and Chemistry*, **21**, 109–114.

Sverdrup LE, Jensen J, Krogh PH and Stenersen J (2002b) Studies on the effect of soil ageing on the toxicity of pyrene and phenanthrene to a soil-dwelling springtail. *Environmental Toxicology Chemistry*, **21**, 489–492.

Sverdrup LE, Nielsen T and Krogh PH (2002c) Soil ecotoxicity of polycyclic aromatic hydrocarbons (PAHs) in relation to soil sorption, lipophilicity and water solubility. *Environmental Science and Technology* (in press).

Sverdrup LE, Krogh PH, Nielsen T and Stenersen J (2002d) Relative sensitivity of three terrestrial invertebrate tests to polycyclic aromatic hydrocarbons. *Environmental Toxicology and Chemistry* (accepted).

Tang J and Alexander M (1999) Mild extractability and bioavailability of polycyclic aromatic hydrocarbons in soil. *Environmental Toxicology and Chemistry*, **18**, 2711–2714.

Tang J, Carroquino MJ, Robertson BK and Alexander M (1998) Combined effect of sequestration and bioremediation in reducing the bioavailability of polycyclic aromatic hydrocarbons in soil. *Environmental Science and Technology*, **32**, 3586–3590.

Terytze K, Bäukle N, Böhmer W, and Müller J (1998) Einschätzung der Konzentrationsprofile polyzyklischer aromatischer Kohlenwasserstoffe (PAK) in Böden des Biosphärenreservats Spreewald. *Zeitschrift für Umweltchemie und Ökotoxikologie*, **10**, 326–332.

Thomas W, Rühling A and Simon H (1984) Accumulation of airborne pollutants (PAH, chlorinated hydrocarbons, heavy metals) in various plant species and humus. *Environmental Pollution A*, **36**, 295–310.

Tschirpke C (1998) Einfluss von DDT auf die Mortalität und Reproduktion von Collembolen. Thesis, Freie Universität Berlin, 68 pp.

Van Brummelen TC, Verweij RA and Van Straalen NM (1991) Determination of benzo[a]-pyrene in isopods (*Porcellio scaber* Latr.) exposed to contaminated food. *Comparative Biochemistry and Physiology*, **100C**, 21–24.

Van Brummelen TC and Stuijfzand SC (1993) Effects of benzo[a]pyrene on survival, growth and energy reserves in the terrestrial isopods *Oniscus asellus and Porcellio scaber*. *Science of the Total Environment*, suppl, 921–930.

Van Brummelen TC, Van Gestel CAM and Verweij RA (1996a) Long-term toxicity of five polycyclic aromatic hydrocarbons for the terrestrial isopods *Oniscus asellus* and *Porcellio scaber*. *Environmental Toxicology and Chemistry*, **15**, 1199–1210.

Van Brummelen, Verweij RA, Wedzinga SA and Van Gestel CAM (1996b) Polycyclic aromatic hydrocarbons in earthworm and isopods from contaminated forest soils. *Chemosphere*, **32**, 315–341.

Van Brummelen TC and Van Straalen NM (1996) Uptake and elimination of benzo[a]pyrene in the terrestrial isopod *Porcellio scaber*. *Archives of Environmental Contamination and Toxicology*, **31**, 277–285.

Van der Kraan C and Van Wingerden WKRE (1996) Use of the eggs of the large marsh grasshopper (*Stethophyma grossum* L) for testing the effects of polycyclic aromatic hydrocarbons (PAH). In Sommeijer MJ and Francke PJ (eds), *Proceedings of the*

Experimental and Applied Entomology section of the Netherlands Entomological Society (NEV), vol. 7. Amsterdam, The Netherlands, pp. 217-222.

Van Straalen NM and Verweij RA (1991) Effects of benzo[a]pyrene on food assimilation and growth efficiency in *Porcellio scaber* (Isopoda). *Archives of Environmental Contamination and Toxicology*, **46**, 134-140.

Warnecke D, Chroszcz G, Schäfer R and Achazi RK (2002) Bodenfauna-Tests. In Hund-Rinke K, Kördel W, Heiden S, Erb R, (eds), *Ökotoxikologische Testbatterien: Ergebnisse eines DBU-geförderten Ringtests*, Erich Schmidt Verlag, Berlin, pp. 187-237.

White JC, Kelsey JW, Hatzinger PB and Alexander M (1997) Factors affecting sequestration and bioavailability of phenanthrene in soils. *Environmental Toxicology and Chemistry*, **16**, 2040-2045.

White JC, Quinones-Rivera A and Alexander M (1998) Effect of wetting and drying on the bioavailability of organic compounds sequestered in soil. *Environmental Toxicology Chemistry*, **17**, 2378-2382.

Wilke W (2000) Polycyclic aromatic hydrocarbons (PAHs) in soil — a review. *Journal of Plant Nutrition and Soil Science*, **163**, 229-248.

Ecotoxicological Studies Focusing on Marine and Freshwater Fish

JERRY F. PAYNE[1], ANNE MATHIEU[2] AND TRACY K. COLLIER[3]

[1]*Science Oceans and Environment Branch, Department of Fisheries and Oceans Ltd, St. John's, NF, Canada*
[2]*Oceans Ltd, St. John's, NF, Canada*
[3]*Northwest Fisheries Science Center, Seattle, WA, USA*

11.1 INTRODUCTION

PAHs originate from both point and diffuse sources (see Chapters 1, 2 and 3). Studies from the 1960s onwards began to report that many coastal harbors, as well as inland rivers and lakes, were contaminated with varying levels of PAHs and more attention began to be placed on assessing their ecotoxicological potential. This assessment included laboratory studies with individual and complex mixtures of PAHs, as well as field studies on feral organisms. Important observations were made on tumors and other pathologies in fish from contaminated environments, with effects being strongly linked in some instances to PAHs. However, in field studies there is invariably the problem of other confounding contaminants, such as organochlorine pesticides and metals, and laboratory studies remain critical for providing insight into the relative importance of various risk factors that may be associated with field observations. Also, unlike many organochlorine compounds and metals that have potential to bioconcentrate in fish tissues, PAHs are rapidly metabolized (e.g. Lemaire *et al.* 1990; see also Chapter 5) and do not bioconcentrate to any substantive extent. Accordingly, there is always the possibility that they are causing adverse effects yet providing little or no chemical signature — the so-called 'hit and run' potential for PAHs.

This overview draws upon information from field and laboratory studies over the past 15 years or so linking various biological effects in fin-fish to PAHs. It

PAHs: An Ecotoxicological Perspective. Edited by Peter E.T. Douben.
© 2003 John Wiley & Sons Ltd

includes information on biochemical, histopathological, genetic, immunological, reproductive, developmental and behavioral effects, which are grouped in relation to field studies (Table 11.1) as well as laboratory studies with complex mixtures of PAHs (Table 11.2) and individual PAHs (Table 11.3). The studies are selective but considered to be fairly representative for each category. Establishing either a strict scientific or 'legal' standard of causal evidence for environmental effects seems improbable. However, considering the combination of field and laboratory studies presently available and using a weight of evidence approach, we suggest that levels of PAHs commonly found in many marine and freshwater environments are important risk factors for various aspects of fish health.

With respect to introductory caution, it should be noted that both weathered petroleum and pyrolytic sources of PAHs are found in association with a variety of N, S and O substituted analogs, phenols, etc., and these compounds, as well as PAHs may be contributing to overall toxicity, even though the term 'PAHs' is commonly used. Parental PAH alone represents a large class of compounds and various studies often measure different numbers of compounds in reporting 'total' PAH or estimate total PAH by fluorescence, making it difficult to make interstudy comparisons between exposures and biological effects. Furthermore, many studies do not measure alkylated forms of PAHs, and this could underestimate toxicity potential at sites primarily contaminated with petroleum hydrocarbons, which are enriched in some alkylated PAHs (e.g. Neff and Anderson 1981).

Due to space limitations, common fish names are used in the tables. Common names can also be more meaningful for a general reader, with the understanding that a specialist can refer to the references.

11.2 FIELD AND LABORATORY STUDIES

11.2.1 BIOCHEMICAL EFFECTS

Biochemical effects have been observed in fish in coastal waters, lakes and rivers in a number of countries. Most observations have been on alteration of phase I and to a much lesser extent phase II enzymes, which play a key role in detoxification and other biochemical processes (e.g. Stegeman and Hahn 1993). Other biochemical effects have occasionally been reported (e.g. changes in hormones, energy reserves and serum enzymes) but most observations in the environment and the major proportion of those noted in Table 11.1 have been on the induction of mixed-function oxygenase (MFO) enzymes, which belong to the phase I group of enzymes. The MFO family of enzymes have iron-containing hemoproteins, cytochrome(s)-P450, as terminal oxidases and terms like induction of MFO enzymes or cytochrome(s)-P450 are often used synonymously. MFO enzymes have been the subject of considerable attention,

TABLE 11.1 Examples of effects observed in feral fish which may be variously linked to PAHs

Species	Biochemical	Histopathologic	Immunological	Genetic	Reproductive	Development	Locations	References
American plaice	X						Canada, Baie des Anglais	Lee et al. (1999)
Atlantic croaker	X			X			USA, Galveston Bay	Willet et al. (1997)
Atlantic tomcod				X			USA, urban and industrialized areas	Stein et al. (1994)
Atlantic tomcod	X						Canada, Miramichi Estuary	Courtenay et al. (1995)
Barbel	X						Italy, River Po	Vigano et al. (1998)
Brown bullhead		X					USA, Black River	Baumann and Harshbarger (1985)
Brown bullhead			X				USA/Canada, different lakes	Pandrangi et al. (1995)
Brown bullhead					X	X	USA, Black and Cuyahoga Rivers	Lesko et al. (1996)
Brown bullhead	X						USA, Black River	Steevens et al. (1996)
Brown bullhead	X	X					USA, Niagara River	Eufemia et al. (1997)
Brown bullhead	X						France, Wetland area of Camargue	Buet et al. (1998)
Brown bullhead		X				X	USA, Schuylkill River	Steyermark et al. (1999)
Brown bullhead	X	X				X	USA, Lower Great Lakes	Arcand-Hoy and Metcalfe (1999)
Brown bullhead	X			X		X	USA, Black River	McFarland et al. (1999)
Carp			X				USA/Canada, lakes	Pandrangi et al. (1995)
Carp		X					USA, West Point Lake	Pritchard et al. (1996)
Carp	X						Czech Republic, ponds	Machala et al. (1997)
Carp	X						France, Wetland area of Camargue	Buet et al. (1998)
Carp	X						USA, Ottawa River	Cormier et al. (2000)
Carp bream	X					X	Russia, Rybinsk reservoir	German and Kozlovskaya (2001)
Cat fish	X			X			USA, Devil's Swamp	Winston et al. (1988)
Cat fish		X					USA, Black River	Baumann and Harshbarger (1995)
Cat fish	X			X			USA, Galveston Bay	Willet et al. (1997)
Chinook salmon	X			X			USA, Puget Sound	Stein et al. (1995)
Chinook salmon			X				USA, Puget Sound	Arkoosh et al. (1998)
Chinook salmon	X						USA, Commencement Bay	Stehr et al. (2000)
Chub	X						Italy, River Po	Vigano et al. (1998)
Cod	X						Norway, Soerfjorden	Goksoyr et al. (1994)
Cod	X						Norway, Soerfjorden	Beyer et al. (1996)
Cod		X	X				Norway, Soerfjorden	Husøy et al. (1996)
Cod	X						Norway, Sunndalsfjord	Naes et al. (1999)

(*continued overleaf*)

TABLE 11.1 (*continued*)

Species	Biochemical	Histopathologic	Immunological	Genetic	Reproductive	Development	Locations	References
Crayfish				X			The Netherlands, River Meuse	Schilderman et al. (1999)
Dab	X						Germany, German Bight	Westernhagen et al. (1999)
Dab	X		X				UK, coastal waters	Lyons et al. (2000)
Deep sea fish	X						Mediterranean Sea	Escartin and Porte (1999)
Eel				X			The Netherlands, freshwater sites	Van Shooten et al. (1995)
Eel	X		X				The Netherlands, freshwater sites	Van der Oost et al. (1996)
Eel	X						France, Wetland area of Camargue	Buet et al. (1998)
Eel	X						UK, Thames estuary	Livingstone et al. (2000)
Eel		X					Canada, St. Lawrence River	Couillard et al. (1997)
Emerald rockcod	X						Antarctica, bays	Miller et al. (1999)
English sole		X					USA, Puget Sound	Malins et al. (1984)
English sole		X					USA, Eagle Harbor	Malins et al. (1985)
English sole		X					Canada, Vancouver Harbor	Goyette et al. (1988)
English sole					X	X	USA, Puget Sound	Casillas et al. (1991)
English sole	X						USA, different estuaries	Collier et al. (1992a)
English sole					X	X	USA, Puget Sound	Collier et al. (1992b)
English sole					X	X	USA, Puget Sound	Johnson et al. (1997)
English sole		X					USA, Pacific and Atlantic coasts	Myers et al. (1998)
European flounder		X					Elbe River	Kohler (1990)
European flounder	X						Norway, Soerfjorden	Beyer et al. (1996)
European flounder		X	X				Norway, Soerfjorden	Husøy et al. (1996)
European flounder		X					The Netherlands, coastal	Vethaak and Wester (1996)
European flounder	X	X					Estonia	Bogovski et al. (1997)
European flounder	X						Germany, German Bight	Westernhagen et al. (1999)
European perch	X						Canada, St. Lawrence River	Hontela et al. (1992)
Fish (various)				X			USA, Buffalo and Detroit Rivers	Maccubbin et al. (1990)
Flounder	X						Chile, polluted bays	Rudolph and Rudolph (1999)
Flounder		X					Chile, polluted bays	George-Nascimento et al. (2000)
Grenadier	X	X					Norway, coastal sites	Forlin et al. (1996)

TABLE 11.1 (*continued*)

Species	Biochemical	Histopathologic	Immunological	Genetic	Reproductive	Development	Locations	References
Herring				X			USA, *Exxon Valdez* oil spill	Hose and Brown (1998)
Med. sea perch	X						France, Mediterranean Sea	Lafaurie *et al.* (1993)
Med. sea perch	X			X			North-west Mediterranean Sea	Burgeot *et al.* (1996)
Milkfish	X						Taiwan, river	Chen *et al.* (1998)
Mullet				X			South-eastern Black Sea	Karakoc *et al.* (1998)
Mummichog		X					USA, Elizabeth River	Volgelbein *et al.* (1990)
Mummichog				X			USA, Elizabeth River	Rose *et al.* (2000)
Nase	X						Italy, River Po	Vigano *et al.* (1998)
Oyster toadfish			X				USA, Elizabeth River	Collier *et al.* (1993)
Perch	X			X			Sweden, Baltic Sea coastline	Ericson *et al.* (1998)
Pike	X						Canada, St. Lawrence River	Hontela *et al.* (1992)
Pike	X						The Netherlands, small Amsterdam lakes	Van der Oost and Heida (1993)
Pikey bream	X						Australia, creeks	Cavanagh *et al.* (2000)
Plaice		X					France, *Amoco Cadiz* spill	Haensly *et al.* (1982)
Rainbow trout	X						Italy, River Po	Vigano *et al.* (1994)
Rainbow trout	X	X					Estonia, River Narva	Tuvikene *et al.* (1999)
Red mullet	X			X			North-west Mediterranean Sea	Burgeot *et al.* (1996)
Red mullet	X						France, Mediterranean Sea	Mathieu *et al.* (1991)
Roach	X						The Netherlands, Amsterdam lakes	Van der Oost and Heida (1993)
Roach	X	X					Estonia, River Narva	Tuvikene *et al.* (1999)
Rock sole						X	USA, Puget Sound	Johnson *et al.* (1998)
Salmon chum	X						USA, Commencement Bay	Stehr *et al.* (2000)
Sardine	X						Spain, north coast	Peters *et al.* (1994)
Sculpin	X			X			Iceland, harbors	Stephensen *et al.* (2000)
Smooth flounder	X						Canada, Miramichi Estuary	Courtenay *et al.* (1995)
Spot			X				USA, Elizabeth River	Weeks and Warinner (1984)
Spot	X						USA, Elizabeth River	Van Veld *et al.* (1990)
Spot			X				USA, Elizabeth River	Faisal *et al.* (1993)

(*continued overleaf*)

TABLE 11.1   (*continued*)

Species	Biochemical	Histopathologic	Immunological	Genetic	Reproductive	Development	Locations	References
Starry flounder	X						USA, different estuaries	Collier *et al.* (1992a)
Starry flounder		X					USA, San Francisco Bay	Stehr *et al.* (1997)
Starry flounder		X					USA, Pacific and Atlantic coasts	Myers *et al.* (1998)
Surf perch	X	X					USA, California petroleum seep	Spies *et al.* (1996)
Tilapia				X			Taiwan, Damsui River	Liu *et al.* (1991)
White croaker		X					USA, Pacific and Atlantic coasts	Myers *et al.* (1998)
White sucker		X					Canada, Lake Ontario	Cairns and Fitzsimons (1988)
White sucker	X	X	X				USA, Sheboygan River	Schrank *et al.* (1997)
White sucker					X		USA/Canada, St. Lawrence River	Ridgway *et al.* (1999)
White sucker	X						USA, Ottawa River	Cormier *et al.* (2000)
Winter flounder		X					USA, coastal sites in New England	Gardner *et al.* (1989)
Winter flounder	X	X			X		USA, north-east coast	Johnson *et al.* (1992)
Winter flounder		X					USA, Jamaica Bay	Augspurger *et al.* (1994)
Winter flounder		X					USA, Boston Harbor	Moore and Stegeman (1994)
Winter flounder				X			USA, urbanized areas	Stein *et al.* (1994)
Winter flounder	X						Canada, Sydney estuary	Vandermeulen *et al.* (1996)
Winter flounder	X	X	X				Canada, St. John's Harbor	French *et al.* (2000)
Yellow perch				X			USA/Canada, different lakes	Pandrangi *et al.* (1995)
Yellow perch					X		Canada, St. Lawrence River	Hontela *et al.* (1995)
Yellow perch	X	X					Estonia, River Narva	Tuvikene *et al.* (1999)

since they are commonly induced in fish and other animals upon exposure to a variety of molecules of environmental importance, including PAHs. The sensitivity of fish to induction by petroleum sources of PAHs, including in small boat harbors, was noted over 25 years ago (Payne 1976) and induction has been widely observed since then in fish from marine and freshwater environments (e.g. Payne *et al.* 1987; Stegeman and Hahn 1993).

TABLE 11.2 Examples of effects observed in fish upon exposure to complex mixtures containing PAHs

Species	Effects							Contamination	References
	Biochemical	Histopathologic	Immunological	Genetic	Reproductive	Development	Behavioral		
American plaice					X	X		Sediments, Baie des Anglais (CAN)	Nagler and Cyr (1997)
Blenny	X							North sea crude oil	Celander et al. (1994)
Bullhead	X							Sediments, Hamilton Harbor (CAN)	Leadly et al. (1999)
Catfish	X			X				Sediments, Black Harbor (USA)	Di Giulio et al. (1993)
Cod	X			X				Mechanically dispersed crude oil	Aas et al. (2000)
Dab	X							Sediments, Liverpool Harbor (UK)	Livingstone et al. (1993)
Eel	X							Po River water (Italy)	Agradi et al. (2000)
Eel	X			X				Petroleum distillate extract	Pacheco and Santos (2001)
English sole		X						Sediment extract, Eagle Harbor (USA)	Schiewe et al. (1991)
English sole				X				Sediments, Eagle Harbor (USA)	French et al. (1996)
Flounder	X							Sediments, Rotterdam Harbor (The Netherlands)	Eggens et al. (1996)
Goby							X	Diesel contaminated sediment	Gregg et al. (1997)
Herring				X				Exxon Valdez oil	Pearson et al. (1995)
Herring	X							Water fractions of crude oil	Thomas et al. (1997)
Herring		X		X		X	X	Weathered Alaska crude oil	Carls et al. (1999)
Herring					X	X		Weathered creosote-treated pilings	Vines et al. (2000)
Medaka		X						Sediments, Great Lakes	Fabacher et al. (1991)
Medaka					X			Outboard motor emission waters	Koehler and Hardy (1999)
Pacific halibut					X			Oil-laden sediments	Moles and Norcross (1998)
Pink salmon					X			Weathered Exxon Valdez crude oil	Heintz et al. (1999)
Pink salmon				X				Exxon Valdez oil	Roy et al. (1999)

(continued overleaf)

TABLE 11.2 (*continued*)

Species	Biochemical	Histopathologic	Immunological	Genetic	Reproductive	Development	Behavioral	Contamination	References
	Effects							Contamination	References
Pink salmon							X	Weathered *Exxon Valdez* crude oil	Wertheimer et al. (2000)
Rainbow trout						X		Effluent from a petroleum refinery	Rowe et al. (1983)
Rainbow trout				X				Sediment, Lake Ontario	Metcalfe et al. (1990)
Rainbow trout								Sediments, Hamilton Harbor (CAN)	Balch et al. (1995)
Rainbow trout	X							Sediments, Skagerrak, Kattegat (Norway)	Magnusson et al. (1996)
Rainbow trout	X					X		Crankcase oil	Hellou et al. (1997)
Rainbow trout	X							Creosote contaminated sediment extracts	Hyoetylaeinen and Oikari (1999)
Rainbow trout				X				Creosote microcosms	Karrow et al. (1999)
Rainbow trout	X							Santa Cruz River water	Petty et al. (2000)
Silverside minnow					X			Combusted crude oil fraction	Al-Yakoob et al. (1996)
Spot							X	Sediments, creek, South Carolina (USA)	Marshall and Coull (1996)
Spot	X	X						Creosote contaminated sediment	Sved et al. (1997)
Spot							X	Sediment contaminated by produced water	Hinkle-Conn et al. (1998)
Sturgeon					X			Weathered coal tar sediments	Kocan et al. (1996)
Surf smelt						X		Sediments, Puget Sound (USA)	Misitano et al. (1994)
Tilapia	X							Sediments, Damsui River (Taiwan)	Ueng et al. (1995)
Tilapia	X							Sediments, Hong Kong	Wong et al. (2001)
Winter flounder	X					X		Petroleum-contaminated sediments	Payne et al. (1988)
Winter flounder				X				Petroleum-contaminated sediments	Payne and Fancey (1989)
Winter flounder			X					Petroleum-contaminated sediments	Khan (1995)
Yellow perch	X	X						Oil sands mining-associated waters	Van den Heuvel et al. (2000)

TABLE 11.3 Examples of effects observed in fish upon exposure to individual PAHs

Species	Biochemical	Histopathologic	Immunological	Genetic	Reproductive	Development	Behavioral	Exposure	Reference
Antarctic rockcod	X							B[a]P*	McDonald et al. (1995)
Arctic charr	X							B[a]P	Wolkers et al. (1996)
Bluegill	X	X						Anthracene	McCloskey and Oris (1993)
Bluegill	X							Anthracene	Choi and Oris (2000)
Brook trout	X			X				B[a]P	Padros et al. (2000)
Brown bullhead	X			X				B[a]P	Ploch et al. (1998)
Brown trout				X				B[a]P	Mitchelmore et al. (1998)
Carp	X							B[a]P, chrysene	Van der Weiden et al. (1994)
Carp	X							MC*	Marionnet et al. (1998)
Catfish								B[a]P, DMBA*	Martin-Alguacil et al. (1991)
Catfish	X			X				B[a]P, BNF*	Ploch et al. (1998)
Catfish	X							B[a]P	Willett et al. (2000)
Catfish						X		Acenaphthene	Dwivedi (2000)
Catfish	X							BNF, DMBA	Weber and Janz (2001)
Chinook salmon			X					B[a]P	Arkoosh et al. (1998)
Chinook salmon	X							BNF	Campbell and Devlin (1996)
Eel	X			X				B[a]P, BNF	Pacheco and Santos (1997)
Eel	X							B[a]P, BNF	Schlezinger and Stegeman (2000)
European flounder	X							B[a]P, phenanthrene	Rocha Monteiro (2000)
European flounder				X				B[a]P	Malmstrom (2000)
Fathead minnow				X	X			Anthracene	Tilghman-Hall and Oris (1991)
Fathead minnow					X			Fluoranthene	Diamond et al. (1995)
Fathead minnow							X	Fluoranthene	Farr et al. (1995)
Fathead minnow		X						Fluoranthene	Weinstein et al. (1997)
Fathead minnow				X				B[a]P	White et al. (1999)
Goby	X							B[a]P	Zheng et al. (2000)
Grouper	X							B[a]P, BNF	Peters (1995)
Guppy		X						DMBA	Hawkins et al. (1989)
Killfish	X			X				B[a]P	Willett et al. (1995)
Mummichog	X							B[a]P	Van Veld et al. (1997)
Mummichog				X				B[a]P	Rose et al. (2001)
Ovale sole	X	X						B[a]P	Au et al. (1999)
Oyster toadfish			X					DMBA	Seeley and Weeks-Perkins (1997)

(*continued overleaf*)

TABLE 11.3 (*continued*)

Species	Effects							Exposure	Reference
	Biochemical	Histopathologic	Immunological	Genetic	Reproductive	Development	Behavioral		
Rainbow trout		X						B[a]P	Hose et al. (1984)
Rainbow trout	X		X					B[a]P	Masfaraud et al. (1992)
Rainbow trout			X					B[a]P	Potter et al. (1994)
Rainbow trout	X	X						MC	Khan and Semalulu (1995)
Rainbow trout				X				Phenanthrene	Passino-Reader et al. (1995)
Rainbow trout	X							B[a]P	Cravedi et al. (1998)
Rainbow trout	X							Retene	Fragoso et al. (1999)
Sea bass		X						B[a]P	Lemaire et al. (1992)
Sea bass	X							BNF	Novi et al. (1998)
Sheepshead minnow		X						B[a]P, DMBA	Hawkins et al. (1991)
Tilapia		X	X					B[a]P	Holladay et al. (1998)
Tilapia		X	X					DMBA	Hart et al. (1998)
Tomcod	X							B[a]P, BNF	Courtenay et al. (1999)
Various species	X		X					B[a]P	Chen et al. (1999)
Zebrafish			X					B[a]P	Hsu et al. (1996)
Zebrafish	X	X					X	Retene	Billiard et al. (1999)

*B[a]P, benzopyrene; MC, methylcholanthrene; DMBA, dimethylbenzanthracene; BNF, β-naphthoflavone.

Presumptive evidence strongly indicates that the MFO enzyme induction observed in fish in many harbors, lakes and rivers worldwide, and represented in Table 11.1, is variably linked to contamination by PAHs. This is further supported by a large number of observations on induction in a variety of species upon exposure to complex mixtures of PAH, as well as individual PAH compounds (see Tables 11.2 and 11.3). Although studies on complex mixtures of PAH provide more information of environmental relevance than studies on single compounds alone, systematic studies on single compounds are critical for shedding light on which compounds or compound types may be causing different biological responses. Most of the observations on biochemical effects in Tables 11.2 and 11.3 refer to induction of MFO enzymes. Also, observations of induction in fish around petroleum development sites in the marine environment (e.g. Stagg *et al.* 1995) and in the vicinity of major oil spills (e.g. Kurelec *et al.* 1977; George *et al.* 1995) provide unambiguous evidence for PAH-linked induction in the environment. Further, both MFO induction and pathological effects have been observed in fish taken in the vicinity of natural petroleum seeps in the Santa Barbara Channel (Spies *et al.* 1996).

Is MFO enzyme induction itself of ecotoxicological importance, beyond being an index of chemical exposure sufficiently high to elicit a biological response? There is a body of literature associating MFO induction with the production of damaging free radicals and adducts, which are important in mutagenic and carcinogenic processes (e.g. Stegeman and Hahn 1993). Induction in fish has also been specifically linked with effects on reproduction, as well as correlated with various organ and cellular disturbances (e.g. Johnson *et al.* 1988; Payne *et al.* 1988; Au *et al.* 1999). Because induction of enzyme systems is also an energy-utilizing process, any energy loss of a non-essential nature could theoretically alter other biochemical and physiological processes in a maladaptive manner. Therefore induction of MFO enzymes, and especially prolonged induction of relatively high levels, can be considered a risk factor for fish health.

11.2.2   HISTOPATHOLOGICAL EFFECTS

Observations on visible signs of disease in marine fish, such as lymphocystis, skin ulceration and fin rot, date back several decades. During the early 1970s considerable interest arose as to whether chemical contamination could play a role in the etiology of these or other fish diseases and extensive studies began to appear from surveys in waters of various countries, including the USA (e.g. Couch 1985; Ziskowski *et al.* 1987), the Netherlands and Belgium (e.g. Moller 1981; Banning 1987), Germany (e.g. Dethlefsen *et al.* 1987) and the UK (e.g. Bucke *et al.* 1983). In addition to external diseases, more emphasis began to be placed on histopathological studies for neoplasms and other tissue and organ abnormalities, which can be powerful indicators of serious chemical injury. Many important observations have been made over the past 25 years linking histopathological effects including carcinogenesis in fish to polluted waters containing elevated levels of PAH. Examples of studies carried out in marine and estuarine waters of the USA, Canada, and Europe, where PAHs likely contributed to a variable degree to some of the effects observed in fish, are noted in Table 11.1. Most of the studies drew attention to the possible importance of PAHs while other studies, although not drawing attention to PAHs, were carried out in areas known to contain elevated levels of the compounds. Although not included as case studies in Table 11.1, some of the effects noted in fish around pulp mills (e.g. Khan and Payne 1997) could also be due in part to PAHs, since such large industrial plants can be expected to release a level of oil and grease into the aquatic environment. Examples of studies carried out in the Great Lakes and in rivers in the USA, where PAHs have been linked to various pathological effects, are also noted in Table 11.1.

Although the evidence linking adverse histopathological effects in fish to pollution is quite strong, it is difficult to resolve cause – effect relationships for various classes of chemicals. However, a few studies of freshwater and marine systems in the USA and Canada have provided very strong evidence for PAHs being a causal factor in fish pathology, including skin and skeletal disorders, liver abnormalities, and neoplasms. The linkage with PAHs is quite robust in these particular studies, since they were carried out at sites heavily contaminated with PAHs from creosote or coke.

As noted previously, creosote is a mixture distilled from coal tar, while coke is the residue left after coal distillation. Both are highly enriched in PAHs (and associated analogs and phenols) and commonly known to be quite mutagenic, carcinogenic, and cytotoxic. The Black River in Ohio was historically contaminated with rather extreme levels of PAHs from a coking plant, and up to 30% of the brown bullheads collected in the vicinity of the plant had neoplasms, including cholangiocarcinomas and hepatocarcinomas (Baumann and Harshbarger 1985). Another site where skin and liver tumors in fish have been linked to discharges from a coking facility is in Hamilton Harbor in Lake Ontario, near a large steel plant (Cairns and Fitzsimons 1988). In the case of creosote, fish collected from a portion of the Elizabeth River in Virginia, where creosote was a major contributor to the total PAH load in the river, displayed cancer and other pathological effects (Vogelbein et al. 1990). High levels of PAHs from creosote have also been associated with neoplasms and other liver disorders in marine fish, e.g., detailed studies carried out on English sole in Puget Sound found a relatively high prevalence of disease in Eagle Harbor which was historically linked with creosote contamination (Malins et al. 1985).

These case studies provide strong evidence for environmental PAHs contributing to carcinogenesis and other disorders in fish. Any case studies where pathological effects decline upon reduction or cessation of inputs of PAHs would provide the strongest evidence. Notable in this regard are observations on the marked decline in liver neoplasms in bullhead from the Black River upon closure of the coking facility (Baumann and Harshbarger 1995), and reductions of several types of liver diseases in English sole following a sediment capping project in Eagle Harbor (Collier and Myers, 1999).

Results obtained in experimental studies with fish chronically exposed to sediments containing petroleum or pyrolytic sources of PAHs, as well as various studies with extracts of PAHs from contaminated sediments, or industrial formulations such as creosote or for instances with specific PAHs, provide additional support for PAHs being a likely cause for some of the pathological effects found in fish in highly contaminated environments. Examples of several such studies are noted in Tables 11.2 and 11.3 and a few are discussed in more detail. Weathered petroleum contains complex mixtures of PAHs and aliphatic hydrocarbons, and whereas chronic exposure of flounders to sediments contaminated with weathered petroleum produced a variety of effects, no such effects were found in flounders similarly exposed to sediments containing a petroleum source of complex aliphatic hydrocarbons, indicating that PAHs were primarily responsible for the effects observed in fish with weathered petroleum (Payne et al. 1995). Epidermal hyperplasia and papillomas developed in brown bullheads upon skin painting with PAH extracts of contaminated sediments from the Buffalo River (Black et al. 1985). Hepatocellular neoplasms and a spectrum of other abnormalities were found in the livers of rainbow trout after microinjection of eggs with extracts of sediments from the Black River (Maccubbin *et al.* 1987). Similar results were obtained by Balch *et al.* (1995) with extracts of sediments from Hamilton Harbor. Other studies have been carried out with fish exposed *in vivo* to extracts of PAHs added to water or sediment. A variety of liver abnormalities and neoplasms were exhibited by medaka upon exposure to extracts of sediments from the Black and Fox Rivers (Fabacher *et al.* 1991). Fin erosion was also a sensitive response in medaka, as well as in spot exposed for short periods to relatively low levels of creosote in sediments (Sved *et al.* 1997).

Concerning individual PAHs, the carcinogenic potency of B[a]P is well known and has produced neoplasms in fish (e.g. Hawkins *et al.* 1991). Ultrastructural studies have also revealed a variety of pathological features in the liver and intestinal tissues of fish exposed to B[a]P (Lemaire *et al.* 1992; Au *et al.* 1999). PAHs that have little or no carcinogenic potential may have considerable potential to cause other pathological effects. Naphthalene is probably best known in this regard, but other PAHs are also important, e.g. severe fin erosion and liver necrosis were some of the effects found in catfish exposed to relatively low concentrations of acenaphthene for short periods (Dwivedi 2000). Also, anthracene, which is known to produce highly cytotoxic by-products in some organisms when they are simultaneously exposed to the chemical and ultraviolet light, damaged the gills of fish upon exposure to quite low levels of the compound (McCloskey and Oris 1993).

Considering the variety of information available from field and laboratory studies, it is reasonable to state that PAHs have the potential to play a role, and possibly sometimes a major role, in the production of pathological effects in feral fish, including skin disorders, organ abnormalities, and neoplasms.

11.2.3 IMMUNOLOGICAL EFFECTS

Immune responses can involve two components, a specific humoral component, such as antibody production, or a non-specific cellular component, such as phagocytosis of bacteria by white blood cells (e.g. Ellis 1989). Alteration of immune systems in fish by contaminants may affect their susceptibility to bacterial, viral, and parasitic infections, decrease their resistance to carcinogenesis, or impair vital functions such as tissue repair. There are few environmental studies which have linked chemically mediated diseases to impairment of immunological functions. Examples of a few such studies are noted in Table 11.1.

Impaired phagocytic functions, including reduced phagocytosis, chemotactic response and chemiluminescence, were found in oyster toadfish, spot and hogchoker taken from a site in the Elizabeth River in Virginia which, as previously noted, is heavily contaminated with PAHs (Weeks and Warinner 1984; Seeley and Weeks-Perkins 1991). Mitogen-stimulated lymphocyte proliferation was also suppressed in spot from the same River (Faisal and Huggett 1993). Hematological changes were noted in white suckers in the lower Sheboyan

River in Wisconsin, which contains high levels of PAHs (Schrank *et al.* 1997). Altered leucocyte numbers (and other effects) were also observed in winter flounders in St. John's Harbor, Newfoundland, in comparison with reference fish taken outside the harbor (French *et al.* 2000). Sediments in this harbor are contaminated throughout with high levels of PAHs but relatively low levels of PCBs and other organochlorines.

A series of recent studies (summarized in Arkoosh and Collier 2002) in the USA suggest that polluted estuaries may be one of the factors contributing to the decline of wild Pacific salmon, an issue of major socioeconomic interest in the USA and Canada. Juvenile salmon taken from contaminated estuaries in Puget Sound and challenged with pathogenic bacteria, exhibited higher mortality than reference fish similarly challenged. It is not known to what degree PAHs are linked with this important field observation of immunotoxicity in salmon, but these studies did show that laboratory exposure of salmon to either a single PAH or a PAHs model mixture could also cause increased mortality following pathogen challenge.

There are few studies dealing with the immunotoxicity to fish of mixtures of PAHs or individual PAHs. Some examples are given in Tables 11.2 and 11.3. Several immunological parameters were evaluated in rainbow trout exposed to creosote and one of the most notable changes was alteration of blood leucocytes. This specific study provided important dose–response information and it is of interest that the LOEC was quite low, around 1 ppb of PAHs in water (Karrow *et al.* 1999).

Weathered petroleum is highly enriched in PAHs, making it a suitable candidate for studies of mixtures of PAHs (see also Section 11.2.2). Macrophage centers represent the primitive analogs of lymph nodes in mammals and are believed to be an integral component of the cellular immune system in fish. Payne and Fancey (1989) recorded a change in the number of macrophage centers in the livers of flounders chronically exposed to a petroleum source of sediment PAHs at levels commonly found in nearshore and inland waters.

Alteration of immunological status can theoretically result in a variety of adverse outcomes, but increased susceptibility to infectious diseases is usually of most concern. Susceptibility to disease is important for any species, especially those identified by agencies as being 'species at risk'. The observation of compromised immune systems in juvenile salmon in Puget Sound, which may be linked to contamination by PAHs, is of note in this regard. Also of interest are laboratory and field observations on alteration of blood cells, which are an important immunological component and can be readily measured.

11.2.4 GENETIC EFFECTS

Damage to genetic material can theoretically lead not only to carcinogenesis and other pathological effects in animals living now, but also produce genetic diseases in future generations. Research on the genotoxic potential of various environmental chemicals, drugs, etc. have been ongoing for decades, with earlier work emphasizing cytogenetic and mutational studies and later investigations incorporating biochemical endpoints, such as specific mutations in DNA, DNA strand breaks, oxidative damage to DNA, or adduct formation (e.g. Pfeifer 1996).

An appreciation of the different types of screening carried out in the environment for genotoxic chemicals is provided by De Flora *et al.* (1991). Notably, a wide variety of individual and complex mixtures of PAHs, as well as environmental extracts enriched in PAHs, have exhibited genetic toxicity in different assays, including bacterial systems. The metabolism of PAHs is extensively covered in Chapter 5, including the formation of DNA adducts. This section briefly reviews other aspects of genetic toxicity, focusing on studies with fish proper.

Genetic toxicity of a more overt and 'visibly' harmful nature can be assessed microscopically with classical cytogenetic techniques. Also, within the past few years an electrophoretic technique based on light microscopy (the Comet assay), which measures DNA fragmentation of individual nuclei, has been used more extensively (e.g. Pfeifer 1996). Classical cytogenetic techniques revealed significant genetic toxicity in herring larvae, in association with the *Exxon Valdez* spill in Alaska, and effects were correlated with levels of PAHs found in mussels in the area (Hose and Brown 1998). Cytogenetic toxicity was also observed in fish larvae exposed as eggs to low levels of petroleum-derived PAHs (Carls *et al.* 1999). The studies with herring eggs are of special interest in this regard, since effects were noted with very low levels of PAHs from weathered oil, which should be essentially free of toxic monoaromatics such as benzene and xylene. DNA fragmentation, as assessed by the Comet technique, can also be interpreted as rather overt DNA damage, and it is of interest that blood cells of bullheads from the Detroit River in the USA and Hamilton Harbor in Canada displayed distinctly higher levels of DNA fragmentation than the cells of fish from reference sites (Pandrangi *et al.* 1995).

Damage to gametic DNA is of special importance, since effects could potentially be transmitted between generations. Studies on the induction of mutations or heritability of pollutant-induced mutations are difficult. However, some studies are beginning to appear pointing to a potential for PAHs to induce mutations in fish and possibly heritability; e.g. exposure of fish larvae to weathered Prudhoe Bay crude oil produced mutations in high frequency at hot spots in oncogenes (Roy *et al.* 1999). Although it is not known whether mutations were actually involved, it is of special interest that White *et al.* (1999), in their studies with fathead minnows exposed to B[a]P, produced heritable reproductive effects in fish removed from the original exposure by at least one generation.

There is a substantial body of evidence from field and laboratory studies indicating that PAHs are a significant risk factor for genetic toxicity in fish. Most studies have dealt with the formation of DNA adducts (Chapter 5), but classical cytogenetic techniques have demonstrated genetic damage in fish larvae, with some of the lowest levels of PAHs known to produce effects, while field studies indicate that environmental levels of PAHs may sometimes be sufficiently high to cause overt damage to DNA in the blood cells of adult fish. Interestingly, there is also recent evidence pointing to a potential for PAHs to produce mutations in reproductive cells, resulting in effects which could be transmitted intergenerationally.

11.2.5 REPRODUCTIVE EFFECTS

Reproductive toxicity is one of the most important types of toxicity because of its potential for producing adverse effects at a population level. Hypotheses about a role for pollutants in the low fertility and high prevalence of embryo mortality in some stocks of salmon in the Great Lakes go back to the 1970s (e.g. Leatherland *et al.* 1998). Organochlorines and especially PCBs have often been implicated in reproductive toxicity, and studies in the coastal waters of Germany and Denmark have provided evidence for effects (Dethlefsen 1988). Establishing cause–effect relationships for hypotheses about reproductive tox-icity is obviously quite difficult, especially in relation to the potential for broad-scale geographical effects. Also, since PAHs are rapidly metabolized and do not necessarily bioaccumulate to any extent, there is always a possibility that they are causing toxic effects yet providing no evidence for the effects. Point sources of contaminants are more suitable for study, and it is of interest that within the past decade or so there have been a number of observations on altered reproductive functions in fish taken close to pulp mills (e.g. McMaster *et al.* 1996). Natural and selected industrial chemicals have also recently been linked to feminizing effects in male fish taken near sewage outfalls (e.g. Purdom *et al.* 1994). Such observations have stimulated considerable interest about the potential for contaminants, including PAHs, to affect fish reproduction.

Are field studies at sites contaminated with relatively high levels of PAHs indicating PAHs to be a risk factor for reproductive toxicity? English sole from sites in Puget Sound exhibited changes in gonadal development, altered levels of plasma steroids and reduced spawning success (Johnson *et al.* 1988; Casillas 1991) but effects were less pronounced in rock sole in the area (Johnson *et al.* 1998). Also, winter flounders from contaminated estuaries in the north-eastern USA displayed little evidence of reproductive dysfunction (Johnson *et al.* 1994). But it was noted that English sole, unlike winter flounders, reside in contaminated estuaries for longer periods throughout their period of vitellogenesis and the greater exposure during this period may account in part for the apparent differential sensitivity between the two flounder species.

A study of freshwater fish in highly degraded habitats found no evidence of serious reproductive dysfunction as displayed by fecundity indices (Lesko *et al.* 1996). Brown bullheads from the Black and Cuyahoga Rivers in Ohio containing sediments with elevated levels of PAHs actually had fecundity indices that were equal to or greater than those of reference populations. However, the potential for producing adverse effects on the fecundity in fish in polluted systems could be compensated for by other factors, such as enhanced food supply due to the absence of competing predators (whose absence could also be linked directly or indirectly to contamination). The authors suggest such a possibility for the lack of effects in bullheads in these highly degraded systems. Also, it should be noted that although fecundity is commonly used as a sign of reproductive health, the quality of eggs or sperm could theoretically be impaired without effects on fecundity or other morphological indices.

The chronic toxicity studies needed to elucidate the effects of chemicals on reproduction in aquatic organisms having a long reproductive cycle require facilities and resources not commonly found in most laboratories. Accordingly, there is limited experimental information to assess the effects of complex mixtures of PAHs on reproduction in fish. Since fecundity indices may often be insensitive indicators of reproductive toxicity, it is also important to place greater emphasis on studies related to the quantity and quality of offspring produced. Nagler and Cyr (1997) recently reported quite novel findings in this regard, with respect to the impairment of sperm quality and effects on egg hatchability in American plaice chronically exposed to sediments highly contaminated with PAHs from the Baie des Anglais in Quebec.

Are there laboratory studies with complex mixtures of PAHs relatively free of other contaminants or individual PAHs to support hypotheses about a potential for PAHs to adversely affect fish reproduction? Certain PAHs closely resemble steroid hormones and exhibit weakened estrogenic or anti-estrogenic responses in some model systems (Nicolas 1999; Navas and Segner 2000). Benzo[a]pyrene decreased circulating levels of 17β-estradiol in fish (Thomas 1988; Singh 1989) and effects on both steroidogenesis and steroid secretion have been noted in *in vitro* studies with ovarian tissue (Afonso *et al.* 1997; Monteiro *et al.* 2000). We measured steroids in the plasma of fish chronically exposed to sediments highly enriched in either petrogenic or pyrolytic sources of PAHs (Idler *et al.* 1995). Steroid concentrations were reduced with both sources of PAHs but it is of interest to note that similar results were obtained with sediments highly contaminated with aliphatic hydrocarbons.

There is some evidence from field and laboratory studies indicating PAHs as a potential risk factor for reproduction in fish. However, since long-term chronic toxicity studies are rather difficult, with reproductive studies being some of the most difficult, experimental evidence is generally lacking about the importance of PAHs to cause reproductive toxicity in fish, especially in relation to dose–response relationships. Also, although measures such as gonad morphometrics and fecundity are commonly used to assess reproductive health in fish, these endpoints may be underestimating potential reproductive effects and more emphasis should also be placed on assessing the quality of sperm and eggs.

11.2.6   DEVELOPMENTAL EFFECTS

Potential for effects on growth and development is also an important issue, with developing organisms usually displaying a greater sensitivity to chemicals than adults. Since adverse developmental effects can be expected with any chemical should dosages be sufficiently high, the environmental relevance of chemical dosage is always a key factor to consider. Notable in this regard are observations that weathered oil, which is essentially free of low molecular weight monoaromatic compounds such as benzene and xylene, produced sublethal effects in herring larvae exposed to extracts of oil in water containing as little as 0.4 ppb PAHs (Carls *et al.* 1999) (Table 11.2). Relatively low concentrations of PAHs in sediment could also affect growth of larvae. Misitano *et al.* (1994) indicated a potential for effects on growth as assessed by DNA content, in larval surf smelt exposed for 96 h to sediment containing 1.5 ppm PAHs. Effects on juvenile fish, which are important in recruitment, would be of special relevance from a fisheries perspective. Growth was reported to be reduced in juvenile flounder chronically exposed to as little as 1.6 ppm of a petroleum source of PAHs in sediment (Moles and Norcross 1998). Provided for general interest in the tables are other examples of 'developmental' effects, but these mainly relate to effects such as slight changes in organ or somatic condition indices and not to developmental effects *per se.*

Although experimental data is limited, there is indication that development may be affected in larval and juvenile fish upon exposure to very low levels of PAH.

11.2.7   BEHAVIORAL EFFECTS

Studies on animal behavior have generally received little attention in aquatic ecotoxicology but alteration of behavior may be an important environmental consideration for fish in some circumstances. Negative effects might include susceptibility to predation, alteration of homing behavior or avoidance of feeding areas. On the other hand, avoidance of contaminated areas could reduce toxicological risk. It is of interest that under Canadian law, avoidance of feeding areas or loss of habitat could have paramountcy over toxicological risks to fish, short of observations on lethality.

There is limited information to assess behavioral effects of PAHs on fish. A few examples are given in Tables 11.2 and 11.3. Presumably of most importance would be sediments impacted with relatively high levels of PAHs such as around oil spill sites. Feeding behavior was affected in goby by suspended sediments contaminated with a diesel oil source of PAHs in the 200 ppm range, with cessation of feeding around 200 ppm (Gregg *et al.* 1997). Alteration of feeding behavior was also noted in spot exposed to a diesel source of PAHs in the 120 ppm range (Hinkle-Conn *et al.* 1998).

Weathered oil remaining in sediments after oil spills has been the subject of considerable attention. It is of interest that pathological effects, including altered swimming behavior, were exhibited by herring larvae hatched from eggs exposed to weathered *Exxon Valdez* oil containing 1 ppb PAHs (Carls *et al.*

1999). As previously noted, this is an extremely low level of PAHs to produce a biological effect.

Considering contamination of water or sediments, contamination of sediments with relatively high levels of PAHs is probably of most importance with respect to any potential for effects on fish behavior. It is not known whether the large quantities of PAHs in water that are expected after large oil spills might affect important behavioral responses in fish, such as alteration of homing behavior to natal rivers by salmon, but compounds with greater solubility and aromaticity, such as xylenes and phenols for instance (or low molecular weight sulfur compounds), are likely of greater importance in this regard.

11.3 CONCLUSION

This overview attempts to summarize information from field and laboratory studies over the past 15 years or so, linking various biological effects in fish to PAHs. It includes information on biochemical, histopathological, genetic, immunological, reproductive, developmental and behavioral effects. The suite of effects are grouped in relation to (a) field studies, (b) laboratory studies with complex mixtures of PAHs and (c) individual PAHs.

All relevant studies could not be included but those listed are a fair representation of biological effects that have been reported. Particular attention has also been given in the case of field studies to observations on different species in various countries. Although select numbers of studies noted in the tables are discussed, the studies altogether provide a body of information for use in a weight-of-evidence approach for assessing the ecotoxicological potential of PAHs for fish. This approach is useful because of the improbability of establishing either a strict scientific or 'legal' standard of causal evidence for regulatory bodies to use in assessing environmental effects associated with mixed contamination. Combining field and laboratory studies and using a weight-of-evidence approach, we suggest that levels of PAHs commonly found in many marine and freshwater environments are causing or contributing to health effects in fish.

Although use of different analytical methods makes it difficult to commonly equate levels of PAHs reported to be present in the environment, it is noted that concentrations of PAHs in sediments in the 10 ppm range have been found in coastal, estuarine and riverine waters of a number of countries, while guidelines indicate potential for effects in fish in the 1 ppm range or lower. Also, concentrations of PAHs in water as low as 1 ppb have been reported to cause serious effects in fish larvae and sublethal effects in fish. This leads to the question of whether contamination by PAHs along major shipping routes, and especially oil tanker routes, or for instance in association with release of large volumes of production waters at petroleum development sites, is adversely affecting populations of fish larvae, including species of commercial importance. It is noted in this regard that concentrations of PAHs as high as in the 100–200 ppb range have been reported to be present along major tanker routes in the Arabian Sea.

Most studies to date have centered around elucidating effects on fish health in association with elevated levels of PAHs in sediments in urban coastal zones or similarly contaminated rivers. Given the apparent extent of contamination in these areas, which may increase, it is important to continue to address potential fish health problems in such regions of concern. This assessment should also place more emphasis on laboratory-based chronic toxicity studies in order to provide better insight into dose–response relationships and thus better information for development of sediment quality guidelines. However, effects have also recently been reported with very low concentrations of PAHs in water. This points to a potential for effects on fish and especially larvae over broad-scale areas away from urban coastal zones and this aspect needs to be addressed much more fully.

REFERENCES

Aas E, Baussant T, Balk L, Liewenborg and Andersen OK (2000) PAHs metabolites in bile, cytochrome P4501A and DNA adducts as environmental risk parameters for chronic oil exposure: a laboratory experiment with Atlantic cod. *Aquatic Toxicology*, **51**(2), 241–258.

Afonso L, Campbell PM, Iwama GK, Devlin RH and Donaldson EM (1997) The effect of the aromatase inhibitor fadrozole and two polynuclear aromatic hydrocarbons on sex steroid secretion by ovarian follicles of coho salmon. *General and Comparative Endocrinology*, **106**(2), 169–174.

Agradi E, Baga R, Cillo F, Ceradini S and Heltai D (2000) Environmental contaminants and biochemical response in eel exposed to Po River water. *Chemosphere*, **41**(10), 1555–1562.

Akcha F, Izuel C, Venier P, Budzinski H, Burgeot T and Narbonne J-F (2000) Enzymatic biomarker measurement and study of DNA adduct formation in benzo[a]pyrene-contaminated mussels, *Mytilus galloprovincialis*. *Aquatic Toxicology*, **49**(4), 269–287.

Al-Yakoob SN, Gundersen D and Curtis L (1996) Effects of the water-soluble fraction of partially combusted crude oil from Kuwait's oil fires (from Desert Storm) on survival and growth of the marine fish *Menidia beryllina*. *Ecotoxicology and Environmental Safety*, **35**(2), 142–149.

Arcand-Hoy LD and Metcalfe CD (1999) Biomarkers of exposure of brown bullheads (*Ameiurus nebulosus*) to contaminants in the lower Great Lakes, North America. *Environmental Toxicology and Chemistry*, **18**(4), 740–749.

Arkoosh, MR and Collier TK (2002) Ecological risk assessment paradigm for salmon: Analyzing immune function to evaluate risk. *Human and Ecological Risk Assessment*, **8**, 265–276.

Arkoosh MR, Casillas E, Huffman P, Clemons E, Evered J, Stein JE and Varanasi U (1998) Increased susceptibility of juvenile chinook salmon from a contaminated estuary to *Vibrio anguillarum*. *Transactions of the American Fisheries Society*, **127**(3), 360–374.

Au DWT, Wu RSS, Zhou BS and Lam PKS (1999) Relationship between ultrastructural changes and EROD activities in liver of fish exposed to benzo[a]pyrene. *Environmental Pollution*, **104**(2), 235–247.

Augspurger TP, Herman RL, Tanacredi JT and Hatfield JS (1994) Liver lesions in winter flounder (*Pseudopleuronectes americanus*) from Jamaica Bay, New York: Indications of environmental degradation. *Estuaries*, **17**(1B), 172–180.

Balch GC, Metcalfe CD and Huestis SY (1995) Identification of potential fish carcinogens in sediment from Hamilton Harbour, Ontario, Canada. *Environmental Toxicology and Chemistry*, **14**(1), 79–91.

Banning P (1987) Long-term recording of some fish diseases using general fishery research surveys in the south-east part of the North Sea. *Diseases of Aquatic Organisms*, **3**, 1–11.

Baumann PC and Harshbarger JC (1985) Frequencies of liver neoplasia in a feral fish population and associated carcinogens. *Marine Environmental Research*, **17**(2–4), 324–327.

Baumann PC and Harshbarger JC (1995) Decline in liver neoplasms in wild brown bullhead catfish after coking plant closes and environmental PAHs plummet. *Environmental Health Perspectives*, **103**(2), 168–170.

Beyer J, Sandvik M, Hylland K, Fjeld E, Egaas E, Aas E, Skare JU and Goksoyr A (1996) Contaminant accumulation and biomarker responses in flounder (*Platichthys flesus*) and Atlantic cod (*Gadus morhua*) exposed by caging to polluted sediments in Soerfjorden, Norway. *Aquatic Toxicology*, **36**(1–2), 75–98.

Billiard SM, Querbach K and Hodson PV (1999) Toxicity of retene to early life stages of two freshwater fish species. *Environmental Toxicity and Chemistry*, **18**(9), 2070–2077.

Black J, Fox H, Black P and Bock F (1985) Carcinogenic effects of river sediment extracts in fish and mice. In Jolley RL, Bull RJ, Davis WP, Katz S, Roberts MH (Jr) and Jacobs VA (eds), *Water Chlorination*, vol. 5. Lewis, Chelsea, MI.

Bogovski S, Veldre I, Sergejev B, Muzyka V and Karlova S (1997) Dynamics of carcinogenic pollution in the Estonian coastal waters. In *Problems of Contemporary Ecology. Temporal Changes in Estonian Nature and Environment. 7. Short Communications of Estonian Seventh Conference in Ecology*, 8–9 May 1997, Tartu, Estonia.

Bucke D, Feist SW and Rolfe M (1983) Fish disease studies in Liverpool Bay and the north-east Irish Sea. *International Council for the Exploration of the Sea, Marine Environmental Quality Committee*, **5**, 9 pp.

Buet A, Roche H, Habert H, Caquet T and Ramade F (1998) Evaluation of a contamination level by organic micropollutants in fishes of the Camargue protected zone. Proposition of an experimental plan to validate pertinent biomarkers. *Ichthyophysiologica Acta*, **21**, 61–76.

Burgeot T, Bocquene G, Porte C, Dimeet J, Santella RM, Garcia de la Parra LM, Pfhol-Leszkowicz A, Raoux C and Galgani F (1996) Bioindicators of pollutant exposure in the north-western Mediterranean Sea. *Marine Ecology Progress Series*, **131**(1–3), 125–141.

Cairns VW and Fitzsimons JD (1988) The occurrence of epidermal papillomas and liver neoplasia in white suckers (*Catostomus commersoni*) from Lake Ontario. *Canadian Technical Report, Fisheries and Aquatic Sciences*, **1607**, 151–152.

Campbell PM and Devlin RH (1996) Expression of CYP1A1 in livers and gonads of Pacific salmon: quantification of mRNA levels by RT-cPCR. *Aquatic Toxicology*, **34**(1), 47–69.

Carls MG, Rice SD and Hose JE (1999) Sensitivity of fish embryos to weathered crude oil: Part 1. Low level exposure during incubation causes malformations, genetic damage and mortality in larval Pacific herring (*Clupea pallasi*). *Environmental Toxicology and Chemistry*, **18**(3), 481–493.

Casillas E, Misitano D, Johnson LL, Rhodes LD, Collier TK, Stein JE, McCain BB and Varanasi U (1991) Inducibility of spawning and reproductive success of female English sole (*Parophrys vetulus*) from urban and non-urban areas of Puget Sound, Washington. *Marine Environmental Research*, **31**, 99–122.

Cavanagh JE, Burns KA, Brunskill GJ, Ryan DAJ and Ahokas JT (2000) Induction of hepatic cytochrome P-450 1A in pikey bream (*Acanthopagrus berda*) collected from agricultural and urban catchments in far north Queensland. *Marine Pollution Bulletin*, **41**(7–12), 377–384.

Celander M, Naef C, Broman D and Forlin L (1994) Temporal aspects of induction of hepatic cytochrome P450 1A and conjugating enzymes in the viviparous blenny (*Zoarces viviparus*) treated with petroleum hydrocarbons. *Aquatic Toxicology*, **29**(3–4), 183–196.

Chapman PM, Barrick RC, Neff JM and Swartz RC (1987) Four independent approaches to developing sediment quality criteria yield similar values for model contaminants. *Environmental Toxicology and Chemistry*, **6**, 723–725.

Chen CM, Ueng TH, Wang HW, Lee SZ and Wang JS (1998) Microsomal monooxygenase activity in milkfish (*Chanos chanos*) from aquaculture ponds near metal reclamation facilities. *Bulletin of Environmental Contamination and Toxicology*, **61**(3), 378–383.

Chen J-P, Xu L-H, Wu Z-B, Zhang Y-Y and Lam PKS (1999) Molecular ecotoxicological indicators of fish intoxicated by benzo[a]pyrene. *China Environmental Science*, **19**(5), 417–420.

Choi J and Oris JT (2000) Evidence of oxidative stress in bluegill sunfish (*Lepomis macrochirus*) liver microsomes simultaneously exposed to solar ultraviolet radiation and anthracene. *Environmental Toxicology and Chemistry*, **19**(7), 1795–1799.

Collier TK, Singh SV, Awasthi YC and Varanasi U (1992a) Hepatic xenobiotic metabolizing enzymes in two species of benthic fish showing different prevalences of contaminant-associated liver neoplasms. *Toxicology and Applied Pharmacology*, **113**(2), 319–324.

Collier TK, Stein JE, Goksor A, Myers MS, Gooch JW, Huggett RJ and Varanasi U (1993) Biomarkers of PAH exposure in oyster toadfish (*Opsanus tau*) from the Elisabeth River, Virginia. *Environmental Sciences*, **2**, 161–177.

Collier TK, Stein JE, Sanborn HR, Hom T, Myers MS and Varanasi U (1992b) Field studies of reproductive success and bioindicators of maternal contaminated exposure in English sole (*Parophrys vetulus*). *Science of the Total Environment*, **116**, 169–185.

Collier TK and Myers MS (1999) Using biomarkers to monitor remedial actions at a subtidal estuarine site severely contaminated with polynuclear aromatic hydrocarbons. In Peakall DB *et al.* (eds), *Biomarkers: A pragmatic Basis for Remediation of Severe Pollution in Eastern Europe*. Kluwer Academic, The Netherlands, pp. 301–304.

Cormier SM, Lin ELC, Millward MR, Schubauer-Berigan MK, Williams DE, Subramanian B, Sanders R, Counts B and Altfater D (2000) Using regional exposure criteria and upstream reference data to characterise spatial and temporal exposures to chemical contaminants. *Environmental Toxicology and Chemistry*, **19**(4), 1127–1135.

Couch JA (1985) Prospective study of infectious and non-infectious diseases in oysters and fishes in three Gulf of Mexico estuaries. *Diseases of Aquatic Organisms*, **1**, 59–82.

Couillard CM, Hodson PV and Castonguay M (1997) Correlations between pathological changes and chemical contamination in American eels, *Anguilla rostrata*, from the St. Lawrence River. *Canadian Journal of Fisheries and Aquatic Sciences*, **54**, 1916–1927.

Courtenay SC, Grunwald CM, Kreamer GL, Fairchild WL, Arsenault JT, Ikonomou M and Wirgin I (1999) A comparison of the dose and time response of CYP1A1 mRNA induction in chemically treated Atlantic tomcod from two populations. *Aquatic Toxicology*, **47**, 43–69.

Courtenay SC, Williams PJ, Vardy C and Wirgin I (1995) Atlantic tomcod (*Microgadus tomcod*) and smooth flounder (*Pleuronectes putnami*) as indicators of organic pollution in the Miramichi Estuary. *Canadian Special Publication of Fisheries and Aquatic Sciences*, **123**, 211–227.

Cravedi JP, Perdu-Durand E and Poupin E (1998) Effects of a polycyclic aromatic hydro-carbon and a PCB on xenobiotic metabolizing activities in rainbow trout (*O. mykiss*). Facteurs de l'environnement et biologie des poissons: colloque Institut Fédératif de Recherche IFR 43. *Bulletin Français de la Pêche et de la Pisciculture*, Paris, 563–570.

De Flora S, Bagnasco M and Zanacchi P (1991) Genotoxic, carcinogenic, and teratogenic hazards in the marine environment, with special reference to the Mediterranean Sea. *Mutation Research*, **258**, 285–320.

Dethlefsen V (1988) Status report on aquatic pollution problems in Europe. *Aquatic Toxicology*, **11**, 259–286.

Dethlefsen V, Waterman B and Hoppenheit M (1987) Diseases of North Sea dab (*Liman-da limanda*) in relation to biological and chemical parameters. *Archiv fūe Fisherei-wissenschaft*, **37**, 107–237.

Di Giulio RT, Habig C and Gallagher EP (1993) Effects of Black Harbor sediments on indices of biotransformation, oxidative stress and DNA integrity in channel catfish. *Aquatic Toxicology*, **26**(1–2), 1–22.

Diamond SA, Oris JT and Guttman SI (1995) Adaptation to fluoranthene exposure in a lab-oratory population of fathead minnows. *Environmental Toxicology and Chemistry*, **14**(8), 1393–1400.

Dwivedi H (2000) Long-term effects of acenaphthene (PAHs) on the liver of catfish *Heteropneustes fossilis*. *Journal of Ecotoxicology and Environmental Monitoring*, **10**(1), 47–52.

Eggens ML, Vethaak AD, Leaver MJ, Horbach GJ, Boon JP and Seinen W (1996) Dif-ferences in CYP1A response between flounder (*Platichthys flesus*) and plaice (*Pleuronectes platessa*) after long-term exposure to harbour dredged spoil in a mesocosm study. *Chemosphere*, **32**(7), 1357–1380.

Eisler R (1987) Polycyclic aromatic hydrocarbon hazards to fish, wildlife, and inverte-brates: a synoptic review. *United States, Fish and Wildlife Service, Biological Report*, **85**(1.11), 81.

Ellis AE (1989) The immunology of teleosts. In Roberts RJ (ed.). *Fish Pathology*. Baillière Tindall, London, pp. 135–152.

Ericson G, Lindesjoo E and Balk L (1998) DNA adducts and histopathological lesions in perch (*Percia fluviatilis*) and northern pike (*Esox lucius*) along a polycyclic aromatic hydrocarbon gradient on the Swedish coastline of the Baltic Sea. *Canadian Journal of Fisheries and Aquatic Sciences*, **55**, 815–824.

Escartin E and Porte C (1999) Hydroxylated PAHs in bile of deep-sea fish. Relationship with xenobiotic metabolizing enzymes. *Environmental Science and Technology*, **33**(16), 2710–2714.

Eufemia NA, Collier TK, Stein JE, Watson DE and Di Gulio RT (1997) Biochemical responses to sediment-associated contaminants in brown bullhead (*Ameiurus nebu-losus*) from the Niagara River ecosystem. *Ecotoxicology*, **6**(1), 13–34.

Fabacher DL, Besser, JM, Schmitt CJ, Harshbarger JC, Peterman PH and Lebo JA (1991) Contaminated sediments from tributaries of the Great Lakes: chemical characterization and carcinogenic effects in medaka (*Oryzias latipes*). *Archives of Environmental Contamination and Toxicology*, **21**(1), 17–34.

Faisal M and Huggett RJ (1993) Effects of polycyclic aromatic hydrocarbons on the lym-phocyte mitogenic responses in spot, *Leiostomus xanthurus*. *Marine Environmental Research*, **35**(1–2), 121–124.

Farr AJ, Chabot CC and Taylor DH (1995) Behavioral avoidance of fluoranthene by fathead minnows (*Pimephales promelas*). *Neurotoxicology and Teratology*, **17**(3), 265–271.

Forlin L, Baden SP, Eriksson S, Granmo A, Lindesjoeoe E, Magnusson K, Ekelund R, Esselin A and Sturve J (1996) Effects of contaminants in roundnose grenadier (*Cory-phaenoides rupestris*) and Norway lobster (*Nephrops norvegicus*) and contaminated levels in mussels (*Mytilus edulis*) in the Sjagerrak and Kattegat compared to the Faroe Islands. *Journal of Sea Research*, **35**(1-3), 209-222.

Fragoso NM, Hodson PV, Kozin IS, Brown RS and Parrott JL (1999) Kinetics of mixed function oxygenase induction and retene excretion in retene-exposed rainbow trout (*Oncorhynchus mykiss*). *Environmental Toxicology and Chemistry*, **18**(10), 2268-2274.

French B, Melvin W, Dawe M and Mathieu A (2000) Habitat degradation and cumulative effects on fish health in St. John's Harbor. Proceedings of the 27th Annual Aquatic Toxicity Workshop. *Canadian Technical Report of Fisheries and Aquatic Sciences*, **2331**, 100.

French BL, Reichert WL, Hom T, Nishimoto M, Sanborn HR and Stein JE (1996) Accumulation and dose-response of hepatic DNA adducts in English sole (*Pleuronectes vetulus*) exposed to a gradient of contaminated sediments. *Aquatic Toxicology*, **36**(1-2), 1-16.

Gardner GR, Pruell RJ and Folmar LC (1989) A comparison of both neoplastic and non-neoplastic disorders in winter flounder (*Pseudopleuronectes americanus*) from eight areas in New England. *Marine Environmental Research*, **28**(1-4), 393-397.

George SG, Wright J and Conroy J (1995) Temporal studies of the impact of the Braer oil spill on inshore feral fish from Shetland, Scotland. *Archives of Environmental Contamination and Toxicology*, **29**(4), 530-534.

George-Nascimento M, Khan RA, Garcias F, Lobos V, Muñoz G and Valdebenito V (2000) Impaired health in flounder, *Paralichthys* (sp), inhabiting coastal Chile. *Bulletin of Environmental Contamination and Toxicology*, **64**(2), 184-190.

German AV and Kozlovskaya VI (2001) Hepatosomatic index and the biochemical composition of the liver in *Abramis brama* in the Sheksna stretch of the Rybinsk reservoir at different levels of toxicant accumulation. *Journal of Ichthyology*, **41**(2), 160-163.

Goksoyr A, Beyer J, Husoy AM, Larsen HE, Westrheim K, Wilhelmsen S and Klungsoyr J (1994) Accumulation and effects of aromatic and chlorinated hydrocarbons in juvenile Atlantic cod (*Gadus morhua*) caged in a polluted fjord (Soerfjorden, Norway). *Aquatic Toxicology*, **29**(1-2), 21-35.

Goyette D, Brand D and Thomas M (1988) Prevalence of idiopathic liver lesions in English sole and epidermal abnormalities in flatfish from Vancouver Harbor, British Columbia. *Environment Canada. Environmental Protection Service. Pacific and Yukon Region*. Vancouver, BC, 48 p.

Gregg JC, Fleeger JW and Carman KR (1997) Effects of suspended, diesel-contaminated sediment on feeding rate in the darter goby, *Gobionellus boleosoma* (Gobiidae). *Marine Pollution Bulletin*, **34**(4), 269-275.

Haensly WE, Neff JM, Sharp JR, Morris AC, Bedgood MF and Boem PD (1982) Histopathology of *Pleuronectes platessa* L. from Aber Wrac'h and Aber Benoit, Brittany, France: long-term effects of AMOCO Cadiz crude oil spill. *Journal of Fish Diseases*, **5**, 365-391.

Hart LJ, Smith SA, Smith BJ, Robertson J, Besteman EG and Holladay SD (1998) Subacute immunotoxic effects of the polycyclic aromatic hydrocarbon 7, 12-dimethylbenzanthracene (DMBA) on spleen and pronephros leukocytic cell counts and phagocytic cell activity in tilapia (*Oreochromis niloticus*). *Aquatic Toxicology*, **41**(1-2), 17-29.

Hawkins WE, Walker WW, Lyttle JS and Overstreet RM (1989) Carcinogenic effects of 7,12-dimethylbenz[a]anthracene on the guppy (*Poecelia reticulata*). *Aquatic Toxicology*, **15**(1), 63-82.

Hawkins WE, Walker WW, Lytle TF, Lytle JS and Overstreet RM (1991) Studies on the carcinogenic effects of benzo[a]pyrene and 7,12-dimethylbenz[a]anthracene on the sheepshead minnow (*Cyprinodon variegatus*). In *Aquatic Toxicology and Risk Assessment* vol. 14. American Society for Testing and Materials, Philadelphia, PA (USA), pp. 97-104.

Heintz RA, Short JW and Rice SD (1999) Sensitivity of fish embryos to weathered crude oil: Part 2. Increased mortality of pink salmon (*Oncorhynchus gorbuscha*) embryos incubating downstream from weathered Exxon Valdez crude oil. *Environmental Toxicology and Chemistry*, **18**(3), 494–503.

Hellou J, Warren W, Andrews C, Mercer G, Payne JF and Howse D (1997) Long-term fate of crankcase oil in rainbow trout: a time–and dose–response study. *Environmental Toxicology and Chemistry*, **16**(6), 1295–1303.

Hinkle-Conn CG, Fleeger JW, Gregg JC and Carman KR (1998) Effects of sediment-bound polycyclic aromatic hydrocarbons on feeding behavior in juvenile spot (*Leiostomus xanthurus*). *Journal of Experimental Marine Biology and Ecology*, **227**(1), 113–132.

Holladay SD, Smith SA, Besteman EG, Deyab ASMI, Gogal RM, Hrubec T, Robertson JL and Ahmed SA (1998) Benzo[a]pyrene-induced hypocellularity of the pronephros in tilapia (*Oreochromis niloticus*) is accompanied by alterations in stromal and parenchymal cells and by enhanced immune cells apoptosis. *Veterinary Immunology and Immunopathology*, **64**(1), 69–82.

Hontela A, Dumont P, Duclos D and Fortin R (1995) Endocrine and metabolic dysfunction in yellow perch, *Perca flavescens*, exposed to organic contaminants and heavy metals in the St. Lawrence River. *Environmental Toxicology and Chemistry*, **14**(4), 725–731.

Hontela A, Rasmussen JB, Audet C and Chevalier G (1992) Impaired cortisol stress response in fish from environments polluted by PAHs, PCBs and mercury. *Archives of Environmental Contamination and Toxicology*, **22**(3), 278–283.

Hose JE and Brown ED (1998) Field applications of the piscine anaphase aberration test: lessons from the Exxon Valdez oil spill. *Mutation Research*, **399**(2), 167–178.

Hose JE, Hannah JB, Puffer HW and Landolt ML (1984) Histologic and skeletal abnormalities in benzo[a]pyrene treated rainbow trout alevins. *Archives of Environmental Contamination and Toxicology*, **13**, 675–684.

Hsu T and Deng F-Y (1996) Studies on the susceptibility of various organs of zebrafish (*Brachydanio rerio*) to benzo[a]pyrene-induced DNA adduct formation. *Chemosphere*, **33**(10), 1975–1980.

Husøy AM, Myers MS and Goksøyr A (1996) Cellular localization of cytochrome P450 (CYP1A) induction and histology in Atlantic cod (*Gadus morhua*) and European flounder (*Platichthys flesus*) after environmental exposure to contaminants by caging in Soerfjorden, Norway. *Aquatic Toxicology*, **36**(1–2), 53–74.

Hyoetylaeinen T and Oikari A (1999) Assessment of the bioactivity of creosote-contaminated sediment by liver biotransformation system of rainbow trout. *Ecotoxicology and Environmental Safety*, **44**(3), 253–258.

Idler DR, So YP, Fletcher GL and Payne JF (1995) Depression of blood levels of reproductive steroids and glucuronides in male winter flounder (*Pleuronectes americanus*) exposed to small quantities of Hibernia crude, used crankcase oil, oily drilling mud and harbour sediments in the 4 months prior to spawning in late May–June. In *Proceedings of the Fifth International Symposium on the Reproductive Physiology of Fish*, Goetz, FW and Thomas P (eds), University of Texas, Austin, TX, 2–8 July 1995, 187.

Johnson LL, Casillas E, Collier TK, McCain BB and Varanasi U (1988) Contaminant effects on ovarian development in English sole (*Parophrys vetulus*) from Puget Sound, Washington. *Canadian Journal of Fisheries and Aquatic Sciences*, **45**, 2133–2146.

Johnson LL, Misitano D, Sol S, Nelson GM, French B, Ylitalo GM and Hom T (1998) Contaminant effects on ovarian development and spawning success in rock sole from Puget Sound, Washington. *Transactions of the American Fisheries Society*, **127**(3), 375–392.

Johnson LL, Sol SY, Lomax DP, Nelson GM, Sloan CA and Casillas E (1997) Fecundity and egg weight in English sole, *Pleuronectes vetulus*, from Puget Sound, Washington:

influence of nutritional status and chemical contaminants. *Fishery Bulletin*, **95**(2), 231–249.

Johnson LL, Stein JE, Collier TK, Casillas E and McCain BB (1992) Bioindicators of contaminant exposure, liver pathology and reproductive development in prespawning female winter flounder (*Pleuronectes americanus*) from urban and nonurban estuaries on the Northeast Atlantic coast. *NOAA Technical Memorandum*, National Marine Fisheries Service, Seattle, WA (USA). Northwest Fisheries Science Centre. 82 p.

Johnson LL, Stein JE, Collier TK, Casillas E and Varanasi U (1994) Indicators of reproductive development in prespawning female winter flounder (*Pleuronectes americanus*) from urban and non-urban estuaries in the north-east United States. *Science of the Total Environment*, **141**, 241–260.

Karakoc FT, Tuli, A, Hewer AT, Gaines AF, Phillips DH and Unsal M (1998) Adduct distributions in piscine DNA: south-eastern Black Sea. *Marine Pollution Bulletin*, **36**(9), 696–704.

Karrow NA, Boermans HJ, Dixon DG, Hontella A, Solomon KR, Whyte JJ and Bols NC (1999) Characterizing the immunotoxicity of creosote to rainbow trout (*Oncorhynchus mykiss*): a microcosm study. *Aquatic Toxicology*, **45**(4), 223–239.

Khan RA (1995) Histopathology in winter flounder, *Pleuronectes americanus*, following chronic exposure to crude oil. *Bulletin of Environmental Contamination and Toxicology*, **54**, 297–301.

Khan RA and Payne JF (1997) A multidisciplinary approach using several biomarkers, including a parasite, as indicators of pollution: a case history from a paper mill in Newfoundland. *Parasitologia*, **39**, 183–188.

Khan RA and Semalulu SS (1995) Ultrastructure and biochemical effects of 3-methylcholanthrene in rainbow trout. *Bulletin of Environmental Contamination and Toxicology*, **54**(5), 731–736.

Kocan RM, Matta MB and Salazar SM (1996) Toxicity of weathered coal tar for shortnose sturgeon (*Acipenser brevirostrum*) embryos and larvae. *Archives of Environmental Contamination and Toxicology*, **31**(2), 161–165.

Koehler ME and Hardy JT (1999) Effects of outboard motor emissions on early development of the killifish *Oryzias latipes*. *Northwest Science*, **73**(4), 277–282.

Kohler A (1990) Identification of contaminant-induced cellular and subcellular lesions in the liver of flounder (*Platichthys flesus*) caught at differently polluted estuaries. *Aquatic Toxicology*, **16**, 271–294.

Kurelec B, Britvic S, Rijavec M, Muller WE and Zahn RK (1977) Benzo[a]pyrene monooxygenase induction in marine fish — molecular response to oil pollution. *Marine Biology (Berlin)*, **44**, 211–216.

Lafaurie M, Mathieu A, Salaun JP, Narbonne JF, Galgani F, Romeo M, Monod JL and Garrigues P (1993) Biochemical markers in pollution assessment: field studies along the north coast of the Mediterranean Sea. *MAP Technical Reports Series*, **71**, 21–42.

Leadly TA, Arcand-Hoy LD, Haffner GD and Metcalfe CD (1999) Fluorescent aromatic hydrocarbons in bile as a biomarker of exposure of brown bullheads (*Ameiurus nebulosus*) to contaminated sediments. *Environmental Toxicology and Chemistry*, **18**(4), 750–755.

Leatherland JF, Ballantyne JS and Van Der Kraak G (1998) Diagnostic assessment of non-infectious disorders of captive and wild fish populations and the use of fish as sentinel organisms for environmental studies. In Leatherland JF and Woo PTK (eds), *Fish Diseases and Disorders, Volume 2: Non-infectious Disorders*. Wallington, UK: CABI Publishing pp. 335–366.

Lee K, Nagler JJ, Fournier M, Lebeuf M and Cyr DG (1999) Toxicological characterisation of sediments from Baie des Anglais on the St. Lawrence estuary. *Chemosphere*, **39**(6), 1019–1035.

Lemaire P, Berhaut J, Lemaire-Gony S and Lafaurie M (1992) Ultrastructural changes induced by benzo[a]pyrene in sea bass (*Dicentrarchus labrax*) liver and intestine: importance of the intoxication route. *Environmental Research*, **57**(1), 59–72.

Lemaire P, Mathieu A, Carriere S, Drai P, Giudicelli J and Lafaurie M (1990) The uptake metabolism and biological half-life of benzo[a]pyrene in different tissues of sea bass, *Dicentrarchus labrax*. *Ecotoxicology and Environmental Safety*, **20**(3), 223–233.

Lesko LT, Smith SB and Blouin MA (1996) The effect of contaminated sediments on fecundity of the brown bullhead in three Lake Erie tributaries. *Journal of Great Lakes Research*, **22**(4), 830–837.

Liu TY, Cheng SL, Ueng TH, Ueng YF and Chi CW (1991) Comparative analysis of aromatic DNA adducts in fish from polluted and unpolluted areas by the super ^{32}P-postlabelling analysis. *Bulletin of Environmental Contamination and Toxicology*, **47**(5), 783–789.

Livingstone DR, Lemaire P, Matthews A, Peters L, Bucke D and Law RJ (1993) Pro-oxidant, antioxidant and 7-ethoxyresorufin *O*-de-ethylase (EROD) activity responses in liver dab (*Limanda limanda*) exposed to sediment contaminated with hydrocarbons and other chemicals. *Marine Pollution Bulletin*, **26**(11), 602–606.

Livingstone DR, Mitchelmore CL, Peters LD, O'Hara SCM, Shaw JP, Chesman BS, Doyotte A, McEvoy J, Ronisz D, Larsson DGJ and Forlin L (2000) Development of hepatic CYP1A and blood vitellogenin in eel (*Anguilla anguilla*) for use as biomarkers in the Thames estuary, UK. *Marine Environmental Research*, **50**(1–5), 367–371.

Long ER, McDonald DD, Smith SL and Calder FD (1995) Incidence of adverse biological effects within ranges of chemical concentrations in marine and estuarine sediments. *Environmental Management*, **19**(1), 81–97.

Lyons BP, Stewart C and Kirby MF (2000) Super ^{32}P-postlabelling analysis of DNA adducts and EROD induction as biomarkers of genotoxin exposure in dab (*Limanda limanda*) from British coastal waters. *Marine Environmental Research*, **50**(1–5), 575–579.

Maccubbin AE, Black JJ and Dunn BP (1990) ^{32}P-postlabelling detection of DNA adducts in fish from chemically contaminated waterways. *Science of the Total Environment*, **94**(1–2), 89–104.

Maccubbin AE, Ersing N, Weinar J and Black JJ (1987) *In vivo* carcinogen bioassays using rainbow trout and medaka fish embryos. In Sandhu SS, DeMarini DM, Moore MM and Mumford JL (eds), *Short-term Bioassays in the Analysis of Complex Environmental Mixtures*. Plenum, New York, pp. 209–223.

MacDonald DD, Ingersoll CG and Berger TA (2000) Development and evaluation of consensus-based quality guidelines for freshwater ecosystems. *Archives of Environmental Contamination and Toxicology*, **39**, 20–31.

Machala M, Petrivalsky M, Nezveda K, Ulrich R, Dusek L, Piacka V and Svobodova Z (1997) Responses of carp hepatopancreatic 7-ethoxyresorufin-*O*-de-ethylase and glutathione-dependent enzymes to organic pollutants — A field study. *Environmental Toxicology and Chemistry*, **16**(7), 1410–1416.

Magnusson K, Ekelund R, Dave G, Granmo A, Forlin L, Wennberg L, Samuelsson MO, Berggren M and Brorstroem-Lunden E (1996) Contamination and correlation with toxicity of sediment samples from the Skagerrak and Kattegat. *Journal of Sea Research*, **35**(1–3), 223–234.

Malins DC, McCain BB, Brown DW, Chan SL, Myers MS, Landahl JT, Prohaska PC, Friedman AJ, Rhodes LD, Burrows DG, Gronlund WD and Hodgins HU (1984) Chemical pollutants in sediments and diseases of bottom-dwelling fish in Puget Sound, Washington. *Environmental Science and Technology*, **18**, 705–713.

Malins DC, Krahn MM, Myers MS, Rhodes LD, Brown DW, Krone CA, McCain BB and Chan, S-L (1985) Toxic chemicals in sediments and biota from a creosote-polluted harbor: relationships with hepatic neoplasms and other hepatic lesions in English sole (*Parophrys vetulus*). *Carcinogenesis*, **6**(10), 1463–1469.

Malmstrom CM (2000) DNA adducts in liver and leukocytes of flounder (*Platichthys flesus*) experimentally exposed to benzo[a]pyrene. *Aquatic Toxicology*, **48**(2–3), 177–184.

Marionnet D, Chambras C, Taysse L, Bosgireaud C and Deschaux P (1998) Modulation of drug-metabolizing systems by bacterial endotoxin in carp liver and immune organs. *Ecotoxicology and Environmental Safety*, **41**(2), 189–194.

Marshall KR and Coull BC (1996) PAHs effects on removal of meiobenthic copepods by juvenile spot (*Leiostomus xanthurus*). *Marine Pollution Bulletin*, **32**(1), 22–26.

Martin-Alguacil N, Babich H, Rosenberg DW and Borenfreund E (1991) *In vitro* response of the brown bullhead catfish cell line, BB, to aquatic pollutants. *Archives of Environmental Contamination and Toxicology*, **20**(1), 113–117.

Martineau D, DeGuise S, Fournier M, Schugart L, Girard C, Lagace A and Beland P (1994) Pathology and toxicology of beluga whales from the St. Lawrence Estuary, Quebec, Canada; past, present and future. *Science of the Total Environment*, **154**, 201–215.

Masfaraud JF, Pfohl-Leszkowic A, Malaveille C, Keith G and Monod G (1992) 7-Ethylresorufin O-de-ethylase activity and level of DNA-adducts in trout treated with benzo[a]pyrene. *Marine Environmental Research*, **34**(1–4), 351–354.

Mathieu A, Lemaire P, Carriere S, Drai P, Giudicelli J and Lafaurie M (1991) Seasonal and sex linked variations in hepatic and extra hepatic biotransformation activities in striped mullet (*Mullus barbatus*). *Ecotoxicology and Environmental Safety*, **22**, 45–57.

Mathieu A, Payne JF, Fancey LL, Santella RM and Young TL (1997) Polycyclic aromatic hydrocarbon–DNA adducts in beluga whales from the Arctic. *Journal of Toxicology and Environmental Health*, **51**(1), 1–4.

McCloskey JT and Oris JT (1993) Effect of anthracene and solar ultraviolet radiation exposure on gill ATPase and selected hematologic measurements in the bluegill sunfish (*Lepomis macrochirus*). *Aquatic Toxicology*, **24**(3–4), 207–218.

McDonald SJ, Kennicutt MC, Liu H and Safe SH (1995) Assessing aromatic hydrocarbon exposure in Atlantic fish captured near Palmer and McMurdo stations, Antarctica. *Archives of Environmental Contamination and Toxicology*, **29**(2), 232–240.

McFarland VA, Inouye LS, Lutz CH, Jarvis AS, Clarke JU and McCant DD (1999) Biomarkers of oxidative stress and genotoxicity in livers of field-collected brown bullhead, *Ameiurus nebulosus*. *Archives of Environmental Contamination and Toxicology*, **37**(2), 236–241.

McMaster ME, Van der Kraak GJ and Munkittrick KR (1996) An epidemiological evaluation of the biochemical basis for steroid hormonal depressions in fish exposed to industrial wastes. *Journal of Great Lakes Research*, **22**(2), 153–171.

Metcalfe CD, Balch GC, Cairns VW, Fitzsimons JD and Dunn BP (1990) Carcinogenic and genotoxic activity of extracts from contaminated sediments in western Lake Ontario. *Science of the Total Environment*, **94**, 125–141.

Miller HC, Mills GN, Bembo DG, Macdonald JA and Evans CW (1999) Induction of cytochrome P4501A (CYP1A) in *Trematomus bernacchii* as an indicator of environmental pollution in Antarctica: assessment by quantitative RT–PCR. *Aquatic Toxicology*, **44**(3), 183–193.

Misitano DA, Casillas E and Haley CR (1994) Effects of contaminated sediments on viability, length, DNA and protein content of larval surf smelt, *Hypomesus pretiosus*. *Marine Environmental Research*, **37**, 1–21.

Mitchelmore CL and Chipman JK (1998) Detection of DNA strand breaks in brown trout (*Salmo trutta*) hepatocytes and blood cells using the single cell gel electrophoresis (comet) assay. *Aquatic Toxicology*, **41**(1–2), 161–182.

Moles A and Norcross BL (1998) Effects of oil-laden sediments on growth and health of juvenile flatfishes. *Canadian Journal of Fisheries and Aquatic Sciences*, **55**(3), 605–610.

Möller H (1981) Fish diseases in German and Danish waters in summer, 1980. *Meeresforschung*, **29**, 1–16.

Monteiro PRR, Reis-Henriques MA and Coimbra J (2000) Plasma steroid levels in female flounder (*Platichthys flesus*) after chronic dietary exposure to single polycyclic aromatic hydrocarbons. *Marine Environmental Research*, **49**, 453–467.

Moore MJ and Stegeman JJ (1994) Hepatic neoplasms in winter flounder *Pleuronectes americanus* from Boston Harbor, Massachusetts, USA. *Diseases in Aquatic Organisms*, **20**, 33–48.

Myers MS, Johnson LL, Olson OP, Stehr CM, Horness BH, Collier TK and McCain BB (1998) Toxicopathic hepatic lesions as biomarkers of chemical contaminant exposure and effects in marine bottomfish species from the Northeast and Pacific coasts, USA. *Marine Pollution Bulletin*, **37**(1–2), 92–113.

Naes K, Hylland K, Oug E, Forlin L and Ericson G (1999) Accumulation and effects of aluminium smelter-generated polycyclic aromatic hydrocarbons on soft-bottom invertebrates and fish. *Environmental Toxicology and Chemistry*, **18**(10), 2205–2216.

Nagler JJ and Cyr DG (1997) Exposure of male American plaice (*Hippoglossoides platessoides*) to contaminated marine sediments decreases the hatching success of their progeny. *Environmental Toxicology and Chemistry*, **16**(8), 1733–1738.

Navas JM and Segner H (2000) Antiestrogenicity of β-naphthoflavone and PAHs in cultured rainbow trout hepatocytes: evidence for a role of the arylhydrocarbon receptor. *Aquatic Toxicology*, **51**(1), 79–92.

Neff JM and Anderson JW (1981) Toxicity of petroleum and specific petroleum hydrocarbons. In *Response of Marine Animals to Petroleum and Specific Petroleum Hydrocarbons*. Applied Science Publishers, London, pp. 5–34.

Nicolas JM (1999) Vitellogenesis in fish and the effects of polycyclic aromatic hydrocarbon contaminants. *Aquatic Toxicology*, **45**, 77–90.

Novi S, Pretti C, Cognetti AM, Longo V, Marchetti S and Gervasi PG (1998) Biotransformation enzymes and their induction by β-naphthoflavone in adult sea bass (*Dicentrarchus labrax*). *Aquatic Toxicology*, **41**(1–2), 63–81.

Pacheco M and Santos MA (1997) Induction of EROD activity and genotoxic effects by polycyclic aromatic hydrocarbons and resin acids on the juvenile eel (*Anguilla anguilla* L.). *Ecotoxicology and Environmental Safety*, **38**(3), 252–259.

Pacheco M and Santos MA (2001) Biotransformation, endocrine, and genetic responses of *Anguilla anguilla* L. to petroleum distillate products and environmentally contaminated waters. *Ecotoxicology and Environmental Safety*, **49**(1), 64–75.

Padros J, Pelletier E, Reader S and Denizeau F (2000) Mutual *in vivo* interactions between benzo[a]pyrene and tributyltin in brook trout (*Salvelinus fontinalis*). *Environmental Toxicology and Chemistry*, **19**(4), 1019–1027.

Pandrangi R, Petras M, Ralph S and Vrzoc M (1995) Alkaline single cell gel (comet) assay and genotoxicity monitoring using bullheads and carp. *Environmental and Molecular Mutagenesis*, **26**(4), 345–356.

Passino-Reader DR, Berlin WH and Hickey JP (1995) Chronic bioassays of rainbow trout fry with compounds representative of contaminants in Great Lakes fish. *Journal of Great Lakes Research*, **21**(3), 373–383.

Payne JF (1976) Field evaluation of benzopyrene hydroxylase induction as a monitor for marine petroleum pollution. *Science*, **191**, 945–946.

Payne JF and Fancey LF (1989) Effect of polycyclic aromatic hydrocarbons on immune responses in fish: change in melanomacrophage centers in flounder (*Pseudopleuronectes americanus*) exposed to hydrocarbon-contaminated sediments. *Marine Environmental Research*, **28**(1–4), 431–435.

Payne JF, Fancey LL, Hellou J, King MJ and Fletcher GL (1995) Aliphatic hydrocarbons in sediments: a chronic toxicity study with winter flounder (*Pleuronectes americanus*) exposed to oil well drill cuttings. *Canadian Journal of Fisheries and Aquatic Sciences*, **52**, 2724–2735.

Payne JF, Fancey LL, Rahimtula A and Porter EL (1987) Review and perspective on the use of mixed-function oxygenase enzymes in biological monitoring. *Comparative Biochemistry and Physiology*, **86** C(2), 233–245.

Payne JF, Kiceniuk J, Fancey LL, Williams U, Fletcher GL, Rahimtula A and Fowler B (1988) What is a safe level of polycyclic aromatic hydrocarbons for fish: subchronic toxicity study on winter flounder (*Pseudopleuronectes americanus*). *Canadian Journal of Fisheries and Aquatic Sciences*, **45**(11), 1983–1993.

Pearson WH, Moksness E and Skalski JR (1995) A field and laboratory assessment of oil spill effects on survival and reproduction on Pacific herring following the Exxon Valdez spill. In *Exxon Valdez Oil Spill: Fate and Effects in Alaskan Waters*. ASTM, Philadelphia, PA, 626–661.

Peters LD (1995) Elevation of 7-ethoxyresorufin *O*-de-ethylase (EROD) activity in immature grouper *Anyperodon leucogrammicus* after exposure to benzo[a]pyrene: cytochrome P4501A1 levels as a biomarker of exposure to polycyclic aromatic hydrocarbons. *Research Bulletin*, **60**, 83–86.

Peters LD, Porte C, Albaiges J and Livingstone DR (1994) 7-Ethoxyresorufin *O*-de-ethylase (EROD) and antioxidant enzyme activities in larvae of sardine (*Sardina pilchardus*) from the North coast of Spain. *Marine Pollution Bulletin*, **28**(5), 299–304.

Petty JD, Jones SB, Huckins JN, Cranor WL, Parris JT, McTague TB and Boyle TP (2000) An approach for assessment of water quality using semipermeable membrane devices (SPMDs) and bioindicator tests. *Chemosphere*, **41**(3), 311–321.

Pfeifer GP (1996) *Technologies for Detection of DNA Damage and Mutations*. Plenum, New York, 441 pp.

Ploch SA, King LC, Kohan MJ and Di Giulio RT (1998) Comparative *in vitro* and *in vivo* benzo[a]pyrene-DNA adduct formation and its relationship to CYP1A in two species of ictalurid catfish. *Toxicology and Applied Pharmacology*, **149**(1), 90–98.

Potter D, Clarius TM, Wright AS and Watson WP (1994) Molecular dosimetry of DNA adducts in rainbow trout (*Oncorhynchus mykiss*) exposed to benzo[a]pyrene by different routes. *Archives of Toxicology*, **69**(1), 1–7.

Pritchard MK, Fournie JW and Blazer VS (1996) Hepatic neoplasms in wild common carp. *Journal of Aquatic Animal Health*, **8**(2), 111–119.

Purdom CE, Hardiman PA, Bye VJ, Eno NC, Tyler CR and Sumpter JP (1994) Estrogenic effects of effluents from sewage treatment works. *Chemistry and Ecology*, **78**, 275–285.

Ridgway LL, Chapleau F, Comba ME and Backus SM (1999) Population characteristics and contaminant burdens of the white sucker (*Catostomus commersoni*) from the St. Lawrence River near Cornwall, Ontario and Massena, New York. *Journal of Great Lakes Research*, **25**(3), 567–582.

Rocha Monteiro PR (2000) Polycyclic aromatic hydrocarbons inhibit *in vitro* ovarian steroidogenesis in the flounder (*Platichthys flesus* L.). *Aquatic Toxicology*, **48**(4), 549–559.

Rose WL, French BL, Reichert WL and Faisal M (2000) DNA adducts in hematopoietic tissues and blood of the mummichog (*Fundulus heteroclitus*) from a creosote-contaminated site in the Elizabeth River, Virginia. *Marine Environmental Research*, **50**(1–5), 581–589.

Rose WL, French BL, Reichert WL and Faisal M (2001) Persistence of benzo[a]pyrene-DNA adducts in hematopoietic tissues and blood of the mummichog, *Fundulus heteroclitus*. *Aquatic Toxicology*, **52**(3–4), 319–328.

Rowe DW, Sprague JB, Heming TA and Brown IT (1983) Sublethal effects of treated liquid effluent from a petroleum refinery. II. Growth of rainbow trout. *Aquatic Toxicology*, **3**(2), 161–169.

Roy NK, Stabile J, Seeb JE, Habicht C and Wirgin I (1999) High frequency of K-ras mutations in pink salmon embryos experimentally exposed to Exxon Valdez oil. *Environmental Toxicology and Chemistry*, **18**(7), 1521–1528.

Rudolph A and Rudolph MI (1999) Activity of benzo[a]pyrene hydroxylase in three marine species. *Bulletin of Environmental Contamination and Toxicology*, **63**(5), 639–645.

Schiewe MS, Weber DD, Myers MS, Jacques FJ, Reichert WL, Krone CA, Malins DC, McCain BB, Chan SL and Varanasi U (1991) Induction of foci of cellular alteration and other hepatic lesions in English sole (*Parophrys vetulus*) exposed to an extract of an urban marine sediment. *Canadian Journal of Fisheries and Aquatic Science*, **48**, 1750–1760.

Schilderman PA, Moonen EJ, Maas LM, Welle I and Kleinjans JC (1999) Use of crayfish in biomonitoring studies of environmental pollution of the River Meuse. *Ecotoxicology and Environmental Safety*, **44**(3), 241–252.

Schlezinger JJ and Stegeman JJ (2000) Induction of cytochrome P450 1A in the American eel by model halogenated and non-halogenated aryl hydrocarbon receptor agonists. *Aquatic Toxicology*, **50**(4), 375–386.

Schrank CS, Cormier SM and Blazer VS (1997) Contaminant exposure, biochemical and histopathological biomarkers in white suckers from contaminated and reference sites in the Sheboygan River, Wisconsin. *Journal of Great Lakes Research*, **23**(2), 119–130.

Seeley KR and Weeks-Perkins BA (1991) Altered phagocytic activity of macrophages in oyster toadfish from a highly polluted subestuary. *Journal of Aquatic Animal Health*, **3**, 224–227.

Seeley KR and Weeks-Perkins BA (1997) Suppression of natural cytotoxic cell and macrophage phagocytic function in oyster toadfish exposed to 7,12-dimethylbenz(a)anthracene. *Fish and Shellfish Immunology*, **7**(2), 115–121.

Shaw GR and Connell DW (2001) DNA adducts as a biomarker of polycyclic aromatic hydrocarbon exposure in aquatic organisms: relationship to carcinogenicity. *Biomarkers*, **6**(1), 64–71.

Singh H (1989) Interaction of xenobiotics with reproductive endocrine functions in a protogynous teleost *Monopterus albus*. *Marine Environmental Research*, **28**, 285–289.

Spies RB, Stegeman JJ, Hinton DE, Woodin B, Smolowitz R, Okihiro M and Shea D (1996) Biomarkers of hydrocarbon exposure and sublethal effects in embiotocid fishes from a natural petroleum seep in the Santa Barbara Channel. *Aquatic Toxicology*, **34**, 195–219.

Stagg, RM, McIntosh A and Mackie P (1995) Elevation of hepatic monooxygenase activity in the dab (*Limanda limanda* L.) in relation to environmental contamination with petroleum hydrocarbons in the northern North Sea. *Aquatic Toxicology*, **33**, 245–264.

Steevens JA, Baumann PC and Jones SB (1996) A comparison of β-adrenoceptors and muscarinic cholinergic receptors in tissues of brown bullhead catfish (*Ameiurus nebulosus*) from the Black River and Old Woman Creek, Ohio. *Environmental Toxicology and Chemistry*, **15**(9), 1551–1554.

Stegeman JJ and Hahn ME (1993) Biochemistry and molecular biology of monooxygenases: current perspectives on forms, functions, and regulation of cytochrome P450 in aquatic species. In Malins DC and Ostrander GK (eds), *Aquatic Toxicology: Molecular, Biochemical, and Cellular Perspectives*. Lewis, Boca Raton, FL, pp. 87–206.

Stehr CM, Myers MS, Burrows DG, Krahn MM, Meador JP, McCain BB and Varanasi U (1997) Chemical contamination and associated liver diseases in two species of fish from San Francisco Bay and Bodega Bay. *Ecotoxicology*, **6**(1), 35–65.

Stehr CM, Brown DW, Hom T, Anulacion BF, Reichert WL and Collier TK (2000) Exposure of juvenile chinook and chum salmon to chemical contaminants in the Hylebos waterway of Commencement Bay, Tacoma, Washington. *Journal of Aquatic Ecosystem Stress and Recovery*, **7**(3), 215–227.

Stein JE, Hom T, Collier TK, Brown DW and Varanasi U (1995) Contaminant exposure and biochemical effects in outmigrant juvenile chinook salmon from urban and non-urban estuaries of Puget Sound, Washington. *Environmental Toxicology and Chemistry*, **14**(6), 1019–1029.

Stein JE, Reichert WL, French BL, and Varanasi U (1993) ^{32}P-postlabelling analysis of DNA adduct formation and persistence in English sole (*Pleuronectes vetulus*) exposed to benzo[a]pyrene and 7H-dibenzo[c,g]carbazole. *Chemico-Biological Interactions*, **88**, 55–69.

Stein JE, Reichert WL and Varanasi U (1994) Molecular epizootiology: assessment of exposure to genotoxic compounds in teleosts. *Environmental Health Perspectives*, **102**(12 suppl.), 19–23.

Stephensen E, Svavarsson J, Sturve J, Ericson G, Adolfsson-Erici M and Forlin L (2000) Biochemical indicators of pollution exposure in shorthorn sculpin (*Myoxocephalus scorpius*), caught in four harbours on the southwest coast of Iceland. *Aquatic Toxicology*, **48**(4), 431–442.

Steyermark AC, Spotila JR, Gillette D and Isseroff H (1999) Biomarkers indicate health problems in brown bullheads from the industrialized Schuylkill River, Philadelphia. *Transactions of the American Fisheries Society*, **128**(2), 328–338.

Sved DW, Roberts MH and Van Veld PA (1997) Toxicity of sediments contaminated with fractions of creosote. *Water Research*, **31**(2), 294–300.

Thomas P (1988) Reproductive endocrine function in female Atlantic croaker exposed to pollutants. *Marine Environmental Research*, **24**, 179–183.

Thomas RE, Carls MG, Rice SD and Shagrun L (1997) Mixed function oxygenase induction in pre- and post-spawn herring (*Clupea pallasi*) by petroleum hydrocarbons. *Comparative Biochemistry and Physiology*, **116C**(2), 141–147.

Tilghman-Hall A and Oris JT (1991) Anthracene reduces reproductive potential and is maternally transferred during long-term exposure in fathead minnows. *Aquatic Toxicology*, **19**(3), 249–264.

Tuvikene A, Huuskonen S, Koponen K, Ritola O, Mauer U and Lindstrom-Seppae P (1999) Oil shale processing as a source of aquatic pollution: monitoring of the biologic effects in caged and feral freshwater fish. *Environmental Health Perspectives*, **107**(9), 745–752.

Ueng YF, Liu TY and Ueng TH (1995) Induction of cytochrome P450 1A1 and monooxygenase activity in tilapia by sediment extract. *Bulletin of Environmental Contamination and Toxicology*, **54**(1), 60–67.

Van den Heuvel MR, Power M, Richards J, MacKinnon M and Dixon DG (2000) Disease and gill lesions in yellow perch (*Perca flavescens*) exposed to oil sands mining-associated waters. *Ecotoxicology and Environmental Safety*, **46**, 334–341.

Van der Oost R and Heida H (1993) Bioaccumulation of organic micropollutants in different aquatic organisms: Sublethal toxic effects on fish. *Marine Environmental Research*, **35**(1–2), 202.

Van der Oost R, Goksoyr A, Celander M, Heida H and Vermeulen NPE (1996) Biomonitoring of aquatic pollution with feral eel (*Anguilla anguilla*). 2. Biomarkers: pollution-induced biochemical responses. *Aquatic Toxicology*, **36**(3–4), 189–222.

Van der Oost R, Van Schooten F-J, Ariese F, Heida H, Satumalay K and Vermeulen NPE (1994) Bioaccumulation, biotransformation and DNA binding of PAHs in feral eel (*Anguilla anguilla*) exposed to polluted sediments: a field survey. *Environmental Toxicology and Chemistry*, **13**(6), 859–870.

Van Der Weiden ME, Hanegraaf FH, Eggens ML, Celander M, Seinen W and Van den Berg M (1994) Temporal induction of cytochrome P450 1A in the mirror carp (*Cyprinus carpio*) after administration of several polycyclic aromatic hydrocarbons. *Environmental Toxicology and Chemistry*, **13**(5), 797–802.

Van Schooten FJ, Maas LM, Moonen EJC, Kleinjans JCS and Van der Oost R (1995) DNA dosimetry in biological indicator species living on PAHs-contaminated soils and sediments. *Ecotoxicology and Environmental Safety*, **30**(2), 171–179.

Van Veld PA, Westbrook DG, Woodin BR, Hale RC, Smith CL, Huggett RJ and Stegeman JJ (1990) Induced cytochrome P-450 in intestine and liver of spot (*Leiostomus xanthurus*) from a polycyclic aromatic hydrocarbon contaminated environment. *Aquatic Toxicology*, **17**(2), 119–132.

Van Veld PA, Vogelbein WK, Cochran MK, Goksoyr A and Stegeman JJ (1997) Route-specific cellular expression of cytochrome P4501A (CYP1A) in fish (*Fundulus heteroclitus*) following exposure to aqueous and dietary benzo[a]pyrene. *Toxicology and Applied Pharmacology*, **142**(2), 348–359.

Vandermeulen JH and Mossman D (1996) Sources of variability in seasonal hepatic microsomal oxygenase activity in winter flounder (*Pleuronectes americanus*) from a coal tar-contaminated estuary. *Canadian Journal of Fisheries and Aquatic Sciences*, **53**(8), 1741–1753.

Vethaak AD and Wester PW (1996) Diseases of flounder *Platichthys flesus* in Dutch coastal and estuarine waters, with particular reference to environmental stress factors. II. Liver histopathology. *Diseases of Aquatic Organisms*, **26**, 99–116.

Vigano L, Arillo A, Bagnasco M, Bennicellli C and Melodia F (1994) Time course of xenobiotic biotransformation enzyme activities of rainbow trout caged in the river Po. *Science of the Total Environment*, **151**(1), 37–46.

Vigano L, Arillo A, Melodia F, Arlati P and Monti C (1998) Biomarker responses in cyprinids of the middle stretch of the River Po, Italy. *Environmental Toxicology and Chemistry*, **17**(3), 404–411.

Vines CA, Robbins T, Griffin FJ and Cherr GN (2000) The effects of diffusible creosote-derived compounds on development in Pacific herring (*Clupea pallasi*). *Aquatic Toxicology*, **51**(2), 225–239.

Vogelbein WK, Fournie JW, Van Veld PA and Huggett RJ (1990) Hepatic neoplasms in the mummichog *Fundulus heteroclitus* from a creosote-contaminated site. *Cancer Research*, **50**(18), 5978–5986.

Weber LP and Janz DM (2001) Effect of β-naphthoflavone and dimethylbenz(a)anthracene on apoptosis and HSP70 expression in juvenile channel catfish (*Ictalurus punctatus*) ovary. *Aquatic Toxicology*, **54**(1–2), 39–50.

Weeks BA and Warinner JE (1984) Effects of toxic chemicals on macrophage phagocytosis in two estuarine fishes. *Marine Environmental Research*, **14**, 327–335.

Weinstein JE, Oris JT and Taylor DH (1997) An ultrastructural examination of the mode of UV-induced toxic action of fluoranthene in the fathead minnow, *Pimephales promelas*. *Aquatic Toxicology*, **39**(1), 1–22.

Wertheimer AC, Heintz RA, Thedinga JF, Maselko JM and Rice SD (2000) Straying of adult pink salmon from their natal stream following embryonic exposure to weathered Exxon Valdez crude oil. *Transactions of the American Fisheries Society*, **129**(4), 989–1004.

Westernhagen H, Kruener G and Broeg K (1999) Ethoxyresorufin *O*-deethylase (EROD) activity in the liver of dab (*Limanda limanda*) and flounder (*Platichthys flesus*) from the German Bight. EROD expression and tissue contamination. *Helgoland Marine Research*, **53**(3–4), 244–249.

White PA, Robitaille S and Rasmussen JB (1999) Heritable reproductive effects of benzo[a]pyrene on the fathead minnow (*Pimephales promelas*). *Environmental Toxicology and Chemistry*, **18**(8), 1843–1847.

Willett KL, Gardinali PR, Lienesch LA and Di Giulio RT (2000) Comparative metabolism and excretion of benzo[a]pyrene in two species of ictalurid catfish. *Toxicological Sciences*, **58**(1), 68–76.

Willett KL, McDonald SJ, Steinberg MA, Beatty KB, Kennicutt MC and Safe SH (1997) Biomarker sensitivity for polynuclear aromatic hydrocarbon contamination in two marine fish species collected in Galveston Bay, Texas. *Environmental Toxicology and Chemistry*, **16**(7), 1472–1479.

Willett K, Steinberg M, Thomsen J, Narasimhan TR, Safe S, McDonald S, Beatty K and Kennicutt MC (1995) Exposure of killifish to benzo[a]pyrene: comparative metabolism, DNA adduct formation and aryl hydrocarbon (Ah) receptor agonist activities. *Comparative Biochemistry and Physiology*, **112B**(1), 93–103.

Winston GW, Shane BS and Henry CB (1988) Hepatic mono-oxygenase induction and promutagen activation in channel catfish from a contaminated river basin. *Ecotoxicology and Environmental Safety*, **16**(3), 258–271.

Wolkers J, Joergensen EH, Nijmeijer SM and Witkamp RF (1996) Time-dependent induction of two distinct hepatic cytochrome P4501A catalytic activities at low temperatures in Arctic charr (*Salvelinus alpinus*) after oral exposure to benzo[a]pyrene. *Aquatic Toxicology*, **35**(2), 127–138.

Wong CKC, Yeung HY, Woo PS and Wong MH (2001) Specific expression of cytochrome P4501A1 gene in gill, intestine and liver of tilapia exposed to coastal sediments. *Aquatic Toxicology*, **54**(1–2), 69–80.

Zheng W, Feng T and Guo X (2000) Effect of benzo[a]pyrene on content of reduced glutathione in ovary of *Boleophthalmus pectinirostris*. *Chinese Journal of Applied and Environmental Biology*, **6**(4), 349–353.

Ziskowski JJ, Despres-Patanjo L, Murchelano RA, Howe AB, Ralph D and Atran S (1987) Disease in commercially valuable fish stocks in the North Atlantic. *Marine Pollution Bulletin*, **18**, 496–504.

Effects of PAHs on Terrestrial and Freshwater Birds, Mammals and Amphibians

HEATH M. MALCOLM AND RICHARD F. SHORE

Centre for Ecology and Hydrology, Monks Wood, UK

12.1 INTRODUCTION

The aim of this chapter is to evaluate the effects of PAHs on terrestrial vertebrates and it focuses on birds, wild mammals and amphibians. Some species of birds that are traditionally associated with marine food chains are included, as they have become increasingly reliant on feeding from terrestrial areas, such as landfills (gulls), or in utilizing inland freshwater bodies (e.g. cormorant, *Phalacrocorax carbo*, and common terns, *Sterna hirundo*). The chapter covers exposure, metabolism and toxicity of PAHs but does not review toxicological studies on laboratory test species. There are limited data on individual PAH compounds and therefore studies in which vertebrates were exposed to various contaminants known to contain PAHs, such as fuel oils, have also been reviewed.

12.2 EXPOSURE AND METABOLISM

There appear to be no data on uptake rates for PAHs by terrestrial vertebrates. However, since PAHs are ubiquitous in the environment (WHO 1998), and have been detected in some body tissues and eggs, it can be concluded that exposure to PAHs does occur in at least some free-living vertebrate species. The main route of exposure of terrestrial vertebrates to PAHs is thought to be via the diet. However, topical exposure is also important in some circumstances, e.g. when incubating, birds become oiled and oil is subsequently transferred from the

PAHs: An Ecotoxicological Perspective. Edited by Peter E.T. Douben.
© 2003 John Wiley & Sons Ltd

feathers to the eggshell. This can result in embryos being exposed *in ovo*. Oiled birds are also orally exposed to PAHs because they ingest oil when preening.

PAHs have been detected in various invertebrate species (Environment Canada 1994; WHO 1998; see also Chapters 8 and 10), some of which are eaten by the vertebrates considered in this chapter. In experimental studies on uptake of specific PAHs by earthworms (*Lumbricus rubellus*), a major food item for many invertebrate-feeding birds and mammals, fluoranthene and to a lesser extent phenanthrene were found to be bioaccumulated when worms were exposed to contaminated soil (Ma *et al.* 1995). Bioaccumulation factors for worms exposed to a nominal initial concentration of 100 mg/kg phenanthrene ranged from 0.027 to 0.097. Bioaccumulation factors were higher during the first 14 days of exposure, and when worms were deprived of a food supply. Bioaccumulation factors for worms similarly exposed to 100 mg/kg fluoranthene ranged from 0.022 to 0.623. In a subsequent field study, Ma *et al.* (1998) calculated bioconcentration factors (Table 12.1) for 11 PAH compounds in earthworms collected from several sites in a floodplain which had experienced long-term PAH contamination. Bioaccumulation factors were higher for the PAH compounds with higher log octanol–water partition coefficients. Van Brummelen *et al.* (1996) calculated bioaccumulation factors for the same earthworm species, with reference to the PAH concentration in the soil compartments, namely, litter, fragmentation material, humus and mineral soil (Table 12.1). Bioaccumulation factors for the litter, fragmentation and humus compartments were approximately equal. Bioaccumulation factors were higher in mineral soil due to the lower PAH concentrations found in this compartment.

PAHs, like organochlorines, are lipophilic and therefore have the potential to be bioaccumulated to hazardous levels by vertebrates (Lu *et al.* 1977). However, bioaccumulation in species at higher trophic levels is likely to be limited. This is because PAHs induce and are rapidly metabolized by mixed-function oxidases

TABLE 12.1 Bioaccumulation factors measured *in situ* in earthworms

PAH	BAF	PAH	BAF
Anthracene	0.080 ± 0.061^a	Benzo[a]anthracene	1.880 ± 0.810^a
Phenanthrene	0.070 ± 0.030^a		$0.027-0.128^b$
	$0.018-0.050^b$	Benzo[b]fluoranthene	1.350 ± 0.552^a
Pyrene	0.330 ± 0.172^a		$0.013-0.072^b$
	$0.024-0.167^b$	Benzo[k]fluoranthene	2.110 ± 0.845^a
Fluoranthene	0.350 ± 0.137^a		$0.022-0.121^b$
	$0.0276-0.142^b$	Benzo[a]pyrene	2.470 ± 1.107^a
Chrysene	2.070 ± 0.980^a		$0.021-0.277^b$
	$0.040-0.279^b$	Benzo[ghi]perylene	9.080 ± 3.778^a
Dibenzo[ah]anthracene	18.020 ± 6.962^a		

[a]Average values ± 95% confidence interval (CI) (Ma *et al*. 1998).
[b]Range of values calculated relative to PAH concentrations in litter, fragmentation, humus and mineral soil (van Brummelen *et al*. 1996).

(Hallett and Brecher 1984). Details are given in Chapter 5. The P450-dependent monooxygenases metabolize PAHs to diol epoxides, which are known to be highly reactive in biological systems, forming adducts with DNA/RNA (Sheffield *et al*. 2001). These metabolites can also bind to other cellular macromolecules to form PAH–diolepoxide–protein adducts, which have been used as biomarkers of PAH exposure in wild vertebrates. For example, there was an increased level of benzo[a]pyrene-diolepoxide adducts in blood albumin and hemoglobin in woodchucks (*Marmota monax*) from an area near an aluminium electrolysis plant (Blondin and Viau 1992). The frequencies of detection of blood albumin adducts was significantly correlated with the benzo[a]pyrene concentration in the vegetation. The ability of PAHs to induce liver monooxygenase activity has also been demonstrated in wood mice (*Apodemus sylvaticus*) sampled from either a pitch-coke plant or an industrial/oil shipping area (Leupold *et al*. 1992). Induction of the enzymes correlated with the different PAH exposures. Mice from the pitch-coke plant had increased 7-ethoxyresorufin *O*-deethylase (EROD) activity, and PROD (pentoxyresorufin-*O*-deethylase) activity was enhanced in mice from the industrial/oil-shipping site. However, EROD activity in wild common shrews (*Sorex araneus*) was not correlated with distance from a blast furnace, despite there being likely differences in exposure to PAHs along a transect away from the furnace (Bosveld *et al*. 1996); PAH concentrations in soil, earthworms and isopods all decreased with increasing distance from the plant (Van Brummelen 1995).

12.3   BIRDS

12.3.1   RESIDUES IN WILD BIRD SPECIES AND THEIR EGGS

Because PAHs are rapidly metabolized by birds, residues are often not detectable in the body organs and tissues and so are not often measured. However, PAHs have been detected in some avian species. For example, the livers of herring gulls (*Larus argentatus*) from two sites in Ontario, Canada, contained concentrations (μg/kg lipid) of anthracene (0.15), fluoranthene (0.082), pyrene (0.076), naphthalene (0.05), fluorine (0.044), acenaphthene (0.038) and benzo[a]pyrene (0.038) (Hallett and Brecher 1984).

Because PAHs are lipophilic, they are more readily detected in eggs. Thus, eggs are potentially good biomonitors of PAH contamination in birds, although concentrations are generally low. This is because birds tend to metabolize and excrete much of their ingested PAH load (Stronkhorst *et al*. 1993) and because metabolism of PAHs also occurs *in ovo*; 94% of a PAH mixture (0.2 mg/kg) injected into white leghorn chicken (*Gallus domesticus*) eggs was metabolized within 14 days of administration (Naf *et al*. 1992). Surprisingly, there appear to have been relatively few studies on PAH levels in the eggs of free-living birds. The concentrations of PAHs reported in eggs from the few studies that have been done are listed in Table 12.2. The majority of these eggs came from either

TABLE 12.2 Concentrations (conc.) of PAH compounds (μg/kg) in avian eggs

Species		Conc.	Species		Conc.
Naphthalene			**Fluoranthene**		
Herring gull	Larus argentatus	12.7[a]	Herring gull	Larus argentatus	< 0.174[a]
Cormorant	Phalacrocorax carbo	29.2[a]	Cormorant	Phalacrocorax carbo	2.89[a]
Chough	Pyrrhocorax pyrrhocorax	2.95[a]	Chough	Pyrrhocorax pyrrhocorax	0.734[a]
Goose	Anser anser	212.2[b]	Goose	Anser anser	50.6[b]
Heron	Ardea cinerea	237[b]	Heron	Ardea cinerea	38.6[b]
Domestic fowl	Gallus gallus	24.4[b]	Domestic fowl	Gallus gallus	41.5[b]
Domestic fowl	Gallus gallus	145.1[b]	Domestic fowl	Gallus gallus	2.87[b]
Mallard	Anas platyrhynchos	276.5[b]	Mallard	Anas platyrhynchos	2309[b]
Pintail	Anas acuta	4.4[b]	Pintail	Anas acuta	73[b]
Shoveler	Anas clypeata	85.1[b]	Shoveler	Anas clypeata	25.5[b]
Tufted duck	Aythya fuligula	45[b]	Tufted duck	Aythya fuligula	57[b]
Pochard	Aythya ferina	2.4[b]	Pochard	Aythya ferina	1.6[b]
Herring gull	Larus argentatus	208.2[b]	Herring gull	Larus argentatus	15.4[b]
Common gull	Larus canus	312.4[b]	Common gull	Larus canus	96.4[b]
Black-headed gull	Larus ridibundus	125[b]	Black-headed gull	Larus ridibundus	6.6[b]
Slavonian grebe	Podiceps auritus	1256[b]	Slavonian grebe	Podiceps auritus	570.4[b]
Common tern	Sterna hirundo	44.2[b]	Common tern	Sterna hirundo	221[b]
Lapwing	Vanellus vanellus	54.2[b]	Lapwing	Vanellus vanellus	26.2[b]
Marsh sandpiper	Tringa stagnatilis	400[b]	Marsh sandpiper	Tringa stagnatilis	330.5[b]
			Common tern	Sterna hirundo	0.1[d]
Acenaphthalene			**Pyrene**		
Herring gull	Larus argentatus	< 0.166[a]	Herring gull	Larus argentatus	< 0.058[a]
Cormorant	Phalacrocorax carbo	< 0.166[a]	Cormorant	Phalacrocorax carbo	< 0.058[a]
Chough	Pyrrhocorax pyrrhocorax	< 0.166[a]	Chough	Pyrrhocorax pyrrhocorax	< 0.058[a]

Acenaphthene

Goose	Anser anser	96.4[b]
Heron	Ardea cinerea	25.5[b]
Domestic fowl	Gallus gallus	1.4[b]
Domestic fowl	Gallus gallus	9.2[b]
Mallard	Anas platyrhynchos	88.4[b]
Pintail	Anas acuta	183.6[b]
Shoveler	Anas clypeata	19.4[b]
Tufted duck	Aythya fuligula	35.4[b]
Pochard	Aythya ferina	0.2[b]
Herring gull	Larus argentatus	174.5[b]
Common gull	Larus canus	60.5[b]
Black-headed gull	Larus ridibundus	2.5[b]
Slavonian grebe	Podiceps auritus	1525[b]
Common tern	Sterna hirundo	24.1[b]
Lapwing	Vanellus vanellus	43.2[b]
Marsh sandpiper	Tringa stagnatilis	125.3[b]
Goose	Anser anser	106.8[b]
Heron	Ardea cinerea	55[b]
Domestic fowl	Gallus gallus	81.7[b]
Domestic fowl	Gallus gallus	21.3[b]
Mallard	Anas platyrhynchos	3218[b]
Pintail	Anas acuta	90[b]
Shoveler	Anas clypeata	89.4[b]
Tufted duck	Aythya fuligula	485[b]
Pochard	Aythya ferina	51[b]
Herring gull	Larus argentatus	19[b]
Common gull	Larus canus	170.8[b]
Black-headed gull	Larus ridibundus	233.8[b]
Slavonian grebe	Podiceps auritus	836.6[b]
Common tern	Sterna hirundo	446.6[b]
Lapwing	Vanellus vanellus	51.3[b]
Marsh sandpiper	Tringa stagnatilis	810.1[b]
Common tern	Sterna hirundo	0.3[d]

Acenaphthylene

Herring gull	Larus argentatus	< 0.174[a]
Cormorant	Phalacrocorax carbo	< 0.174[a]
Chough	Pyrrhocorax pyrrhocorax	< 0.174[a]
Goose	Anser anser	6.5[b]
Heron	Ardea cinerea	27.6[b]
Domestic fowl	Gallus gallus	1.4[b]
Domestic fowl	Gallus gallus	24[b]
Mallard	Anas platyrhynchos	261[b]
Pintail	Anas acuta	96[b]
Shoveler	Anas clypeata	183.8[b]
Tufted duck	Aythya fuligula	30.2[b]
Pochard	Aythya ferina	0.6[b]
Herring gull	Larus argentatus	58.9[b]
Common gull	Larus canus	60.5[b]

Benzo[a]anthracene

Herring gull	Larus argentatus	< 0.174[a]
Cormorant	Phalacrocorax carbo	5.98[a]
Chough	Pyrrhocorax pyrrhocorax	< 0.174[a]

Chrysene

Herring gull	Larus argentatus	< 0.678[a]
Cormorant	Phalacrocorax carbo	< 0.678[a]
Chough	Pyrrhocorax pyrrhocorax	< 0.678[a]

Benzo[b]fluoranthene

Herring gull	Larus argentatus	2.23[a]
Cormorant	Phalacrocorax carbo	4.17[a]
Chough	Pyrrhocorax pyrrhocorax	0.934[a]

(continued overleaf)

TABLE 12.2 (continued)

Species		Conc.	Species		Conc.
Black-headed gull	*Larus ridibundus*	1.8[b]	**Benzo[k]fluoranthene**		
Slavonian grebe	*Podiceps auritus*	1100[b]	Herring gull	*Larus argentatus*	1.17[a]
Common tern	*Sterna hirundo*	20[b]	Cormorant	*Phalacrocorax carbo*	2.23[a]
Lapwing	*Vanellus vanellus*	38.4[b]	Chough	*Pyrrhocorax pyrrhocorax*	1.53[a]
Marsh sandpiper	*Tringa stagnatilis*	902.5[b]			
Fluorene			**Benzo[a]pyrene**		
Herring gull	*Larus argentatus*	19.1[a]	Herring gull	*Larus argentatus*	< 0.116[a]
Cormorant	*Phalacrocorax carbo*	< 0.194[a]	Cormorant	*Phalacrocorax carbo*	4.92[a]
Chough	*Pyrrhocorax pyrrhocorax*	16.9[a]	Chough	*Pyrrhocorax pyrrhocorax*	< 0.116[a]
Goose	*Anser anser*	32.1[b]	Goose	*Anser anser*	19.6[b]
Heron	*Ardea cinerea*	7.8[b]	Heron	*Ardea cinerea*	9.6[b]
Domestic fowl	*Gallus gallus*	2.5[b]	Domestic fowl	*Gallus gallus*	1.4[b]
Domestic fowl	*Gallus gallus*	112.8[b]	Domestic fowl	*Gallus gallus*	24[b]
Mallard	*Anas platyrhynchos*	25.2[b]	Mallard	*Anas platyrhynchos*	50.4[b]
Pintail	*Anas acuta*	26.8[b]	Pintail	*Anas acuta*	33[b]
Shoveler	*Anas clypeata*	127.7[b]	Shoveler	*Anas clypeata*	31.9[b]
Tufted duck	*Aythya fuligula*	23.9[b]	Tufted duck	*Aythya fuligula*	10.8[b]
Pochard	*Aythya ferina*	0.7[b]	Pochard	*Aythya ferina*	0.5[b]
Herring gull	*Larus argentatus*	120[b]	Herring gull	*Larus argentatus*	47.4[b]
Common gull	*Larus canus*	36.1[b]	Common gull	*Larus canus*	10.2[b]
Black-headed gull	*Larus ridibundus*	5[b]	Black-headed gull	*Larus ridibundus*	2.6[b]
Slavonian grebe	*Podiceps auritus*	390.4[b]	Slavonian grebe	*Podiceps auritus*	30.4[b]
Common tern	*Sterna hirundo*	22.4[b]	Common tern	*Sterna hirundo*	29[b]
Lapwing	*Vanellus vanellus*	135.5[b]	Lapwing	*Vanellus vanellus*	12.8[b]
Marsh sandpiper	*Tringa stagnatilis*	120.1[b]	Marsh sandpiper	*Tringa stagnatilis*	29.2[b]

Phenanthrene

Common name	Scientific name	Value
Herring gull	Larus argentatus	1.45[a]
Herring gull	Larus argentatus	1[c]
Cormorant	Phalacrocorax carbo	< 0.194[a]
Chough	Pyrrhocorax pyrrhocorax	< 0.194[a]
Goose	Anser anser	91.2[b]
Heron	Ardea cinerea	23[b]
Domestic fowl	Gallus gallus	29.6[b]
Domestic fowl	Gallus gallus	53.3[b]
Mallard	Anas platyrhynchos	425.4[b]
Pintail	Anas acuta	124.5[b]
Shoveler	Anas clypeata	47.9[b]
Tufted duck	Aythya fuligula	379[b]
Pochard	Aythya ferina	1.2[b]
Herring gull	Larus argentatus	81.2[b]
Common gull	Larus canus	122[b]
Black-headed gull	Larus ridibundus	99.9[b]
Slavonian grebe	Podiceps auritus	1340[b]
Common tern	Sterna hirundo	270.4[b]
Common tern	Sterna hirundo	0.4[d]
Lapwing	Vanellus vanellus	188.4[b]
Marsh sandpiper	Tringa stagnatilis	603.9[b]

Anthracene

Common name	Scientific name	Value
Herring gull	Larus argentatus	< 0.136[a]
Cormorant	Phalacrocorax carbo	< 0.136[a]
Chough	Pyrrhocorax pyrrhocorax	< 0.136[a]
Goose	Anser anser	25.6[b]

Dibenz[ah]anthracene

Common name	Scientific name	Value
Herring gull	Larus argentatus	< 0.166[a]
Cormorant	Phalacrocorax carbo	< 0.166[a]
Chough	Pyrrhocorax pyrrhocorax	< 0.166[a]

Benzo[ghi]perylene

Common name	Scientific name	Value
Herring gull	Larus argentatus	< 0.194[a]
Cormorant	Phalacrocorax carbo	< 0.194[a]
Chough	Pyrrhocorax pyrrhocorax	0.858[a]
Goose	Anser anser	8.4[b]
Heron	Ardea cinerea	5.2[b]
Domestic fowl	Gallus gallus	1.4[b]
Domestic fowl	Gallus gallus	2.7[b]
Mallard	Anas platyrhynchos	17.2[b]
Pintail	Anas acuta	14[b]
Shoveler	Anas clypeata	9.8[b]
Tufted duck	Aythya fuligula	7.6[b]
Pochard	Aythya ferina	0.3[b]
Herring gull	Larus argentatus	8.2[b]
Common gull	Larus canus	7.3[b]
Black-headed gull	Larus ridibundus	2.9[b]
Slavonian grebe	Podiceps auritus	15.8[b]
Common tern	Sterna hirundo	14.2[b]
Lapwing	Vanellus vanellus	7.7[b]
Marsh sandpiper	Tringa stagnatilis	25.1[b]

Indeno[1,2,3-cd]pyrene

Common name	Scientific name	Value
Herring gull	Larus argentatus	< 0.155[a]
Cormorant	Phalacrocorax carbo	< 0.155[a]

(continued overleaf)

TABLE 12.2 *(continued)*

Species		Conc.	Species		Conc.
Heron	*Ardea cinerea*	30.8[b]	Chough	*Pyrrhocorax pyrrhocorax*	< 0.155[a]
Domestic fowl	*Gallus gallus*	9[b]	Goose	*Anser anser*	7.2[b]
Domestic fowl	*Gallus gallus*	1.6[b]	Heron	*Ardea cinerea*	4.6[b]
Mallard	*Anas platyrhynchos*	419[b]	Domestic fowl	*Gallus gallus*	1.4[b]
Pintail	*Anas acuta*	7.2[b]	Domestic fowl	*Gallus gallus*	3.4[b]
Shoveler	*Anas clypeata*	37.2[b]	Mallard	*Anas platyrhynchos*	19.4[b]
Tufted duck	*Aythya fuligula*	41.1[b]	Pintail	*Anas acuta*	95[b]
Pochard	*Aythya ferina*	0.4[b]	Shoveler	*Anas clypeata*	8.4[b]
Herring gull	*Larus argentatus*	5.8[b]	Tufted duck	*Aythya fuligula*	6.5[b]
Common gull	*Larus canus*	28.2[b]	Pochard	*Aythya ferina*	0.2[b]
Black-headed gull	*Larus ridibundus*	133.8[b]	Herring gull	*Larus argentatus*	4.6[b]
Slavonian grebe	*Podiceps auritus*	701.3[b]	Common gull	*Larus canus*	9[b]
Common tern	*Sterna hirundo*	36.5[b]	Black-headed gull	*Larus ridibundus*	3.2[b]
Lapwing	*Vanellus vanellus*	66.6[b]	Slavonian grebe	*Podiceps auritus*	12.6[b]
Marsh sandpiper	*Tringa stagnatilis*	375[b]	Common tern	*Sterna hirundo*	10.9[b]
			Lapwing	*Vanellus vanellus*	4.9[b]
			Marsh sandpiper	*Tringa stagnatilis*	19.8[b]

[a] Eggs from various sites on the UK coast, concentrations expressed as lipid weight. Mean percentage lipid was 6.5% (herring gull), 3.4% (cormorant) and 5.2% (chough) (Shore *et al.* 1999).

[b] Eggs from Lake Baikal region, concentrations expressed as dry weight. Percentage water content of eggs was not specified (Lebedev *et al.* 1998).

[c] Eggs from German Coast, concentrations expressed as wet weight (Jacob and Grimmer 1994).

[d] Eggs from Scheldt Estuary, concentrations expressed as wet weight (Stronkhorst *et al.* 1993).

the UK coastline (Shore *et al*. 1999) or Lake Baikal in Russia (Lebedev *et al*. 1998). The PAHs present in eggs may be the result not only of maternal transfer but also of direct application to the egg shell of small quantities of oil adhering to the plumage and feet of oiled incubating birds (Birkhead *et al*. 1973).

12.3.2   ACUTE TOXICITY

The acute toxicity of four PAH compounds to red-winged blackbirds (*Agelaius phoeniceus*) exposed via gavage were reported by Schafer *et al*. (1983). The LD_{50} values for acenaphthene, fluorine, anthracene, and phenanthrene were 101, 101, 111, and 113 mg/kg bw, respectively. The LD_{50} for anthracene in house sparrows (*Passer domesticus*) was 244 mg/kg bw. The toxicity of crude oil fractions to birds has also been determined, and the effects linked to the chemical composition of the fraction. Herring gull nestlings given a single (12 ml) oral dose of Prudoe Bay crude oil, or its fractions, had reduced growth, and increased adrenal and nasal gland weights within 8 days of exposure (Peakall *et al*. 1982). The fraction that produced the greatest effect contained a methylated series of chrysenes, benzanthracenes, phenylanthracenes, binaphthyls, and traces of benzopyrenes. Growth rates were not suppressed following exposure to the fraction containing alkylated naphthalenes, biphenyls, anthracenes, phenanthrenes, and fluorenes.

12.3.3   REPRODUCTIVE TOXICITY

PAHs are known to cause reproductive toxicity in birds and can be the result of exposure of either adult birds or eggs; e.g. chronic exposure for up to 5 months to benzo[a]pyrene, administered as weekly 10 mg/kg bw injections, resulted in complete infertility in female pigeons (Hough *et al*. 1993). The authors believed that this reproductive toxicity was a result of changes in the levels of estrogens. Toxicity studies on eggs demonstrated that *in ovo* application of 1 - 5 μl of various PAHs resulted in embryotoxicity, including embryo death and teratogenicity (Hoffman 1979).

The toxicity of crude oils, rather than specific PAHs, to bird embryos has been well documented (for review, see Hoffman 1990). Toxicity studies have usually involved applying oil directly to the egg surface or injecting the oil into the egg. Originally, toxicity was thought to result from the physical blocking of the eggshell pores, thus preventing the uptake of oxygen essential for normal development. However, during the late 1970s, the cause of embryonic death in eggs exposed to various petroleum products, including fuel oils, was established to be chemical toxicity and not anoxia. The fractions of the oils reported to be the most embryotoxic were those containing PAH compounds, whereas treatment with the aliphatic compounds present in crude oil had no significant effect on the developing embryos (Walters *et al.* 1987). The embryotoxic potential of the aromatic fraction was also correlated with the degree of P450 induction in the embryo liver and kidney, and with hydrocarbon hydroxylase activities.

Dose–response studies with crude oil have demonstrated that transfer of minute amounts of crude oil to the eggshell can result in toxic effects. An LD_{50} of 1.3 and 2.2 μl/egg has been reported for Prudhoe Bay and Hibernia crude oil, respectively, following application to the shell of white leghorn chicken eggs on day 8 of incubation (Lee *et al.* 1986). Gross pathological effects, including liver necrosis, renal lesions, extensive edema, growth retardation and teratogenicity, have also been reported in the embryos of white leghorn chickens and mallards following the application to eggs of microliter volumes of either fuel oils or PAH mixtures (Hoffman and Gay 1981; Matsumoto and Kashimoto 1986; Matsumoto 1988; Couillard and Leighton 1989, 1990). Other experimental studies have shown that, as with individual PAH compounds, exposure to crude oil results in induction of mixed-function oxidases in the embryo. Application of 5 μl Prudhoe Bay crude oil to chick egg shells on day 11 of development maximally induced various enzymes, including cytochrome P450-levels (by four-fold), naphthalene hydroxylase (six-fold), benzo[a]pyrene 3-hydroxylase (14-fold) and EROD (24-fold); induction of glutathione-S-transferase was not observed. When disulfiram (which reduces hepatic cytochrome P450 levels in rodents, and hence inhibits metabolism of PAHs) was given 1 h prior to application of Prudhoe Bay Crude Oil to the eggshell, embryo mortality was reduced from 60% to 20% (Lee *et al.* 1986), thereby demonstrating that metabolism of PAHs by the mixed-function oxidase system can actually enhance toxicity in some circumstances.

The degree to which the embryotoxic effects are manifested depends on various factors. These include the amount of oil transferred to the eggs, the chemical composition of the oil, and the stage of incubation at which the contamination occurs (Lewis and Malecki 1984). Toxic effects are more likely to be manifested if the exposure is during the very early stages of incubation (Hoffman 1979; Lewis and Malecki 1984). Reduced growth, as indicated by reduced body size, weight and leg length, was reported in white leghorn embryos following injection of benzo[a]pyrene into the yolk sac. Significant effects were reported in eggs injected with 50 μg/egg on days 4 or 6, but only at 100 μg/egg on day 9 (Anwer and Mehrotra 1988). The difference in toxicity at different stages in development may be due to several factors. As the mass of the embryo increases during incubation, the administered dose, relative to embryo body weight, decreases. However, the timing of exposure relative to developmental stage is also important. The PAH compounds shown to be the most toxic in the 72 h chick embryo test are also known to be ligands to the Ah receptor and to induce EROD activity in chick embryo liver (Brunstrom 1992). The Ah receptor is expressed in the chick embryo liver early in development, and chick embryos are extremely sensitive to Ah receptor ligands. This may explain why the chicken appears to be more sensitive to CYP1A (EROD)-inducing potencies of various chemicals, including benzo[k]fluoranthene, than other species examined (Brunstrom and Halldin 1998).

In addition to the factors listed above, toxicity can also be affected by interspecies differences in sensitivity to PAHs. The embryotoxicity of a mixture of 18 PAH compounds was examined in white leghorn chickens, mallard, common eider (*Somateria mollissima*) and turkey (*Meleagris gallopavo*) (Brunstrom *et al.* 1990). A dose of 2.0 mg/kg egg increased embryo mortality above control levels in all four species but the mallard was more sensitive than the other species, as it was the only one in which mortality was significantly increased by a dose of 0.2 mg/kg. Benzo[k]fluoranthene was the most toxic of the individual PAHs tested, with significant mortality reported in turkey, duck and eider exposed to 0.2 mg/kg, and chickens exposed to 2.0 mg/kg. Previous laboratory studies have shown chicken embryos to be the most sensitive of the four species when exposed to other contaminants, such as PCBs, but this is clearly not the case for PAHs.

The toxicity of crude oils to birds has also been demonstrated in field studies, although there is a paucity of data for terrestrial species. Fry *et al.* (1986) administered Santa Barbara crude oil to wedge-tailed shearwaters (*Puffinus pacificus*) either orally, via a gelatin capsule, or topically, via application to breast plumage. Application of 2 ml to breast plumage reduced breeding success by reducing egg laying and hatching success. The effects in the orally-exposed birds were less pronounced. There were no effects on the growth of chicks from exposed birds, but survival was reduced. The number of exposed birds returning to the colony the following year was reduced.

12.4 MAMMALS

12.4.1 RESIDUES IN TERRESTRIAL MAMMALS

As with birds, there have been few studies on exposure of terrestrial mammals to PAHs. Various PAH compounds have been detected in the livers and kidneys of a small number of species, especially from areas known to have been contaminated with oil. Eleven PAHs were detected in the livers of deer mice (*Peromyscus maniculatus*) from a contaminated area in South Carolina and New Jersey, USA. Concentrations ranged from 0.05 mg/kg for benzo[b]fluoranthene to 4.56 mg/kg for benz[a]anthracene (not specified whether concentrations were expressed on a wet, dry, or lipid weight basis). Other PAHs detected (mg/kg) were acenaphthylene (1.91–3.92), acenaphthene (0.10–0.14), fluorine (0.22–0.34), benz[a]anthracene (0.55–4.56), chrysene (0.01–0.32), benzo[b]fluoranthene (0.05–2.64), benzo[k]fluoranthene (0.05–0.07). Dibenz[ah]anthracene and indeno[1,2,3-cd]pyrene were below the limits of detection (Dickerson *et al.* 1994).

Hepatic concentrations of naphthalene and methylnaphthalene were not significantly elevated in Heermann's kangaroo rats (*Dipodomys heermanni*) collected from an area contaminated by an oil field blowout, compared to those from control areas. In contrast, concentrations of other compounds, including terpanes, steranes and monoaromatic and triaromatic steranes were elevated in the kangaroo rats from the oiled area (Kaplan *et al.* 1996). Concentrations of the other 23 PAH compounds measured were below the limits of detection (0.3 μg/kg; not specified whether expressed on a wet, dry, or lipid weight basis) in all samples analyzed.

12.4.2   TOXICITY

Various studies have demonstrated the acute toxicity and reproductive toxicity of PAHs to laboratory mammals (Environment Canada 1994; WHO 1998). Experimental studies in wild species have also demonstrated the potential toxic effects of PAHs on other endpoints, some of which may be particularly important for free-living mammals, e.g. food consumption was reduced by 2–30% in deer mice exposed to a diet containing 2% PAH compounds for 3 days (Schafer and Bowles 1985). Suppression of immune function, as measured by antibody formation, has also been observed in the same species, when animals were exposed to PAH concentrations a magnitude below the concentrations required for induction of EROD activity (Dickerson *et al.* 1994). The ED_{50}s for immunosuppression for methylcolanthrene, dibenz[a]anthracene and dimethylbenz[a]anthracene were 0.14, 0.048, and 0.026 mg/kg/day, respectively. The corresponding ED_{50} values for EROD induction were 1.24, 0.85, and > LD_{50} dose (value not stated), respectively.

In addition to clinical signs of toxicity, PAH-mediated toxicity is indicated by the presence of DNA adducts. High levels of adducts are associated with increased carcinogenicity. Although increased concentrations of adducts have been reported in wild mammals (Blondin and Viau 1992), there do not appear to be any studies that have linked adducts with the presence of tumors in free-living mammals or associated effects on long-term survival. It is doubtful if increased carcinogenicity is likely to have a significant impact on populations of free-living small mammals and other normally short-lived species.

There do not appear to be any published ecological studies investigating the specific effects of PAHs on wild mammal populations. However, Gashev (1992) investigated the effect of oil spills on small mammal abundance and diversity. Small mammal populations in lightly or moderately contaminated sites recovered to control levels in 8–9 years, while those in heavily contaminated sites took up to 14 years to reach 50% of control levels. The reproductive success of the red-backed vole (*Clethrionomys rutilus*) did not differ between control and contaminated sites, but the proportion of males and juveniles, the number of embryos per female, and the embryo resorption rate, were all higher in voles from the contaminated site. All effects were attributed to the oil contamination, either by direct toxicity, of which PAHs may have been a contributory factor, or indirect effects, such as a reduction of food availability.

12.5 AMPHIBIANS

There is relatively little information on how amphibians are affected by PAHs, but they appear to be similar to higher vertebrates in their physiological responses to PAH insult, e.g. the biochemical responses of newt (*Pleurodeles waltl*) larvae exposed to benzo[a]pyrene are similar to those described in higher vertebrates, viz. micronuclei formation, production of DNA adducts, and induction of liver EROD activity (Marty *et al*. 1998). However, amphibians may be at particular risk of synergistic interactions between PAH toxicity and other environmental stressors. Studies have apparently shown that the toxicity of PAHs to spotted salamander (*Ambystoma maculatum*), northern leopard frog (*Rana pipiens*) and the African clawed frog (*Xenopus laevis*) increased synergistically with the presence of UV light (Hatch and Burton 1998). The 96 h LC_{50} values for leopard frogs and clawed frogs exposed to fluoranthene in the presence of UV light were 366 and 193 μg/L, respectively. A 288 h LC_{50} value of 247 μg/L was reported for the salamander. However, the LC_{50} values for these species when not exposed to UV light were not specified, and so the true synergistic nature of PAH insult in UV exposure is difficult to evaluate.

Fluoranthene in particular has been identified as a potential problem to some amphibians, as it may pose a hazard to ranid larvae at concentrations well below the upper limit of the water solubility of the compound. Photoinduced toxicity, including necrosis and histopathalogical changes in the skin, has been reported in bullfrog larvae (*Rana catesbeiana*) exposed to 10 μg/l fluoranthene (Walker *et al*. 1998). Furthermore, Northern leopard frogs (*Rana pipiens*) exposed to fluoranthene concentrations of 2–10 μg/l accumulated sufficient fluoranthene within 48 h to be lethal when combined with UVA exposure at levels simulating exposure to the midday sun in northern latitudes (Monson *et al*. 1999).

12.6 CONCLUSION

Although there have been relatively few studies on exposure to and the effects of PAHs in terrestrial and freshwater vertebrates, it is clear that such exposure does occur. Laboratory and field studies indicate that although PAHs are lipophilic, they do not tend to be accumulated at high concentrations by wild vertebrates, as the compounds are rapidly metabolized. One of the main direct toxic effects in adult wild birds and mammals exposed to PAHs may, therefore, be the formation of adducts and the associated increased risk of carcinogenicity. This is only likely to have potential ecological significance in long-lived species, where long-term survival and, potentially, lifetime productivity, may be adversely affected. Whether exposure to PAHs can compromise the ability of adult wild birds and mammals to combat other environmental stressors, such as disease, is uncertain, although PAHs are known to an have immunosuppressive effects in vertebrates (WHO 1998). In contrast to higher vertebrates, there is some evidence that certain PAHs may be potentially toxic to adult (and larval) amphibians. However, examination of the available environmental data (WHO 1998) suggests that the aqueous concentrations of PAHs that caused toxic effects in the laboratory are greater than those commonly found in natural water bodies.

Laboratory studies have clearly demonstrated that specific PAHs and crude oils that are rich in PAHs have embryotoxic effects in birds. This is arguably one of the most likely ecologically significant effects that PAHs may exert on birds, although marine rather than freshwater and terrestrial species are likely to be at greater risk of becoming oiled and so exposed to PAHs. The concentrations of PAHs measured to date in the eggs of various free-living birds (Table 12.2) have generally been below the concentrations reported to cause embryotoxicity in laboratory studies (Lebedev et al. 1998; Shore et al. 1999). However, assessments of the potential embryotoxic effects of PAHs are based on single-compound tests, but simultaneous exposure to different PAHs can result in additive toxicity (Brunstrom 1992). Therefore, comparison of residues in eggs with experimental doses of single PAHs may under-emphasize the likelihood of adverse effects, although the embryotoxicity of PAH mixtures is poorly understood.

Given the paucity of information on the levels and effects of PAHs on the reproductive success in wild birds, further investigative studies are warranted. These should focus on key species that, because of their diet, behavior or ecology, are most at risk of exposure. Such studies need to both quantify the individual compounds to which adult birds and their embryos *in ovo* are exposed, and to employ techniques, such as CALUX assays (Murk et al. 1998), that give a measure of the toxic equivalence of mixtures of PAHs (and other Ah receptor-binding xenobiotics) to which birds may be simultaneously exposed.

REFERENCES

Anwer J and Mehrotra NK (1988) Teratogenic effects of benzo[a]pyrene in developing chick embryo. *Toxicology Letters*, **40**, 195–201.

Birkhead TR, Lloyd C and Corkhill P (1973) Oiled seabirds successfully cleaning their plumage. *British Birds*, **66**, 535–537.

Blondin O and Viau C (1992) Benzo[a]pyrene–blood protein adducts in wild woodchucks used as biological sentinels of environmental polycyclic aromatic hydrocarbons contamination. *Archives of Environmental Contamination and Toxicology*, **23**, 310–315.

Bosveld ATC, de Bie PAF and Weggemans J (1996) Effect of polycyclic aromatic hydrocarbons on hepatic EROD induction in experimentally dosed and environmentally exposed shrews (*Crocidura russula/Sorex araneus*). *Organohalogen Compounds*, **29**, 94-97.

Brunstrom B (1992) Embryolethality and induction of 7-ethoxyresorufin *O*-deethylase in chick embryos by polychlorinated biphenyls and polycyclic aromatic hydrocarbons having Ah receptor affinity. *Chemico-Biological Interactions*, **81**, 69-77.

Brunstrom B and Halldin K (1998) EROD induction by environmental contaminants in avian embryo livers. *Comparative Biochemistry and Physiology C-Pharmacology Toxicology & Endocrinology*, **121**, 213-219.

Brunstrom B, Broman D and Naf C (1990) Embryotoxicity of polycyclic aromatic hydrocarbons (PAHs) in three domestic avian species, and of PAHs and co-planar polychlorinated biphenyls (PCBs) in the common eider. *Environmental Pollution*, **67**, 133-143.

Couillard CM and Leighton FA (1989) Comparative pathology of Prudhoe Bay crude oil and inert shell sealants in chicken embryos. *Fundamental and Applied Toxicology*, **13**, 165-173.

Couillard CM and Leighton FA (1990) Sequential study of the pathology of Prudhoe Bay crude oil in chicken embryos. *Ecotoxicology and Environmental Safety*, **19**, 17-23.

Dickerson R, Hooper M, Gard N, Cobb G and Kendall R (1994) Toxicological foundations of ecological risk assessment: biomarker development and interpretation based upon laboratory and wildlife species. *Environmental Health Perspectives*, **102**, 65-69.

Environment Canada (1994) *Canadian Environmental Protection Act. Priority Substances List Assessment Report: Polycyclic Aromatic Hydrocarbons*. Ministry of Supply & Services, Ottawa, Canada, p. 61.

Fry DM, Swenson J, Addiego LA, Grau CR and Kang A (1986) Reduced reproduction of wedge-tailed shearwaters exposed to weathered Santa Barbara crude oil. *Archives of Environmental Contamination and Toxicology*, **15**, 453-463.

Gashev SN (1992) Effects of oilspills on the fauna and ecology of small mammals from the central Ob' Region. *Soviet Journal of Ecology*, **23**, 99-106.

Hallett DJ and Brecher RW (1984) Cycling of polynuclear aromatic hydrocarbons in the Great Lakes ecosystem. In Niagru JO and Simmons MS (eds), *Toxic Contaminants in the Great Lakes*. Wiley, New York, pp. 213-238.

Hatch AC and Burton GA (1998) Effects of photoinduced toxicity of fluoranthene on amphibian embryos and larvae. *Environmental Toxicology and Chemistry*, **17**, 1777-1785.

Hoffman DJ (1979) Embryotoxic and teratogenic affects of crude oils on mallard embryos on day one of development. *Bulletin of Environmental Contamination and Toxicology*, **22**, 632-637.

Hoffman DJ (1990) Embryotoxicity and teratogenicity of environmental contaminants to bird eggs. *Reviews of Environmental Contamination and Toxicology*, **115**, 39-89.

Hoffman DJ and Gay ML (1981) Embryotoxic effects of benzo[a]pyrene, chrysene, and 7,12-dimethylbenz[a]anthracene in petroleum hydrocarbon mixtures in mallard ducks. *Journal of Toxicology and Environmental Health*, **7**, 775-787.

Hough JL, Baird MB, Sfeir GT, Pacini CS, Darrow D and Wheelock C (1993) Benzo[a]pyrene enhances atherosclerosis in white carneau and show racer pigeons. *Arteriosclerosis and Thrombosis*, **13**, 1721-1727.

Jacob J and Grimmer G (1994) Environmental sample bank. PAH analysis in different matrices. In *Biochemical Institute for Environmental Carcinogens — Annual Report 1993*. Biochemical Institute for Environmental Carcinogens, Grosshansdorf, p. 50.

Kaplan I, Lu ST, Lee RP and Warrick G (1996) Polycyclic aromatic hydrocarbon biomarkers confirm selective incorporation of petroleum in soil and kangaroo rat liver samples near an oil well blowout site in the western San Joaquin Valley, California. *Environmental Toxicology and Chemistry*, **15**, 696-707.

Lebedev AT, Poliakova OV, Karakhanova NK, Petrosyan VS and Renzoni A (1998) The contamination of birds with organic pollutants in the Lake Baikal region. *Science of the Total Environment*, **212**, 153–162.

Lee YZ, Obrien PJ, Payne JF and Rahimtula AD (1986) Toxicity of petroleum crude oils and their effect on xenobiotic metabolizing enzyme activities in the chicken embryo *in ovo*. *Environmental Research*, **39**, 153–163.

Leupold D, Jacob J, Raab C, Grimmer C and Becker W (1992) Induction of cytochrome P450-dependent mono-oxygenases in the liver of wood mice (*Apodemus sylvaticus*), its influence on the metabolite profiles of PAH and implications for monitoring of environmental xenobiotics. *Fresenius Journal of Analytical Chemistry*, **343**, 149–150.

Lewis SJ and Malecki RA (1984) Effects of egg oiling on larid productivity and population dynamics. *Auk*, **101**, 584–592.

Lu PY, Metcalf RL, Plummer N and Mandel D (1977) The environmental fate of three carcinogens: benzo[a]pyrene, benzidine, and vinyl chloride evaluated in laboratory model ecosystems. *Archives of Environmental Contamination and Toxicology*, **6**, 129–142.

Ma WC, Immerzeel J and Bodt J (1995) Earthworm and food interactions on bioaccumulation and disappearance in soil of polycyclic aromatic hydrocarbons: studies on phenanthrene and fluoranthene. *Ecotoxicology and Environmental Safety*, **32**, 226–232.

Ma WC, van Kleunen A, Immerzeel J and de Maagd PGJ (1998) Bioaccumulation of polycyclic aromatic hydrocarbons by earthworms: assessment of equilibrium partitioning theory in *in situ* studies and water experiments. *Environmental Toxicology and Chemistry*, **17**, 1730–1737.

Marty J, Djomo JE, Bekaert C and Pfohl-Leszkowicz A (1998) Relationships between formation of micronuclei and DNA adducts and EROD activity in newts following exposure to benzo[a]pyrene. *Environmental and Molecular Mutagenesis*, **32**, 397–405.

Matsumoto H (1988) Toxicity to chicken embryos of organic extracts from airborne particulates separated into five sizes. *Bulletin of Environmental Contamination and Toxicology*, **41**, 44–49.

Matsumoto H and Kashimoto T (1986) Embryotoxicity of organic extracts from airborne particulates in ambient air in the chicken embryo. *Archives of Environmental Contamination and Toxicology*, **15**, 447–452.

Monson PD, Call DJ, Cox DA, Liber K and Ankley GT (1999) Photoinduced toxicity of fluoranthene to northern leopard frogs (*Rana pipiens*). *Environmental Toxicology and Chemistry*, **18**, 308–312.

Murk AJ, Leonards PEG, van Hattum B, Luit R, van der Weiden MEJ and Smit M (1998) Application of biomarkers for exposure and effect of polyhalogenated aromatic hydrocarbons in naturally exposed European otters (*Lutra lutra*). *Environmental Toxicology and Pharmacology*, **6**, 91–102.

Naf C, Broman D and Brunstrom B (1992) Distribution and metabolism of polycyclic aromatic hydrocarbons (PAHs) injected into eggs of chicken (*Gallus domesticus*) and common eider duck (*Somateria mollissima*). *Environmental Toxicology and Chemistry*, **11**, 1653–1660.

Peakall DB, Hallett DJ, Bend JR, Foureman GL and Miller DS (1982) Toxicology of Prudhoe Bay crude oil and its aromatic fractions to nestling herring gulls. *Environmental Research*, **27**, 206–215.

Schafer EW and Bowles WA (1985) Acute oral toxicity and repellency of 933 chemicals to house and deer mice. *Archives of Environmental Contamination and Toxicology*, **14**, 111–129.

Schafer EW, Bowles WA and Hurlbut J (1983) The acute oral toxicity, repellency and hazard potential of 998 chemicals to one or more species of wild and domestic birds. *Archives of Environmental Contamination and Toxicology*, **12**, 355–382.

Sheffield SR, Sawicka-Kapusta K, Cohen JB and Rattner BA (2001) Rodentia and Lagomorpha. In Shore RF and Rattner BA (eds), *Ecotoxicology of Wild Mammals*. Wiley, Chichester, pp. 215–314.

Shore RF, Wright J, Horne JA and Sparks TH (1999) Polycyclic aromatic hydrocarbon (PAH) residues in the eggs of coastal-nesting birds from Britain. *Marine Pollution Bulletin*, **38**, 509–513.

Stronkhorst J, Ysebaert TJ, Smedes F, Meininger PL and Dirksen S (1993) Contaminants in eggs of some waterbird species from Scheldt Estuary, SW Netherlands. *Marine Pollution Bulletin*, **26**, 572–578.

van Brummelen T (1995) *Distribution and Ecotoxicology of PAHs in Forest Soil*. PhD Thesis, University of Amsterdam, p. 175.

van Brummelen T, Verweij RA, Wedzinga SA and van Gestel CAM (1996) Polycyclic aromatic hydrocarbons in earthworms and isopods from contaminated forest soils. *Chemosphere*, **32**, 315–341.

Walker SE, Taylor DH and Oris JT (1998) Behavioral and histopathological effects of fluoranthene on bullfrog larvae (*Rana catesbiana*). *Environmental Toxicology and Chemistry*, **17**, 734–739.

Walters P, Khan S, Obrien PJ, Payne JF and Rahimtula AD (1987) Effectiveness of a Prudhoe Bay crude oil and its aliphatic, aromatic and heterocyclic fractions in inducing mortality and aryl-hydrocarbon hydroxylase in chick embryo *in ovo*. *Archives of Toxicology*, **60**, 454–459.

WHO (1998) *Environmental Health Criteria 202: Selected Non-heterocyclic Polycyclic Aromatic Hydrocarbons*. World Health Organization, Geneva, p. 883.

13

Effects of PAHs on Marine Birds, Mammals and Reptiles

PETER H. ALBERS[1] AND THOMAS R. LOUGHLIN[2]
[1]*USGS Patuxent Wildlife Research Center, Laurel, MD, USA*
[2]*National Marine Mammal Laboratory, Alaska Fisheries Science Center, Seattle, WA, USA*

13.1 INTRODUCTION

Polycyclic aromatic hydrocarbons (PAHs) are found in all abiotic and biotic components of the marine environment. True PAHs have two to seven benzene rings and are composed solely of carbon and hydrogen, with alternating single and double bonds between carbons in the ring structure. In their reports of PAH presence or biological effects, investigators sometimes group true PAHs with compounds consisting of aromatic and non-aromatic rings, or compounds with N, S, or O within the ring (heterocycle) or substituted for an attached hydrogen. Descriptions of the sources, fate, and chemical characteristics of PAHs are presented in other chapters of this book and in a recent review of PAHs by Eisler (2000); treatment of these subjects will be kept to a minimum in this chapter.

Assessments of marine pollution often include PAHs in the mixture of substances, usually chlorinated pesticides, polychlorinated biphenyls (PCBs), dioxins, furans, and metals, which are considered hazardous to plant or animal life. However, a recent workshop on marine mammals and persistent ocean contaminants concluded that PAHs were a 'less widely recognized' marine contaminant (Marine Mammal Commission 1999). The quantity of published literature on effects of petroleum on marine mammals and birds (not reptiles) is large, but the emphasis is on combined physical and toxic effects of whole crude or refined petroleum, rather than just the toxic effects of specific fractions. Because petroleum is a major environmental source of the two- to six-ring PAHs, selective reports of the toxic effects of petroleum will be incorporated into subsequent sections.

PAHs: An Ecotoxicological Perspective. Edited by Peter E.T. Douben.
© 2003 John Wiley & Sons Ltd

The objectives of this chapter are to: (a) describe the exposure of marine mammals, birds, and reptiles to PAHs; and (b) evaluate what is known about the biological consequences of such exposure.

13.2 EXPOSURE, METABOLISM, AND TOXICITY

Sources of PAHs in the marine environment are numerous and have been covered in Chapters 2 and 3. PAHs are subjected to physical, chemical, and biological degradation (see also Chapter 6); the large PAHs (four or more benzene rings) are most resistant to microbial degradation and are most likely to settle into sediments (Colwell and Walker 1977; Neff 1985; James 1989). These substances also are ingested by macroinvertebrates and vertebrates.

Concentrations of PAHs in water, sediment, or tissue are reported for individual PAHs, subsets of PAHs, or total PAHs; and sometimes as crude oil or chrysene equivalents. Sometimes it is unclear what is meant by 'total PAHs'. Commonly, groups of 10–40 parent PAHs or their alkylated homologs are quantified and used as the basis for a 'subset' or 'total' PAH characterization.

Concentrations of PAHs in marine and estuarine waters are not as well documented as PAHs in fresh waters (see also Chapters 1 and 8). The mean and range of total particulate PAHs for marine (11; 2–25 ng/L) and estuarine (170; 33–618 ng/L) waters off the mouth of the River Seine (435; 158–687 ng/L) illustrate the diluting effects of oceans on pollutants in a river draining highly populated areas (Fernandes *et al.* 1997). In contrast, the "mixing zone" of the Rivers Ob and Yenisei and the Kara Sea within the Arctic Circle in Russia had ≤ 0.1 ng/L total particulate PAHs in marine water and ≤ 0.9 ng/L total particulate PAHs in mixing zone water (Fernandes and Sicre 1999).

Concentrations of PAHs in slightly contaminated marine and estuarine sediments would be < 1.0 μg/g (ppm) (e.g. Fernandes and Sicre 1999; Savinov *et al.* 2000; Yang 2000) and concentrations in highly contaminated sediment could exceed 50 μg/g (Roberts *et al.* 1989; Eisler 2000).

Information on tissue concentrations for marine mammals, birds, and reptiles is limited compared to fish and invertebrates. Based on published reports for all marine vertebrates, including fish, concentrations of 'total' PAHs in the range of 1–4 mg/kg wet weight (ww) in liver, kidney, brain, or muscle are possible for mammals, birds, and reptiles inhabiting moderately to highly contaminated environments (Mulcahy and Ballachy 1994; Eisler 2000).

Mammals, birds, reptiles, fish, and many macroinvertebrates (crustaceans, polychaetes, echinoderms, insects) have well-developed mixed-function oxidase (MFO) systems which enable them to efficiently metabolize and excrete some of the hydrocarbons ingested during feeding, grooming, and respiration (Lee 1977; McEwan and Whitehead 1980; Engelhardt 1983; James 1989; Rattner *et al.* 1989; Jenssen *et al.* 1990; Eisler 2000). As with microbes, large aromatic hydrocarbons are the most difficult group of hydrocarbons to excrete, regardless of MFO capability (Varanasi *et al.* 1989; Eisler 2000). Accumulation of PAHs by mammals, birds, and reptiles is mostly associated with body lipid content and activity patterns or distributions that coincide with high concentrations of PAHs (McElroy *et al.* 1989; Eisler 2000). Trophic level increases in accumulation have not been observed in aquatic ecosystems (McElroy *et al.* 1989; Broman *et al.* 1990).

The mechanism of toxicity for PAHs seems to be interference with cellular membrane function and enzyme systems associated with the membrane (Neff 1985). Although unmetabolized PAHs can have toxic effects, a major concern in animals is the ability of reactive metabolites, such as epoxides and dihydrodiols, of some PAHs to bind to cellular proteins and DNA. The resulting biochemical disruptions and cell damage lead to mutations, developmental malformations, tumors, and cancer (Santodonato *et al.* 1981; Varanasi 1989; Eisler 2000). Four-, five-, and six-ring PAHs have greater carcinogenic potential than the two-, three-, and seven-ring PAHs (Neff 1985; Eisler 2000; see also Chapter 5). The addition of alkyl groups to the base PAH structure often produces carcinogenicity or enhances existing carcinogenic activity (e.g. 7,12-dimethylbenzo[a]anthracene). Some halogenated PAHs are mutagenic without metabolic activation (Fu *et al.* 1999) and the toxicity and, possibly, the carcinogenicity of PAHs can be increased by exposure to solar ultraviolet radiation (Ren *et al.* 1994; Arfsten *et al.* 1996). Cancerous and precancerous neoplasms have been induced in aquatic organisms in laboratory studies, and cancerous and non-cancerous neoplasms have been found in demersal fish from heavily polluted sites (Neff 1985; Baumann 1989; Chang *et al.* 1998; Eisler 2000). However, sensitivity to PAH-induced carcinogenesis differs substantially among animals (Neff 1985; Eisler 2000).

In water, the toxicity of individual PAHs increases as molecular weight (MW) increases up to MW 202 (fluoranthene, pyrene). Beyond MW 202, a rapid decline in solubility reduces PAH concentrations to less than lethal levels, regardless of their intrinsic toxicity. However, sublethal effects can result from exposure to these very low concentrations of high MW compounds (Neff 1985). Except for the vicinity of chemical or petroleum spills, environmental concentrations of PAHs in water are usually several orders of magnitude below levels that are acutely toxic to aquatic organisms. Sediment PAH concentrations can be much higher than water concentrations but the limited bioavailability of these PAHs greatly reduces their toxic potential (Eisler 2000). In general, caution should be employed when assessing the aquatic 'toxicity' of biogenic or anthropogenic PAHs because bioavailability (solubility, sediment sequestration, mechanism of exposure) and chemical modification determine how much toxicity is realized.

13.3 MAMMALS

13.3.1 GENERAL

Marine mammals have been a subject of international concern since the virus-induced epizootics affecting seals and dolphins in the Atlantic Ocean and

Mediterranean Sea during the late 1980s and early 1990s (Ross *et al.* 1996) and the *Exxon Valdez* oil spill in Alaska, USA in 1989 (Loughlin 1994a). Marine mammal strandings in Europe and North America are promptly investigated by scientists and often covered by media representatives. Internet websites devoted to reports of stranded mammals and the results of subsequent investigations, e.g. Blaylock (2002) and Cowan (2002), have been established.

During the last decade, research attention has been focused on organochlorine and metal contaminants as possible causes of deteriorated health of marine mammals (Reijnders and de Ruiter-Dijkman 1995; Lee *et al.* 1996; Marine Mammal Commission 1999; Muir *et al.* 1999). Some investigators have reported evidence to implicate these contaminants as a cause of health problems, e.g. impaired immunological responses in harbor seals (*Phoca vitulina*) caused by halogenated aromatic hydrocarbons in the diet (De Swart *et al.* 1994; Ross *et al.* 1996), and a proposed biochemical mechanism for reproductive failure and respiratory inflammation (Troisi *et al.* 2001). Other investigations have cast doubt on the causal nature of contaminants, e.g. the absence of correlations between concentrations of contaminants and epidermal diseases in bottlenose dolphins (*Tursiops truncatus*), and the concurrent correlations between the diseases and oceanographic conditions (Wilson *et al.* 1999). Although PAHs are ubiquitous in the environment, their effects on marine mammal health are less studied than the effects of the more persistent organochlorines and metals (Marine Mammal Commission 1999).

13.3.2   EFFECTS OF PAHs

The metabolism and effects of some PAHs have been well documented in laboratory rodents and domestic mammals but poorly documented in wild mammals (Eisler 2000). Target organs for PAH toxic action are skin, small intestine, kidney, and mammary gland; tissues of the hematopoietic, lymphoid, and immune systems; and gametic tissue. Non-alkylated PAHs are rapidly metabolized, hence accumulation is less likely than for alkylated PAHs (Eisler 2000). Because PAH toxicity occurs at levels that can also induce DNA alterations and cancer, concerns about the carcinogenic and mutagenic potential predominate. Consequently, PAHs that are carcinogenic or mutagenic are most studied (Eisler 2000; Ingram *et al.* 2000). Partially aromatic PAHs, alkylated fully- and partially-aromatic PAHs, and metabolites of non-alkylated fully-aromatic PAHs have enhanced potential to induce cancerous and non-cancerous neoplasms in most of the epithelial tissues of laboratory and domestic mammals. Species differences in sensitivity to carcinogenesis appear to be largely a function of differences in levels of MFO activities (Eisler 2000).

13.3.3   FIELD STUDIES

The best-studied marine mammal, with regard to PAHs and health, is the beluga whale (*Delphinapterus leucas*) of Canada (Table 13.1). During an 8 year period

TABLE 13.1 Reports of PAHs or their DNA adducts in marine mammals subjected to chronic or acute environmental exposure to PAHs. Sample sizes are in parentheses

Species	PAH Occurrence	Location	Reference[f]
Chronic exposure			
Sperm whale (*Physeter catodon*)	Blubber (4); 15 PAHs; $\mu = 99$ ng/g (ppb) dw[a]	Belgium	1
Beluga whale	Liver (1); B[a]P[b]; 0/1[c] Kidney (1); B[a]P; 0/1 Blubber (1); B[a]P; 0/1 Lung (1); B[a]P; 0/1	Canada (Gulf of St. Lawrence)	2
Beluga whale	Liver (5); 16 PAHs; range = 111–303 ng/g ww Brain (5); 16 PAHs; 49–222 ng/g Muscle (5); 16 PAHs; 70–230 ng/g	Canada (Gulf of St. Lawrence)	3
Beluga whale	Liver (16); PAH adducts; 15/16 Kidney (4); PAH adducts; 4/4 Brain (12); PAH adducts; 9/12	Canada (Arctic)	4
Beluga whale	Liver (18); aromatic DNA adducts; 4/4 6/6 8/8	Canada (Gulf of St. Lawrence) (MacKenzie Delta) (East Hudson Bay)	5
Beluga whale	Liver (9); B[a]P DNA adducts; 6/9 Brain (11); DNA adducts; 10/11	Canada (St. Lawrence Estuary)	6
Beluga whale	Liver (4); B[a]P DNA adducts; 0/4 Brain (4); DNA adducts; 0/4 Brain (3); DNA adducts; 3/3	Canada (MacKenzie Estuary) (Gulf of St. Lawrence)	7
Beluga whale	Brain (3); B[a]P DNA adducts; 3/3	Canada (Gulf of St. Lawrence)	8
Harp seal (*Pagophilus groenlandicus*)	Muscle (10); 16 PAHs; median[d] = nd[e]–4.4 ng/g ww Liver (10); 16 PAHs; nd–4.7 ng/g Kidney (10); 16 PAHs; nd–4.3 ng/g Blubber (10); 16 PAHs; nd–23.5 ng/g	Canada (Labrador)	9
Harp seal	Muscle (28); total PAH; range[d] = nd–830 ng/g dw chrysene equivalents	Canada (Labrador, Newfoundland)	10

(continued overleaf)

TABLE 13.1 (*continued*)

Species	PAH Occurrence	Location	Reference[f]
Harbor seal	Liver (15); >3 PAHs; range = nd–<30 ng/g ww Blubber (15); >3 PAHs; nd–<30 ng/g	USA (New York, Massachusetts)	11
Harbor porpoise	Muscle (26); 9 PAHs; μ range = 0.6–8.8 ng/g ww; μ total PAHs = 32.4 mg/g	UK	12
Seal (four spp.) Whale (three spp.) Dolphin (two spp.) Porpoise (one spp.)	Muscle (21); total PAH; range = 1100–1210 ng/g dw chrysene equivalents	Canada (Labrador, Newfoundland)	13
Acute exposure			
Harbor seal	Muscle (10); 39 PAHs; μ = 2 ng/g ww Liver (19); 39 PAHs; 6 ng/g Kidney (4); 39 PAHs; nd Brain (11); 39 PAHs; 1 ng/g Blubber (17); 39 PAHs; 191 ng/g Heart (3); 39 PAHs; nd Lung (3); 39 PAHs; nd Blood (4); 39 PAHs; nd	USA (Prince William Sound, Alaska)	14
Sea otter	Muscle, liver, kidney, brain, intestine (10); 39 PAHs; μ = 1780 ng/g ww across all tissues	USA (Prince William Sound, Alaska)	15
Sea otter	Liver (18); total PAH; μ = 95 ng/g ww (est.) Lung; total PAH: severe congestion (5); μ = 21 ng/g; mild or moderate congestion (9); μ = 4 ng/g	USA (Prince William Sound, Alaska)	16
Minke whale	Liver (1); 39 PAHs; 202 ng/g ww Blubber (1); 39 PAHs; 105 ng/g	USA (Prince William Sound, Alaska)	17
Harbor porpoise	Liver (3); 39 PAHs; μ = 118.3 ng/g ww Blubber (2); 39 PAHs; μ = 297.5 ng/g	USA (Prince William Sound, Alaska)	17

[a] dw, dry weight, ww, wet weight.
[b] B[a]P, benzo[a]pyrene.
[c] Frequency of detection.
[d] Males and females combined.
[e] nd, not detected.
[f] 1, Holsbeek *et al.* 1999; 2, Martineau *et al.* 1985; 3, Beland *et al.* 1991; 4, Mathieu *et al.* 1997; 5, Ray *et al.* 1991; 6, Martineau *et al.* 1994; 7, Shugart 1990; 8, Martineau *et al.* 1988; 9, Zitko *et al.* 1998; 10, Hellou *et al.* 1991; 11, Lake *et al.* 1995; 12, Law and Whinnett, 1992; 13, Hellou *et al.* 1990; 14, Frost *et al.* 1994; 15, Mulcahy and Ballachey, 1994; 16, Williams *et al.* 1995; 17, Loughlin, 1994b.

(1983 – 1990) a total of 45 whales found dead along the shores of the St. Lawrence Estuary were aged and necropsied (Beland *et al.* 1993; Martineau *et al.* 1994). The beluga whales were afflicted with many more pathological abnormalities than a reference collection of five Arctic beluga whales (shot), and 13 cetaceans and 17 pinnipeds found dead (approximately half from trauma) in the St. Lawrence Estuary. Although a sample of organisms found dead would be expected to have more abnormalities than a sample wherein half of the organisms died from trauma, the difference between samples was profound (Beland *et al.* 1993). The incidence of neoplasms in all beluga whales was 40% (Beland *et al.* 1993) and malignant neoplasms were found in nine of 38 mature whales, an incidence of 24% (Martineau *et al.* 1994). The calculated cancer rate for the whale population of approximately 500 animals was greater than for man and all domestic animals except the dog (Martineau *et al.* 1994). Parent PAHs were sought, unsuccessfully, in tissue of one whale by Martineau *et al.* (1985); but Beland *et al.* (1991) found total PAHs (16) of 111 – 303 ng/g in liver, 49 – 222 ng/g in brain, and 70 – 230 ng/g in muscle. Mostly, investigators have measured the quantity of PAH or aromatic metabolites as DNA adducts in liver, brain, and kidney tissue. The presence of benzo[a]pyrene (B[a]P) adducts in St. Lawrence beluga whales (Martineau *et al.* 1988, 1994), but not in four Arctic beluga whales (Shugart 1990), implicated PAHs as a possible causal agent for the health problems of St. Lawrence whales. However, other reports have shown that PAH adducts occur in detectable quantities in Arctic beluga whales (Mathieu *et al.* 1997) and uncharacterized aromatic adducts occur in similar concentrations in the Arctic and the St. Lawrence Estuary (Ray *et al.* 1991). The presence of substantial amounts of other contaminants (PCBs, chlorinated pesticides, metals) in the St. Lawrence Estuary and in tissues of the beluga whale (Beland *et al.* 1993; Martineau *et al.* 1994) further complicates efforts to link the high incidence of neoplasms to a particular class of chemical pollutants.

In contrast to the chronic nature of PAH exposure for beluga whales, marine mammals were acutely exposed to PAHs during and after the *Exxon Valdez* oil spill in Prince William Sound, Alaska, in March 1989 (Table 13.1). Oiled sea otters (*Enhydra lutris*) that were found dead in early April 1989 contained two to eight times as much petroleum hydrocarbons as otters collected from unoiled areas of Prince William Sound. Necropsies performed a year after collection revealed a suite of lesions characteristic of exposure to petroleum, but the role of PAHs in the production of the lesions was not determined (Mulcahy and Ballachey 1994). Necropsies performed on sea otters that died in rehabilitation centers revealed that death was primarily due to hypothermia, shock, and stress; toxicity was a secondary factor (Williams *et al.* 1995). Concentrations of PAHs and saturated hydrocarbons in liver, kidney, and lung were poorly correlated with the degree of oiling and were correlated with the amount of tissue damage only in the lung. The liver contained the highest concentrations of PAHs, averaging approximately 95 ng/g wet weight and ranging up to 235 ng/g.

Harbor seals were collected from oiled areas of Prince William Sound from April through July 1989, as well as from the Gulf of Alaska, unoiled areas of Prince William Sound, and a reference area during 1989–1990 (Frost *et al.* 1994). Most of the seals were alive at the time of collection. Tissue concentrations of PAHs were much lower than for sea otters collected immediately after the spill. Fluorescent bile metabolites of PAHs, as phenanthrene equivalents, were 20 times higher in oiled areas of Prince William Sound than in unoiled areas and 70 times higher than in a reference area. A high incidence of brain lesions was detected in seals from oiled areas of Prince William Sound and the Gulf of Alaska, but the role of PAHs in lesion production was not determined (Frost *et al.* 1994).

A minke whale (*Balaenoptera acutorostrata*) and three harbor porpoises (*Phocoena phocoena*) found dead in Prince William Sound in October and June 1989, respectively, had concentrations of PAHs in liver and blubber that were intermediate between those of the sea otters and harbor seals (Loughlin 1994b). The cause of death could not be determined for these cetaceans nor for any of a number of other dead whales and harbor porpoises found in the path of crude oil from the *Exxon Valdez*.

The amount of PAHs in an acutely exposed animal is a function of the magnitude, duration, and mode of exposure; the chemical composition and pattern of degradation of the spilled petroleum; and the metabolic removal of the PAHs during and after the exposure event. Also, the toxic effects of PAHs from petroleum are difficult to separate from the toxic effects of other petroleum components or the physiological consequences of fur and epidermal fouling by the whole crude or refined oil (Williams *et al.* 1995). Consequently, concentrations of PAHs in tissue have limited usefulness as indices of acute PAH exposure. Frost *et al.* (1994) and Mulcahy and Ballachey (1994) suggested a PAH evaluation procedure consisting of a combination of tissue analysis for PAHs, bile analysis for PAH metabolites, and a pathological assessment.

In addition to the previously-described field studies of characterized PAH contamination, intentional or opportunistic baseline and trend investigations of PAHs in tissues of collected or dead marine mammals have been performed in North America and Europe. Polycyclic aromatic hydrocarbons sometimes are the focus of the investigation (Hellou *et al.* 1990, 1991; Law and Whinnett 1992), but at other times they are part of a suite of contaminants being studied (Lake *et al.* 1995; Zitko *et al.* 1998; Holsbeek *et al.* 1999).

13.4 BIRDS

13.4.1 GENERAL

The effect on marine birds of pollution has been studied through field and laboratory investigations for many years. Unlike marine mammals, diseases that could be linked to anthropogenic pollutants have not caused large numbers of deaths. Organochlorine and metal concentrations are frequently measured in the soft tissues, eggs, and feathers of birds (e.g. Donaldson *et al.* 1997; Debacker *et al.* 1997; Olafsdottir *et al.* 1998; Anthony *et al.* 1999; Muir *et al.* 1999; Burger and Gochfeld 2000); but PAHs are seldom the primary objective of chemical analyses and are usually a 'secondary' class of contaminants when multiple contaminants are studied. Large petroleum discharges, such as the *Amoco Cadiz* (1978), *Exxon Valdez* (1989), and Gulf War (1991) oil spills, have generated considerable scientific and public concern about the plight of marine birds.

13.4.2   EFFECTS OF PAHs ON EMBRYOS

The significant toxicity of crude and refined petroleum to the embryos of marine birds has been demonstrated by field experiments involving petroleum applications to the shells of incubated eggs (Hoffman 1990). Laboratory experiments have shown that the PAH fraction of crude and refined oils is responsible for the lethal and sublethal effects on mallard (*Anas platyrhynchos*) and chicken (*Gallus domesticus*) embryos caused by eggshell oiling (Hoffman 1990). Further, the toxicity of PAHs to bird embryos is a function of the quantity and molecular structure of the PAHs (Matsumoto and Kashimoto 1986). Brunstrom *et al.* (1990) injected a mixture of 18 PAHs into eggs of the chicken, turkey (*Meleagris gallopavo*), domestic mallard, and common eider (*Somateria mollissima*), and found the mallard to be the most sensitive species and benzo[k]fluoranthene (four rings) and indeno[1,2,3-cd]pyrene (five rings) to be the most toxic of the PAHs tested. The most toxic PAHs were found to have additive effects on death of embryos and the cause of toxicity was proposed to be a mechanism controlled by the Ah receptor (Brunstrom 1991). Naf *et al.* (1992) injected a mixture of 16 PAHs into chicken and common eider eggs and reported that > 90% was metabolized by day 18 of incubation (chicken), with the greatest concentration of PAHs in the gallbladder of both species. Mayura *et al.* (1999) injected fractionated PAH mixtures from coal tar into the yolk of chicken eggs and found that death, liver lesions and discoloration, and edema increased as the proportion of five- and over five-ring aromatics, compared to the proportion of two- to four-ring aromatics, increased in the mixture.

13.4.3   EFFECTS OF PAHs ON NESTLINGS AND ADULTS

Adult and nestling marine, non-marine, and domestic birds have been dosed, fed, or externally oiled in a variety of field and laboratory experiments designed to evaluate the effects of petroleum on physiology, growth, reproduction, and behavior (Albers 2003). However, only a few of these investigations have attempted to determine the role of PAHs in the causation of adverse effects. Male mallard ducks fed a mixture of 10 PAHs for 7 months in a chronic ingestion study had greater hepatic stress responses and higher testis weights than male mallards fed a mixture of 10 alkanes (Patton and Dieter 1980). Nestling herring gulls (*Larus argentatus*) were administered single doses of crude oils or their aromatic or aliphatic fractions (Miller *et al.* 1982; Peakall *et al.* 1982). Retardation of nestling weight gain and increased adrenal and nasal gland weights was attributed to the PAHs with four or more rings. Immune function and MFO activity of European starlings (*Sturnus vulgaris*) were altered by oral or subcutaneous doses of 7,12-dimethylbenzo[a]anthracene, a four-ring PAH (Trust *et al.* 1994). The coefficient of variation of nuclear DNA volume of lesser scaup (*Athya affinis*) red blood cells was positively correlated with the concentration of total (41) PAHs in scaup carcasses from a highly polluted canal in East Chicago, Indiana, USA (Custer *et al.* 2000).

13.4.4 FIELD INVESTIGATIONS

Reports of concentrations of PAHs in tissues of wild marine birds or studies of the hazard of environmental PAHs to marine birds are less common than studies of the effects of petroleum (Table 13.2). The geometric mean concentrations (1–4 ng/g ww) of phenanthrene, fluoranthene, and pyrene in herring gull eggs from the 'polluted' Scheldt Estuary in The Netherlands (Stronkhorst *et al.* 1993) were approximately an order of magnitude more than concentrations of the same PAHs in eggs of the herring gull, cormorant (*Phalacrocorax carbo*), shag (*Phalacrocorax aristotelis*), and chough (*Pyrrhocorax pyrrhocorax*) collected from several 'background' sites around the UK (Shore *et al.* 1999). In both instances, the concentrations were less than the approximate eggshell or injection doses of individual PAHs required to induce toxicity in chicken and mallard embryos. Significant differences in PAH accumulation were found among the four UK species. Broman *et al.* (1990) described the distribution, biotransformation, and flux of 19 PAHs in seston (microbes, phytoplankton, zooplankton), blue mussels (*Mytilus edulis* L.), and common eiders. Concentrations of PAHs decreased from seston to mussels to eiders — an indication of trophic-level increases in PAH metabolism. Also, concentrations of PAHs in the eiders were higher in the gallbladder than in fat and liver. Kayal and Connell (1995) reported 75 ng/g and 85 ng/g ww, respectively, of nine PAHs in pooled muscle tissue from Australian pelicans (*Pelecanus conspicillatus*) and silver gulls (*Larus novaehollandiae*) collected in an estuary downstream from Brisbane, Australia. Concentrations of PAHs in crabs and fish were greater than in the birds, hence trophic accumulation was not occurring. Late-winter cormorants (*Phalacrocorax auritus*) from the polluted Houston Ship Canal in Galveston Bay, Texas, USA, had mean carcass concentrations for eight individual PAHs of 20–590 ng/g ww (1080 ng/g for all eight) (King *et al.* 1987). In

TABLE 13.2 Reports of PAHs in eggs or tissues of marine birds subjected to chronic environmental exposure to PAHs. Sample sizes are in parentheses

Species	PAH Occurrence	Location	Reference[f]
Eggs			
Common tern	Phenanthrene (6); $\mu = 4$ ng/g ww[a] Fluoranthene (6); 1 ng/g Pyrene (6); 3 ng/g	The Netherlands (Scheldt Estuary)	1[g]
Herring gull	16 PAHs (15); median = 29 ng/g lipid	UK	2
Cormorant	16 PAHs (5); 71 ng/g		
Shag	16 PAHs (4); 228.1 ng/g		
Chough	16 PAHs (6); 30.6 ng/g		
Tissue			
Seston	19 PAHs (10); $\mu = 5488$ ng/g dwo	Sweden (Baltic Sea)	3
Blue mussel	19 PAHs (6); 480 ng/g dwo		
Common eider gallbladder	19 PAHs (3); 148 ng/g dw		
Australian pelican	Muscle[b]; 9 PAHs (3); $\mu = 75$ ng/g ww	Australia (Brisbane Estuary)	4
Silver gull	Muscle[c]; 9 PAHs (3); 85 ng/g		
Cormorant	Carcass[d] (10); 8 PAHs; $\mu = 1080$ ng/g ww	USA (Houston Ship Channel, Galveston Bay, Texas)	5
Redhead	Carcass[e] (15); 7 PAHs; $\mu = 28$ ng/g ww	USA (Laguna Madre, Texas and Chandeleur Islands, Louisiana)	6
Canvasback	Skin and fat (42); fluorene; $\mu = 10$ ng/g ww Skin and fat (42); naphthalene; 18 ng/g Skin and fat (42); phenanthrene; 22 ng/g	USA (San Francisco Bay, California)	7
Common eider	Carcass[f] (1); 16 PAHs; 4.5 ng/g ww	Sweden (Baltic Sea)	8

[a]dw, dry weight; ww, wet weight; dwo, dry weight organic.
[b]Composite of two birds.
[c]Composite of five birds.
[d]Minus feathers, skin, bill, feet, and gastrointestinal tract.
[e]Minus feathers, feet, wings, gastrointestinal tract, head, and liver.
[f]Composite of five carcasses minus feathers, bones, feet, head, and intestinal contents.
[g]1, Stronkhorst *et al*. 1993; 2, Shore *et al*. 1999; 3, Broman *et al*. 1990; 4, Kayal and Connell 1995; 5, King *et al*. 1987; 6, Michot *et al*. 1994; 7, Miles and Ohlendorf 1993; 8, Naf *et al*. 1992.

contrast, late-winter male redhead ducks (*Aythya americana*) from the Laguna Madre and the Chandeleur Islands (west and east of Galveston Bay) had mean carcass concentrations for seven individual PAHs of < 10–40 ng/g ww (28 ng/g for all seven) (Michot *et al.* 1994). Also, Miles and Ohlendorf (1993) reported mean concentrations of 10, 18, and 22 ng/g ww for fluorene, naphthalene, and phenanthrene, respectively, in the skin and fat of wintering canvasbacks (*Aythya valisineria*) in San Francisco Bay, California, USA; and a composite sample of juvenile common eider carcasses from the Baltic Sea contained 4.5 ng/g ww of 16 PAHs (Naf *et al.* 1992).

13.5 REPTILES

13.5.1 GENERAL

Compared to mammals and birds, marine reptiles are poorly studied with regard to environmental contaminants or disease. Responses of reptiles to petroleum or PAH exposure are not well characterized, and most of the available information is for turtles. Marine turtles have been a focus of international attention during the last 20 years; concerns about organochlorine and petroleum pollution, plastic debris, and an increasing incidence of fibropapillomas are most often cited (Hutchinson and Simmonds 1992; Herbst 1994; Landsberg *et al.* 1999).

13.5.2 EFFECTS OF PAHs ON EMBRYOS

Atlantic Ridley (*Lepidochelys kempi*) and loggerhead (*Caretta caretta*) embryos died or developed abnormally when the eggs were exposed to oiled sand; weathered oil was less harmful to the embryos than fresh oil (T.H. Fritts and M.A. McGehee, Fish and Wildlife Service, Denver Wildlife Research Center, Denver, Colorado, USA, unpublished report).

13.5.3 EFFECTS OF PAHs ON JUVENILES AND ADULTS

Deaths of reptiles associated with acute or chronic exposure to petroleum in fresh or salt water have occasionally been reported. Several species of reptiles were killed by a spill of bunker C fuel oil in the St. Lawrence River (E.S. Smith, New York Department of Environmental Conservation, Albany, New York, USA, unpublished report). Sea snakes were possibly killed by crude oil in the Arabian Gulf (Pierce 1991). Petroleum could have been a contributing factor in the deaths of sea turtles off the coast of Florida, USA (Witham 1978) in the Gulf of Mexico following the *Ixtoc I* oil spill (Hall *et al.* 1983), and in the Arabian Gulf after the Gulf War oil spills (Symens and Al Salamah 1993), but the causes of death were not determined.

Experimental exposure studies with petroleum or PAHs are few. Exposure of juvenile loggerhead turtles to crude oil slicks revealed effects on respiration, skin, blood characteristics and chemistry, and salt gland function (Vargo et al. 1986). Juvenile loggerhead turtles were exposed to artificially weathered crude oil for 4 days followed by an 18 day recovery period; blood abnormalities and severe skin and mucosal changes from exposure were reversed during the recovery period (Lutcavage et al. 1995).

13.5.4   FIELD INVESTIGATIONS

Ingestion of oil and plastic objects has been reported for green (*Chelonia mydas*), loggerhead, and Atlantic Ridley turtles (Witham 1978; Hall et al. 1983; Gramentz 1988). Tissues or eggs of marine reptiles have seldom been analyzed for petrogenic or anthropogenic PAHs. Twenty nests of loggerhead turtles from the Gulf coast of Florida, USA, were sampled for unhatched eggs, but no 'reference' eggs were collected before hatching (Alam and Brim 2000). The eggs were pooled within each nest and analyzed for 43 PAHs. Four PAHs were detected with a frequency of detection of 5, 10, 40, and 45%; concentrations were 60-377 ng/g dw. Adipose and liver tissue from a single leatherback turtle (*Dermochelys coriacea*) found drowned in a fishing net off the west coast of Britain were analyzed for 18 PAHs (Godley et al. 1998). Concentrations were mostly below the limit of detection (0.1 ng/g ww); detected PAHs were 0.2-6 ng/g in adipose tissue ($n = 5$; total $= 12$ ng/g) and 0.2 to 2.2 ng/g in liver ($n = 6$; total 5.5 ng/g).

13.6   CONCLUSIONS

The primary concern about chronic exposure to PAHs is the potential for reactive metabolites of some PAHs to cause damage to DNA, RNA, and cellular proteins. However, most reports of PAH concentrations in tissues of wild animals have limited usefulness for determinations of comparative exposure or biological hazard because of the variation in identity and number of PAHs sought, tissues analyzed, and exposure scenario; and differences in species metabolic capabilities.

Although some PAHs have been well studied in laboratory and domestic mammals, the effects of acute or chronic exposure to individual PAHs, or mixtures of PAHs, are not well known for wild mammals and reptiles, and only partially understood for birds. Furthermore, the complex and jointly ubiquitous nature of PAHs, halogenated hydrocarbons, and metal contaminants makes it difficult to identify biological responses caused by PAHs. Separation of PAH effects from the effects of whole crude or refined petroleum is also difficult. Consequently, investigations of the biological effects of mixtures of PAHs, alone or in combination with other contaminants, and determinations of 'background' PAH concentrations, are needed to accurately assess the potential hazard to wild mammals, birds, and reptiles of exposure to PAHs in polluted marine environments. No tissue or environmental criteria have been promulgated by any regulatory agency of the USA for protection of aquatic or terrestrial wildlife from the toxic effects of PAHs.

No efforts have been made to relate PAHs to effects on populations or communities of birds and reptiles, although attempts to relate petroleum spills to these levels of organization for birds are common in the scientific and non-scientific literature. Efforts to relate PAHs to effects on populations or communities of marine mammals are few, most notably the beluga whale population in the St. Lawrence Estuary of Canada. The scientific literature contains a limited number of efforts, most of which occurred as a result of the *Exxon Valdez* oil spill in Alaska, USA, to address the effects of petroleum spills on populations of marine mammals.

REFERENCES

Alam SK and Brim MS (2000) Organochlorine, PCB, PAH, and metal concentrations in eggs of loggerhead sea turtles (*Caretta caretta*) from north-west Florida, USA. *Journal of Environmental Science and Health*, **B35**, 705–724.

Albers PH (2003) Petroleum and individual polycyclic aromatic hydrocarbons. In Hoffman DJ, Rattner BA, Burton GA Jr and Cairns J Jr (eds), *Handbook of Ecotoxicology*, 2nd edition. Lewis, Boca Raton, FL, pp. 341–371.

Anthony RG, Miles AK, Estes JA and Isaacs FB (1999) Productivity, diets, and environmental contaminants in nesting bald eagles from the Aleutian Archipelago. *Environmental Toxicology and Chemistry*, **18**, 2054–2062.

Arfsten DP, Schaeffer DJ and Mulveny DC (1996) The effects of near ultraviolet radiation on the toxic effects of polycyclic aromatic hydrocarbons in animals and plants: a review. *Ecotoxicology and Environmental Safety*, **33**, 1–24.

Baumann PC (1989) PAH, metabolites, and neoplasia in feral fish populations. In Varanasi U (ed.), *Metabolism of Polycyclic Aromatic Hydrocarbons in the Aquatic Environment*. CRC Press, Boca Raton, FL, pp. 269–289.

Beland P, DeGuise S and Plante R (1991) *Toxicology and Pathology of St. Lawrence Marine Mammals. Final Report for Wildlife Toxicology Fund, World Wildlife Fund.* St. Lawrence National Institute of Ecotoxicology, Montreal, Canada, 95 pp.

Beland P, DeGuise S, Girard C, Lagace A, Martineau D, Michaud R, Muir DCG, Norstrom RJ, Pelletier E, Ray S and Shugart LR (1993) Toxic compounds and health and reproductive effects in St. Lawrence beluga whales. *Journal of Great Lakes Research*, **19**, 766–775.

Blaylock RA (2002) Marine mammal mortalities linked to environmental conditions. http://capita.wustl.edu/NEW/blaylock.html, pp. 1–4.

Broman D, Naf C, Lundbergh I and Zebuhr Y (1990) An *in situ* study on the distribution, biotransformation and flux of polycyclic aromatic hydrocarbons (PAHs) in an aquatic food chain (seston — *Mytilus edulis* L. — *Somateria mollissima* L.) from the Baltic: an ecotoxicological perspective. *Environmental Toxicology and Chemistry*, **9**, 429–442.

Brunstrom B (1991) Embryolethality and induction of 7-ethoxyresorufin *O*-de-ethylase in chick embryos by polychlorinated biphenyls and polycyclic aromatic hydrocarbons having Ah receptor affinity. *Chemical–Biological Interactions*, **81**, 69–77.

Brunstrom B, Broman D and Naf C (1990) Embryotoxicity of polycyclic aromatic hydrocarbons (PAHs) in three domestic avian species, and of PAHs and co-planar polychlorinated biphenyls (PCBs) in the common eider. *Environmental Pollution*, **67**, 133–143.

Burger J and Gochfeld M (2000) Metal levels in feathers of 12 species of seabirds from Midway Atoll in the northern Pacific Ocean. *Science of the Total Environment*, **257**, 37–52.

Chang S, Zdanowicz VS and Murchelano RA (1998) Associations between liver lesions in winter flounder (*Pleuronectes americanus*) and sediment chemical contaminants from north-east United States estuaries. *ICES Journal of Marine Science*, **55**, 954–969.

Colwell RR and Walker JD (1977) Ecological aspects of microbial degradation of petroleum in the marine environment. *Critical Reviews in Microbiology*, **5**, 423–445.

Cowan DF (2002) Pathology reports. Texas Marine Mammal Stranding Network. http://www.tmmsn.org/pathology/pathology.html

Custer TW, Custer CM, Hines RK and Sparks DW (2000) Trace elements, organochlorines, polycyclic aromatic hydrocarbons, dioxins, and furans in lesser scaup wintering on the Indiana Harbor Canal. *Environmental Pollution*, **110**, 469–482.

Debacker V, Holsbeek L, Tapia G, Gobert S, Joiris CR, Jauniaux T, Coignoul F and Bouquegneau J-M (1997) Ecotoxicological and pathological studies of common guillemots, *Uria aalge*, beached on the Belgian coast during six successive wintering periods (1989–90 to 1994–95). *Diseases of Aquatic Organisms*, **29**, 159–168.

De Swart RL, Ross PS, Vedder LJ, Timmerman HH, Heisterkamp S, Van Loveren H, Vos JG, Reijnders PJH and Osterhaus DME (1994) Impairment of immune function in harbor seals (*Phoca vitulina*) feeding on fish from polluted waters. *Ambio*, **23**, 155–159.

Donaldson GM, Braune BM, Gaston AJ and Noble DG (1997) Organochlorine and heavy metal residues in breast muscle of known-age thick-billed murres (*Uria lomvia*) from the Canadian Arctic. *Archives of Environmental Contamination and Toxicology*, **33**, 430–435.

Eisler R (2000) *Handbook of Chemical Risk Assessment*, vol. 2. Lewis, Boca Raton, FL, pp. 1343–1411.

Engelhardt FR (1983) Petroleum effects on marine mammals. *Aquatic Toxicology*, **4**, 199–217.

Fernandes MB and Sicre M-A (1999) Polycyclic aromatic hydrocarbons in the Arctic: Ob and Yenisei estuaries and Kara Sea shelf. *Estuarine, Coastal and Shelf Science*, **48**, 725–737.

Fernandes MB, Sicre MA, Boireau A and Tronczynski J (1997) Polyaromatic hydrocarbon (PAH) distributions in the Seine River and its estuary. *Marine Pollution Bulletin*, **34**, 857–867.

Frost KJ, Lowry LF, Sinclair EH, Verhoef J and McAllister DC (1994) Impacts on distribution, abundance, and productivity of harbor seals. In Loughlin TR (ed.), *Marine Mammals and the Exxon Valdez*. Academic Press, New York, pp. 97–117.

Fu PP, Von Tungeln LS, Chiu L-H and Own ZY (1999) Halogenated polycyclic aromatic hydrocarbons: a class of genotoxic environmental pollutants. *Environmental Carcinogens and Ecotoxicology Reviews*, **C17**, 71–109.

Godley BJ, Gaywood MJ, Law RJ, McCarthy CJ, McKenzie C, Patterson IAP, Penrose RS, Reid RJ and Ross HM (1998) Patterns of marine turtle mortality in British waters (1992–1996) with reference to tissue contaminant levels. *Journal of the Marine Biological Association of the United Kingdom*, **78**, 973–984.

Gramentz D (1988) Involvement of loggerhead turtle with the plastic, metal, and hydrocarbon pollution in the central Mediterranean. *Marine Pollution Bulletin*, **19**, 11–13.

Hall RJ, Belisle AA and Sileo L (1983) Residues of petroleum hydrocarbons in tissues of sea turtles exposed to the *Ixtoc I* oil spill. *Journal of Wildlife Diseases*, **19**, 106–109.

Hellou J, Stenson G, Ni I-H and Payne JF (1990) Polycyclic aromatic hydrocarbons in muscle tissue of marine mammals from the north-west Atlantic. *Marine Pollution Bulletin*, **21**, 469–473.

Hellou J, Upshall C, Ni IH, Payne JF and Huang YS (1991) Polycyclic aromatic hydrocarbons in harp seals (*Phoca proenlandica*) from the north-west Atlantic. *Archives of Environmental Contamination and Toxicology*, **21**, 135–140.

Herbst LH (1994) Fibropapillomatosis in marine turtles. *Annual Review of Fish Diseases*, **4**, 389–425.

Hoffman DJ (1990) Embryotoxicity and teratogenicity of environmental contaminants to bird eggs. *Reviews of Environmental Contamination and Toxicology*, **115**, 39–89.

Holsbeek L, Joiris CR, Debacker V, Ali IB, Roose P, Nellissen J-P, Gobert S, Bouquegneau J-M and Bossicart M (1999) Heavy metals, organochlorines and polycyclic aromatic hydrocarbons in sperm whales stranded in the southern North Sea during the 1994/1995 winter. *Marine Pollution Bulletin*, **38**, 304–313.

Hutchinson J and Simmonds M (1992) Escalation of threats to marine turtles. *Oryx*, **26**, 95–102.

Ingram AJ, Phillips JC and Davies S (2000) DNA adducts produced by oils, oil fractions and polycyclic aromatic hydrocarbons in relation to repair processes and skin carcinogenesis. *Journal of Applied Toxicology*, **20**, 165–174.

James MO (1989) Microbial degradation of PAH in the aquatic environment. In Varanasi U (ed.), *Metabolism of Polycyclic Aromatic Hydrocarbons in the Aquatic Environment*. CRC Press, Boca Raton, FL, pp. 69–91.

Jenssen BM, Ekker M and Zahlsen K (1990) Effects of ingested crude oil on thyroid hormones and on the mixed function oxidase system in ducks. *Comparative Biochemistry and Physiology*, **95C**, 213–216.

Kayal S and Connell DW (1995) Polycyclic aromatic hydrocarbons in biota from the Brisbane River Estuary, Australia. *Estuarine, Coastal and Shelf Science*, **40**, 475–493.

King KA, Stafford CJ, Cain BW, Mueller AJ and Hall HD (1987) Industrial, agricultural, and petroleum contaminants in cormorants wintering near the Houston Ship Channel, Texas, USA. *Colonial Waterbirds*, **10**, 93–99.

Lake CA, Lake JL, Haebler R, McKinney R, Boothman WS and Sadove SS (1995) Contaminant levels in harbor seals from the north-eastern United States. *Archives of Environmental Contamination and Toxicology*, **29**, 128–134.

Landsberg JH, Balazs GH, Steidinger KA, Baden DG, Work TM and Russell DJ (1999) The potential role of natural tumor promoters in marine turtle fibropapillomatosis. *Journal of Aquatic Animal Health*, **11**, 199–210.

Law RJ and Whinnett JA (1992) Polycyclic aromatic hydrocarbons in muscle tissue of harbour porpoises (*Phocoena phocoena*) from UK waters. *Marine Pollution Bulletin*, **24**, 550–553.

Lee JS, Tanabe S, Umino H, Tatsukawa R, Loughlin TR and Calkins DC (1996) Persistent organochlorines in Steller sea lion (*Eumetopias jubatus*) from the bulk of Alaska and the Bering Sea, 1976–1981. *Marine Pollution Bulletin*, **32**, 535–544.

Lee RF (1977) Fate of oil in the sea. In Four PL (ed.), *Proceedings of the 1977 Oil Spill Response Workshop*. FWS/OBS/77-24. United States Fish and Wildlife Service, Washington, DC, pp. 43–54.

Loughlin TR (ed.) (1994a) *Marine Mammals and the Exxon Valdez*. Academic Press, New York, 395 pp.

Loughlin TR (1994b) Tissue hydrocarbon levels and the number of cetaceans found dead after the spill. In Loughlin TR (ed.), *Marine Mammals and the Exxon Valdez*. Academic Press, New York, pp. 359–370.

Lutcavage ME, Lutz PL, Bossart GD and Hudson DM (1995) Physiologic and clinico-pathologic effects of crude oil on loggerhead sea turtles. *Archives of Environmental Contamination and Toxicology*, **28**, 417–422.

Marine Mammal Commission (1999) *Marine Mammals and Persistent Ocean Contaminants: Proceedings of the Marine Mammal Commission Workshop*, Keystone, Colorado, 12–15 October 1998, Bethesda, MD, pp. 6–16, 87–92.

Martineau D, Lagace A, Masse R, Morin M and Beland P (1985) Transitional cell carcinoma of the urinary bladder in a beluga whale (*Delphinapterus leucas*). *Canadian Veterinary Journal*, **26**, 297–302.

Martineau D, Lagace A, Beland P, Higgins R, Armstrong D and Shugart LR (1988) Pathology of stranded beluga whales (*Delphinapterus leucas*) from the St. Lawrence Estuary, Quebec, Canada. *Journal of Comparative Pathology*, **98**, 287–311.

Martineau D, DeGuise S, Fournier M, Shugart L, Girard C, Lagace A and Beland P (1994) Pathology and toxicology of beluga whales from the St. Lawrence Estuary, Quebec, Canada. Past, present and future. *Science of the Total Environment*, **154**, 201–215.

Mathieu A, Payne JF, Fancey LL, Santella RM and Young TL (1997) Polycyclic aromatic hydrocarbon–DNA adducts in beluga whales from the Arctic. *Journal of Toxicology and Environmental Health*, **51**, 1–4.

Matsumoto H and Kashimoto T (1986) Embryotoxicity of organic extracts from airborne particulates in ambient air in the chicken embryo. *Archives of Environmental Contamination and Toxicology*, **15**, 447–452.

Mayura K, Huebner HJ, Dwyer MR, McKenzie KS, Donnelly KC, Kubena LF and Phillips TD (1999) Multi-bioassay approach for assessing the potency of complex mixtures of polycyclic aromatic hydrocarbons. *Chemosphere*, **38**, 1721–1732.

McElroy AE, Farrington JW and Teal JM (1989) Bioavailability of polycyclic aromatic hydrocarbons in the aquatic environment. In Varanasi U (ed.), *Metabolism of Polycyclic Aromatic Hydrocarbons in the Aquatic Environment*. CRC Press, Boca Raton, FL, pp. 1–39.

McEwan EH and Whitehead PM (1980) Uptake and clearance of petroleum hydrocarbons by the glaucous-winged gull (*Larus glaucescens*) and the mallard duck (*Anas platyrhynchos*). *Canadian Journal of Zoology*, **58**, 723–726.

Michot TC, Custer TW, Nault AJ and Mitchell CA (1994) Environmental contaminants in redheads wintering in coastal Louisiana and Texas. *Archives of Environmental Contamination and Toxicology*, **26**, 425–434.

Miles AK and Ohlendorf HM (1993) Environmental contaminants in canvasbacks wintering on San Francisco Bay, California. *California Fish and Game*, **79**, 28–38.

Miller DS, Hallett DJ and Peakall DB (1982) Which components of crude oil are toxic to young seabirds? *Environmental Toxicology and Chemistry*, **1**, 39–44.

Muir D, Braune B, DeMarch B, Norstrom R, Wagemann R, Lockhart L, Hargrave B, Bright D, Addison R, Payne J and Reimer K (1999) Spatial and temporal trends and effects of contaminants in the Canadian Arctic marine ecosystem: a review. *Science of the Total Environment*, **230**, 83–144.

Mulcahy DM and Ballachey BE (1994) Hydrocarbon residues in sea otter tissues. In Loughlin TR (ed.), *Marine Mammals and the Exxon Valdez*. Academic Press, New York, pp. 313–330.

Naf C, Broman D and Brunstrom B (1992) Distribution and metabolism of polycyclic aromatic hydrocarbons (PAHs) injected into eggs of chicken (*Gallus domesticus*) and common eider duck (*Somateria mollissima*). *Environmental Toxicology and Chemistry*, **11**, 1653–1660.

Neff JM (1985) Polycyclic aromatic hydrocarbons. In Rand GM and Petrocilli SR (eds), *Fundamentals of Aquatic Toxicology*. Hemisphere, New York, pp. 416–454.

Olafsdottir K, Skirnisson K, Gylfadottir G, Johannesson T (1998) Seasonal fluctuations of organochlorine levels in the common eider (*Somateria mollissima*) in Iceland. *Environmental Pollution*, **103**, 153–158.

Patton JF and Dieter MP (1980) Effects of petroleum hydrocarbons on hepatic function in the duck. *Comparative Biochemistry and Physiology*, **65C**, 33–36.

Peakall DB, Hallett DJ, Bend JR, Foureman GL and Miller DS (1982) Toxicity of Prudhoe Bay crude oil and its aromatic fractions to nestling herring gulls. *Environmental Research*, **27**, 206–215.

Pierce V (1991) The effects of the Arabian Gulf oil spill on wildlife. In Junge RE (ed.), *1991 Proceedings of the American Association of Zoo Veterinarians*. Calgary, Canada, pp. 370–375.

Rattner BA, Hoffman DJ and Marn CN (1989) Use of mixed-function oxygenases to monitor contaminant exposure. *Environmental Toxicology and Chemistry*, **8**, 1093–1102.

Reijnders PJH and de Ruiter-Dijkman E (1995) Toxicological and epidemiological significance of pollutants in marine mammals. In Blix AS, Walloe L and Ulltang O (eds), *Whales, Seals, Fish and Man*. Elsevier Science BV, Amsterdam, pp. 575–587.

Ray S, Dunn BP, Payne JF, Fancey L, Helbig R and Beland P (1991) Aromatic DNA-carcinogen adducts in beluga whales from the Canadian Arctic and Gulf of St. Lawrence. *Marine Pollution Bulletin*, **22**, 392–396.

Ren L, Huang X-D, McConkey BJ, Dixon DG and Greenberg BM (1994) Photo-induced toxicity of three polycyclic aromatic hydrocarbons (fluoranthene, pyrene and naphthalene) to the duckweed *Lemna gibba* L. G-3. *Ecotoxicology and Environmental Safety*, **28**, 160–171.

Roberts MH Jr, Hargis WJ Jr, Strobel CJ and DeLisle PF (1989) Acute toxicity of PAH contaminated sediments to the estuarine fish, *Leiostomus xanthurus*. *Bulletin of Environmental Contamination and Toxicology*, **42**, 142–149.

Ross P, DeSwart R, Addison R, Van Loveren H, Vos J and Osterhaus A (1996) Contaminant-induced immunotoxicity in harbour seals: wildlife at risk? *Toxicology*, **112**, 157–169.

Santodonato S, Howard P and Basu D (1981) Health and Ecological Assessment of Polynuclear Aromatic Hydrocarbons. *Journal of Environmental Pathology and Toxicology*, **5**(Special Issue), 119–122, 191–300.

Savinov VM, Savinova TN, Carroll J, Matishov GG, Dahle S and Naes K (2000) Polycyclic aromatic hydrocarbons (PAHs) in sediments of the White Sea, Russia. *Marine Pollution Bulletin*, **40**, 807–818.

Shore RF, Wright J, Horne JA and Sparks TH (1999) Polycyclic aromatic hydrocarbon (PAH) residues in the eggs of coastal-nesting birds from Britain. *Marine Pollution Bulletin*, **38**, 509–513.

Shugart LR (1990) Detection and quantitation of benzo[a]pyrene-DNA adducts in brain and liver tissues of beluga whales (*Delphinapterus leucas*) from the St. Lawrence and Mackenzie estuaries. In *For the Future of the Beluga*. University of Quebec Press, Sillery, Canada, pp. 219–223.

Stronkhorst J, Ysebaert TJ, Smedes F, Meininger PL, Dirksen S and Boudewijn TJ (1993) Contaminants in eggs of some waterbird species from the Scheldt Estuary, SW Netherlands. *Marine Pollution Bulletin*, **26**, 572–578.

Symens P and Al Salamah MI (1993) The impact of the Gulf War oil spills on wetlands and waterfowl in the Arabian Gulf. In Moser M and van Vessem J (eds), *Wetland and Waterfowl Conservation in South and West Asia*. Special Publication No. 25. International Waterfowl and Wetlands Research Bureau, Slimbridge, UK, pp. 24–28.

Troisi GM, Haraguchi K, Kaydoo DS, Nyman M, Aguilar A, Borrell A, Siebert U and Mason CF (2001) Bioaccumulation of polychlorinated biphenyls (PCBs) and dichlorodiphenylethane (DDE) methyl sulfones in tissues of seal and dolphin morbillivirus epizootic victims. *Journal of Toxicology and Environmental Health A*, **62**, 1-8.

Trust KA, Fairbrother A and Hooper MJ (1994) Effects of 7,12-dimethylbenzo[a]anthracene on immune function and mixed-function oxygenase activity in the European starling. *Environmental Toxicology and Chemistry*, **13**, 821-830.

Varanasi U, Stein JE and Nishimoto M (1989) Biotransformation and disposition of PAH in fish. In Varanasi U (ed.), *Metabolism of Polycyclic Aromatic Hydrocarbons in the Aquatic Environment*. CRC Press, Boca Raton, FL, pp. 93-149.

Varanasi U (1989) Metabolic activation of PAH in subcellular fractions and cell cultures from aquatic and terrestrial species. In Varanasi U (ed.), *Metabolism of Polycyclic Aromatic Hydrocarbons in the Aquatic Environment*. CRC Press, Boca Raton, FL, pp. 203-251.

Vargo S, Lutz PL, Odell DK, Van Vleet ES and Bossart GD (1986) Effects of oil on marine turtles. *Minerals Management Service Outer Continental Shelf Report*, MMS 86-0070, Vienna, Virginia USA, 12 pp.

Wilson B, Arnold H, Bearzi G, Fortuna CM, Gaspar R, Ingram S, Liret C, Pribanic S, Read AJ, Ridoux V, Schneider K, Urian KW, Wells RS, Wood C, Thompson PM and Hammond PS (1999) Epidermal diseases in bottlenose dolphins: impacts of natural and anthropogenic factors. *Proceedings of the Royal Society of London B*, **266**, 1077-1083.

Witham R (1978) Does a problem exist relative to small sea turtles and oil spills? In *Proceedings of the Conference on Assessment of Ecological Impacts of Oil Spills*. American Institute of Biological Sciences, Keystone, Colorado USA, pp. 630-632.

Williams TM, O'Connor DJ and Nielsen SW (1995) The effects of oil on sea otters: histopathology, toxicology, and clinical history. In Williams TM and Davis RW (eds), *Emergency Care and Rehabilitation of Oiled Sea Otters: A Guide for Oil Spills Involving Fur-bearing Marine Mammals*. University of Alaska Press, Fairbanks, AK, pp. 3-22.

Yang G-P (2000) Polycyclic aromatic hydrocarbons in the sediments of the South China Sea. *Environmental Pollution*, **108**, 163-171.

Zitko V, Stenson G and Hellou J (1998) Levels of organochlorine and polycyclic aromatic compounds in harp seal beaters (*Phoca groenlandica*). *Science of the Total Environment*, **221**, 11-29.

PAH Interactions with Plants: Uptake, Toxicity and Phytoremediation

BRUCE M. GREENBERG

University of Waterloo, Waterloo, ON, Canada

14.1 INTRODUCTION

Most plant species are sensitive to polycyclic aromatic hydrocarbons (PAHs) to some degree. PAH impairment of plant growth and/or development will limit primary productivity, constraining total biological activity in an ecosystem. Accumulation of PAHs by plants represents an entry point of hazardous compounds into the food web, initiating a biomagnification process (Jones *et al*. 1989; Salanitro *et al*. 1997; Thomas *et al*. 1998). Plants can be used as sentinel species for the detection of PAH contamination in the environment (Stephenson *et al*. 1997; ASTM 1998). Further, a number of plant bioindicators have been developed that can be used for rapid assessment of negative impacts of PAHs (Byl and Klaine 1991; Plewa 1991; Huang *et al*. 1997a; Gensemer *et al*. 1999; Marwood *et al*. 2001). As an added benefit, bioindicators reveal a great deal about the underlying mechanisms of toxicity (Huang *et al*. 1997a; Babu *et al*. 2001). Finally, plants can grow in some contaminated soils, generating large amounts of biomass for the remediation of PAHs (Salt *et al*. 1998; Huang *et al*. 2001).

This chapter covers several aspects of the phytotoxicity of PAHs. PAH bioaccumulation by plants is examined, along with the general phytotoxicity of PAHs. Next, plant assays of PAH impacts and PAH mechanisms of plant toxicity are reviewed. Finally, advances in the phytoremediation of PAHs are presented.

14.2 UPTAKE AND TOXICITY

14.2.1 UPTAKE

Plants can assimilate PAHs from soil, water, or air. This is because roots have a high capacity for hydrophobic compounds (Salanitro *et al*. 1997; Schwab *et al*.

PAHs: An Ecotoxicological Perspective. Edited by Peter E.T. Douben.
© 2003 John Wiley & Sons Ltd

1998; Binet *et al.* 2000; Huang *et al.* 2001). In aquatic systems, assimilation of contaminants by plants is rapid and efficient, even from non-dissolved phases, such as sediment. Because plant tissues have a higher affinity for organics than the aqueous phase, the bioconcentration factors (BCFs) can be very high (Duxbury *et al.* 1997; Thomas *et al.* 1998). Plants can also assimilate organics following aerial deposition on the leaves (Thomas *et al.* 1998). Contaminants received in this manner can be highly toxic, and represent an important entry point of organic compounds into the food chain.

Because PAHs are lipophilic, they tend to accumulate in plants, especially in membrane bilayers (Duxbury *et al.* 1997; Thomas *et al.* 1998). Indeed, plants grown in areas with high PAH loads in the soil or air have high bioconcentrations of PAHs (Jones *et al.* 1989).

PAHs, without being activated, are generally only phytotoxic at high concentrations (Huang *et al.* 1993); e.g. in a study by Chaineau *et al.* (1997), soils contaminated with fuel oils inhibited seed germination and growth in lettuce, barley, clover, and maize at high concentrations in soil (0.3–4 g/kg). For bean, wheat, and sunflower, even higher concentrations were required (4–9 g/kg). The smaller PAHs seemed to be the most toxic.

14.2.2   TOXICITY

PAHs absorb sunlight strongly, and can act as photosensitizers (Krylov *et al.* 1997). If sunlight is present, PAH toxicity to aquatic and terrestrial plants increases dramatically (Huang *et al.* 1993; Ren *et al.* 1996; Krylov *et al.* 1997). Obviously, plants provide a relevant model for the study of toxicity of photoactive contaminants. Photoinduced toxicity of PAHs is due to two processes: photosensitization and photomodification reactions (Krylov *et al.* 1997). Photosensitization reactions usually proceed via formation of reactive oxygen species (ROS), especially $^1O_2$. $^1O_2$ formed within a biological organism is highly damaging (Foyer *et al.* 1994; Allan and Fluhr 1997). Photomodification of PAHs, which generally occurs via oxidation of the parent compound, results in a mixture of compounds with high toxicity (Mallakin *et al.* 1999). This is similar to increased PAH hazards following cytochrome P450 activation (Plewa 1991). Interestingly, it was found that humic acids ameliorate PAH toxicity to duckweed (*Lemna gibba*), probably due to binding of the PAHs (Gensemer *et al.* 1999). Therefore, PAHs are an example of an environmental toxicant where one environmental factor (light) can enhance risk, while another (binding) can lower risk.

Synergistic toxicity of PAHs with other classes of contaminants has been investigated. The combined effects of fly ash and SO_2 on cucumber were examined (Tung *et al.* 1995). At concentrations where neither fly ash nor SO_2 alone had an effect, the compounds combined caused significant chlorosis. The active organic compounds in the fly ash were probably PAHs. Recently, we found that a metal (Cu^{2+}) combined with an oxygenated PAH (1,2-dihydroxyanthroquinone) had synergistic toxicity (Babu *et al.* 2001). This toxicity was due to Cu^{2+} catalyzed transfer of electrons to O_2 to form ROS such as superoxide and hydrogen peroxide.

14.3   ASSAYS FOR DETECTION OF PAH PHYTOTOXICITY

14.3.1   WHOLE PLANT ASSAYS

Numerous plant toxicity tests have been developed (Kapustka 1997; Stephenson *et al.* 1997). They include germination, growth, root elongation, reproduction, and life-cycle tests. Nearly all have been used for assessment of PAH toxicity. Germination is often used as a simple and rapid assay. However, it lacks sensitivity because seed coats can be impermeable to PAHs (Kapustka 1997). Plant growth is usually a better toxicity assay, although it is more time-consuming and can require a great deal of growth-chamber or greenhouse space. Reproduction, yield and full life-cycle assays may be the most important endpoints, but they are cumbersome and time-consuming to perform.

The number of plant species that have been used in PAH toxicity testing are extensive (Wang 1991; Kapustka 1997; Stephenson *et al.* 1997). An overriding standard method of toxicity testing with plants was recently developed (ASTM 1998). Stephenson *et al.* (1997) compared numerous plant species for toxicity assessment, and proposed *Brassica napus* (canola) and *Zea mays* (corn) as good broad range species for toxicity testing. Due to the variety of conditions that plants are grown under (e.g. soil vs. hydroponics, different lighting conditions, etc.), it is often difficult to make comparisons between studies. Thus, generalizations about plant sensitivity to specific PAHs should be viewed with caution. Where possible, multi-species and multi-endpoint tests should be performed. It is also important to consider whether the assay was conducted in soil or aqueous medium. Further, the light quality will greatly influence the toxic strength of the PAHs being tested.

Toxicity of PAHs in soil to plants has been examined. The toxicity of PAHs generally drops with increasing organic content of the soil (Maliszewska-Kordybach and Smreczak 2000), because hydrophobic PAHs become less bioavailable. PAHs were also found to inhibit alfalfa root nodulation in soils, but only when the PAHs were bioavailable (Wetzel and Werner 1995). Plants have also been used as sentinel species, e.g. the hazards of airborne PAHs (at road sides) were detected with lichens (Owczarek *et al.* 2001). Plants have also been used to test soils for residual PAH toxicity following various clean-up processes (Baud-Grasset *et al.* 1993).

Aquaculture (hydroponics) has been used for PAH toxicity studies (Huang *et al.* 1993; Wang 1994; Ren *et al.* 1996). It greatly simplifies the application of test compounds to plants. However, there is debate about whether these tests are environmentally relevant, because many PAHs are hydrophobic. However, the sensitivity of terrestrial plants in aquaculture is similar to the sensitivity of the aquatic plant *Lemna* (duckweed) (Wang 1991; Huang *et al.* 1993; Ren *et al.* 1996). In both cases, toxicity is observed in the parts per million range, with a similar ranking of the order of toxicity of different PAHs. Thus, for toxicity ranking purposes and determination of toxicity equivalents (TEQs), hydroponic tests with terrestrial plants and *Lemna* are environmentally relevant. *Lemna* is, of course, fully relevant for toxicity assessment of aquatic environments. Further, freshwater algae and macrophytes have been used extensively for studies on PAH toxicity (Wang 1991; Gala and Geisy 1994; Marwood *et al.* 1999, 2001). We have shown that Lake Erie phytoplankton can be used to assess the toxicity of both intact and oxygenated PAHs (Marwood *et al.* 1999). These tests showed that photoinduced phyototoxicity of PAHs is environmentally relevant to natural populations of phytoplankton. It has also been shown that natural populations of phytoplankton can acclimate to PAH stress, especially if photoinduced toxicity is not a factor (Kelly *et al.* 1999).

Apoptosis (or programmed cell death) is a complicating factor in phytotoxicity (Allan and Fluhr 1997; Babu *et al.* 2001). This is an important natural process for plant cells that are destined to become structural elements or to become senescent. These cells must die before they can carry out their assigned function. Any contaminant that can trigger apoptosis would be highly phytotoxic. At a cellular level, apoptosis is triggered by high cytosolic Ca^{2+} and mechanistically proceeds via a ROS (e.g. singlet O_2, superoxide, hydrogen peroxide, or hydroxyl radical; Foyer *et al.* 1994; Allan and Fluhr 1997). Strikingly, PAHs and oxidized PAH contaminants can cause rapid production of ROS (Krylov *et al.* 1997; Babu *et al.* 2001).

In summary, there are a large number of plant assays for PAH toxicity. Appropriate plants and tests should be employed for comprehensive environmental assessment, e.g. if risk assessment is being done on a soil site, then a terrestrial plant grown in soil should be employed. For ranking the relative hazards of toxicants, an aquatic plant system might be more accurate because chemical delivery can be better controlled.

14.3.2   PLANT PROCESS ASSAYS

Bioindicators generally detect a biochemical aspect of toxicant action (e.g. membrane damage, enzyme inhibition, and DNA damage). They can provide rapid and direct indications of toxicant impact in the environment. The specificity of bioindicators gives them great sensitivity, and they often do not rely on knowledge of the history of chemical exposure. Therefore, if plants are sampled from a PAH-contaminated site and compared at a bioindicator level to plants from a reference site, one can determine whether the plants are stressed. However, bioindicators tend to be toxicant-specific, because not all compounds will inhibit the same biological processes. Therefore, it is important to choose bioindicators that are relevant to the mechanisms of action of PAHs.

Bioindicators have the important benefit of providing knowledge about the mechanism of toxicity. This can lead to an understanding of the biological receptor targeted by a particular chemical class, and extrapolations to related molecules; e.g. for PAH-photoinduced toxicity, the physiology of cellular damage via ROS ($^1O_2$) is well understood (Foyer *et al*. 1994; Krylov *et al*. 1997; Babu *et al*. 2001). Non-specific peroxidation of lipids and proteins in membranes occurs in the presence of ROS, and these types of effects can be assayed as indicators of toxicity.

Inhibition of photosynthesis is a key mechanism of toxicant action in plants (Huang *et al*. 1997a; Marwood *et al*. 2001). At the whole-plant level, this can result in impaired plant growth, lower yields and loss of competitive advantage (Gensemer *et al*. 1999; Marwood *et al*. 2001). There are several reliable methods for measuring impacts on photosynthesis, including chlorosis, carbon fixation, and electron transport (Huang *et al*. 1997a; Marwood *et al*. 2001). In particular, assays of photosystem II (PSII) electron transport can be extremely sensitive measures of damage to almost any point in the photosynthetic apparatus (Marwood *et al*. 2001). This is because PSII is the first step in photosynthesis, and inhibition of most metabolic activities in the chloroplast downstream from PSII will lead to a block in PSII electron transport. When PSII does not utilize the light it absorbs, this energy instead results in oxidative damage (Huang *et al*. 1997a; Babu *et al*. 2001). Photosynthetic electron transport can be measured by a few techniques (Babu *et al*. 2001; Marwood *et al*. 2001). *In vivo*, PSII can be measured by chlorophyll *a* (Chl *a*) fluorescence or O_2 evolution, and primary productivity can be assessed by CO_2 fixation. *In vitro*, PSII activity can be quantified with the aid of electron acceptors.

Gas exchange measurements (e.g. O_2 evolution and CO_2 fixation) are well-established techniques used in environmental assessment; e.g. periphyto-photosynthesis was measured in using CO_2 fixation as an indicator of effluent toxicity (Lewis 1992). It was found to relate well to animal bioindicators of toxicity. Also the phytotoxic effects of industrial effluents and sewage were measured based on transpiration (Gadallah 1996). Both of these effluents were thought to be toxic, at least partly due to the PAHs in the mixtures.

Chl *a* fluorescence is a powerful bioindicator and has been used for toxicant assessment (Huang *et al*. 1997a; Marwood *et al*. 2001). Chl *a* fluorescence was used as a bioindicator of photoinduced toxicity of PAHs to plants (Huang *et al*. 1997a; Gensemer *et al*. 1999; Marwood *et al*. 2001). The primary site of action was found to be inhibition of electron transport downstream of PSII (Huang *et al*. 1997a; Babu *et al*. 2001). This was followed by inhibition of PSII, probably due to excitation pressure on PSII once the downstream electron transport was blocked. A linkage between inhibition of photosynthesis and inhibition of plant growth was established (Huang *et al*. 1997a; Marwood *et al*. 2001), helping to validate Chl *a* fluorescence as a bioindicator of PAH impacts on plants.

Tradescantia plants exposed to diesel fuel (containing PAHs) at 0.1 to 100 mg/kg in soil were assayed for toxicity via three bioindicators (Green *et al*.

1996). They were plant morphology, photosynthesis (Chl *a* fluorescence), and a micronucleus assay (DNA synthesis). There was a good correlation between inhibition of growth and photosynthesis. Further, complex mixtures of PAHs have been shown to compromise membrane integrity in plants (McCann and Solomon 2000). This is consistent with the narcotic action of PAHs.

Photosynthesis is an enzymatic assay of contaminant impact. Not surprisingly, other enzymatic assays have been used as bioindicators, e.g. intact and photomodified PAHs have been found to upregulate ROS-scavenging enzymes (manuscript in preparation). This is consistent with the mechanism of action of PAHs in light. Note that the P450IA1 (EROD) assay is not as effective in plants as in animals. This is because plants, while having a broad range of robust and inducible P450 activities, do not have a strong P450IA1 system (Gentile *et al.* 1991; Plewa 1991).

The effects of contaminants can be sensitively and specifically assayed by gene expression. Chaperonins and heat shock proteins are upregulated in response to chemical stress (Schoffl *et al.* 1998). Their function is to stabilize proteins when denaturing conditions exist. It is not surprising they are activated, because PAHs elevate the levels of ROS, which in turn cause protein denaturation (Babu *et al.* 2001). This is an area of research that is at an early stage and warrants much greater attention.

Clearly, plant bioindicators have great potential to test the impacts of environmental contaminants on plants. They are rapid, selective and sensitive. They can also be used without prior knowledge of the history of exposure. Bioindicators can reveal a great deal about the mechanism of toxicant action, which allows extrapolation to related contaminants.

14.4   QSARs FOR PAH PHYTOTOXICOLOGY

QSARs correlate the physicochemical properties of molecules to observable biological responses. They are useful for understanding the mechanisms of action of groups of related chemicals and for predicting the environmental risks associated with those chemicals. In developing a QSAR model, especially for plants, it is essential to consider the attributes of the environmental compartment in which the contaminant of interest resides, as this will dictate which physicochemical properties are likely to be most influential in toxicity. For instance, because solar radiation is ubiquitous in the environment and can enhance the phytotoxicity of certain chemicals, it represents a modifying factor that can be used in QSAR modeling (see Krylov *et al.* 1997).

QSARs for toxicity of PAHs and chloroaromatic compounds to plants have been described. Without modifying environmental factors, toxicity was related to lipophilicity, determined as log K_{ow} (Hulzebos *et al.* 1991). This relationship was strongest when the toxicity assays were carried out in aquaculture. Thus, as with animals, water solubility and lipophilicity (as described by K_{ow}) is a factor that can be used in PAH QSAR modeling. This is important for PAHs because, in the absence of modifying factors, toxicity is driven by narcosis due to accumulation of PAHs in biological membranes.

A QSAR encompassing modifying factors for environmental perturbations has been generated. It was demonstrated that light can be successfully incorporated as a modifying factor in QSARs (Huang *et al.* 1997b; Krylov *et al.* 1997). Photo-induced toxicity of PAHs occurs via photosensitization reactions (e.g. generation of $^1O_2$) and by photomodification (photooxidation and/or photolysis) of the chemicals to more toxic species. A QSAR developed for toxicity of 16 PAHs to *L. gibba* showed that photosensitization and photomodification additively contribute to toxicity (Krylov *et al.* 1997).

14.5 PHYTOREMEDIATION OF PAHs

For bioremediation of hydrophobic material to be effective, the available biomass for concentrating the contaminants must be large. For this reason, phytoremediation of PAHs has recently received considerable attention (McIntire and Lewis, 1997; Salt *et al.* 1998; Huang *et al.* 2001). Plants have extensive root systems that explore a large volume of soil and assimilate contaminants over a wide area. Roots can also enhance microbial activity by supplying substrate and nutrients. Phytoremediation has been successfully used to remediate a variety of contaminants in soil and groundwater; e.g. hybrid poplar trees have been used for the removal of herbicides such as atrazine (Burken and Schnoor, 1997). Many other plants have been used to take up and/or degrade various organic contaminants in soils, including PAHs (Siciliano *and* Germida, 1997; McIntire and Lewis, 1997; Huang *et al.* 2001).

The first aspect of PAH–plant interactions that must be understood for phytoremediation is the bioaccumulation of PAHs by plants. The uptake and depuration kinetics of three representative PAHs, anthracene (ANT), phenanthrene (PHE) and benzo[a]pyrene (B[a]P), and their photomodified products, were determined for *L. gibba* (Duxbury *et al.* 1997). Like many plants, *L. gibba* had a high capacity for intact ANT, PHE and B[a]P. Net assimilation of all three PAHs was always higher when the chemicals were delivered with DMSO, rather than from sand. PAHs were rapidly assimilated in the light, albeit net assimilation for PAHs was generally lower than in darkness. For terrestrial plants in soil, phytoaccumulation of PAHs is primarily into the roots with little partitioning to other parts of the plant (Schwab *et al.* 1998; Binet *et al.* 2000).

Given that plants take up PAHs rapidly and with high bioconcentration factors, phytoremediation holds a great deal of promise. One study showed both uptake and degradation of pyrene by plants (Liste and Alexander 2000). Phytoremediation has also been performed with petroleum hydrocarbons containing a high content of PAHs. Plants were able to achieve approximately 50% PAH removal from soil (Pradham *et al.* 1998). The only real issue in these cases was the speed of removal and achieving enough root biomass due to PAH phytotoxicity.

Although using plants for remediation of persistent organic contaminants holds advantages over other methods, many limitations exist for current application on a large scale (Salt *et al.* 1998; Huang *et al.* 2001). For instance, when PAH concentrations in the soil are high, many plants will not grow well enough to provide sufficient biomass for successful remediation. In many cases, contaminated soils are poor in nutrients, which will limit plant growth, slowing the remediation process. Further, contaminated soils often do not contain the appropriate microorganisms for the efficient degradation of the contaminants or for the promotion of plant growth. Therefore, until now phytoremediation processes have been, in general, slow and the time scale for complete remediation is often unacceptably long.

To improve remediation, multiple techniques that complement different aspects of contaminant removal have been applied to soils in combination (Huang *et al.* 2001). This has resulted in an enhanced multiprocess phytoremediation system which has improved and accelerated the overall remediation, resulting in removal of 95% of total PAHs. The remediation system includes: physical (volatilization), photochemical (photooxidation), and microbial degradation and plant growth (phytoremediation) processes. The techniques applied to realize these processes are landfarming (aeration and light exposure), microbial remediation (introduction of contaminant degrading bacteria), and phytoremediation (plant growth with plant growth-promoting rhizobacteria). This system has been very effective for removing persistent, strongly bound PAHs from soil (Huang *et al.* 2001). It appears that the combination of these components may be a viable solution for remediating persistent organic contaminants from soils.

14.6 CONCLUDING REMARKS

PAH impacts on plants have been examined using whole-organism measures of toxicity. Germination is often used as a simple and fast assay; however, it lacks sensitivity. Although, plant growth and reproduction are usually better toxicity assays, they are more time-consuming. The primary mechanism of PAH toxicity to plants involves a photoinduction process. There are a wide range of plant responses to PAHs, and plant sensitivity to PAHs can be enhanced by co-contamination with metals. Bioindicators can provide a rapid and direct indication of toxicant impact in natural environments. In developing QSAR models for plants, it is essential to consider the attributes of the environmental compartment in which the contaminant of interest resides, as this will dictate which physicochemical properties are likely to be the most influential in toxicity. Finally phytoremediation holds great promise for the removal of PAHs from contaminated sites.

ACKNOWLEDGMENTS

I want to thank the members of my research group for their assistance in preparing this manuscript. This work was supported by grants from the National

Science and Engineering Research Council (NSERC), the Canadian Networks of Toxicology Centres (CNTC), Imperial Oil and CRESTech.

REFERENCES

Allan AC and Fluhr R (1997) Two distinct sources of elicited reactive oxygen species in tobacco epidermal cells. *Plant Cell*, **9**, 1559–1572.

ASTM (1998) Standard guide for conducting terrestrial plant toxicity tests. In *Annual Book of ASTM Standards*, vol. 11.05. E1963-98. American Society for Testing and Materials, West Conshohocken, PA, 20 pp.

Babu TS, Marder JB, Tripuranthakan S, Dixon DG and Greenberg BM (2001) Synergistic effects of a photo-oxidized PAH and copper on photosynthesis and plant growth: evidence that active oxygen formation is a mechanism of copper toxicity. *Environmental Toxicology and Chemistry*, **20**, 1351–1358.

Baud-Grasset F, Baud-Grasset S and Safferman SI (1993) Evaluation of the bioremediation of a contaminated soil with phytotoxicity tests. *Chemosphere*, **26**, 1365–1374.

Binet P, Portal J and Leyval C (2000) Fate of polycyclic aromatic hydrocarbons (PAH) in the rhizosphere and mycorrhizosphere of ryegrass. *Plant and Soil*, **227**, 207–213.

Burken JG and Schnoor JL (1997) Uptake and metabolism of atrazine by poplar trees, *Environmental Science and Technology*, **31**, 1399–1406.

Byl TD and Klaine SJ (1991) Peroxidase activity as an indicator of sublethal stress in the aquatic plant *Hydrilla verticillata* (royal). In Gorsuch JW, Lower WR, Wang W and Lewis MA (eds), *Plants for Toxicity Assessment*, vol. 2. American Society for Testing and Materials, ASTM STP 1115, West Conshohocken, PA, pp. 101–106.

Chaineau CH, Morel JL and Oudot J (1997) Phytotoxicity and plant uptake of fuel oil hydrocarbons. *Journal of Environmental Quality*, **26**, 1478–1483.

Duxbury CL, Dixon DG and Greenberg BM (1997) The effects of simulated solar radiation on the bioaccumulation of polycyclic aromatic hydrocarbons by the duckweed *Lemna gibba*. *Environmental Toxicology and Chemistry*, **16**, 1739–1748.

Foyer CH, Lelandais M and Kunert KJ (1994) Photooxidative stress in plants. *Physiologia Plantarum*, **92**, 696–717.

Gadallah MAA (1996) Phytotoxic effects of industrial and sewage waste waters on growth, chlorophyll content, transpiration rate and relative water content of potted sunflower plants. *Water Air and Soil Pollution*, **89**, 33–47.

Gala WR and Giesy JP (1994) Flow cytometric determination of the photoinduced toxicity of anthracene to the green alga *Selenastrum capricornutum*. *Environmental Toxicology and Chemistry*, **13**, 831–840.

Gensemer RW, Dixon DG and Greenberg BM (1999) Using chlorophyll a fluorescence induction to detect the onset of anthracene photoinduced toxicity in *Lemna gibba*, and the mitigating effects of humic acid. *Limnology and Oceanography*, **44**, 878–888.

Gentile JM, Johnson P and Robbins S (1991) Activation of aflatoxin and benzo[a]pyrene by tobacco cells in the plant cell/microbe coincubation assay. In Gorsuch JW, Lower WR, Wang W and Lewis MA (eds), *Plants for Toxicity Assessment*, vol. 2. ASTM STP 1115. American Society for Testing and Materials, West Conshohocken, PA, pp. 318–325.

Green BT, Wiberg CT, Woodruff JL, Miller EW, Poage VL, Childress DM, Feulner JA, Prosch SA, Runkel JA, Wanderscheid RL, Wierma MD, Yang X, Choe HT and Mercurio SD (1996) Phytotoxicity observed in *Tradescantia* correlates with diesel fuel contamination in soil. *Environmental and Experimental Botany*, **36**, 313–321.

Huang XD, Dixon DG and Greenberg BM (1993) Impacts of UV irradiation and photooxidation on the toxicity of polycyclic aromatic hydrocarbons to the higher plant *Lemna gibba* L. G3 (duckweed). *Environmental Toxicology and Chemistry*, **12**, 1067–1077.

Huang XD, McConkey BJ, Babu TS and Greenberg BM (1997a) Mechanisms of photoinduced toxicity of anthracene to plants: inhibition of photosynthesis in the aquatic higher plant *Lemna gibba* (duckweed). *Environmental Toxicology and Chemistry*, **16**, 1707–1715.

Huang XD, Krylov SN, Ren L, McConkey BJ, Dixon DG and Greenberg BM (1997b) Mechanistic QSAR model for the photoinduced toxicity of polycyclic aromatic hydrocarbons: II. An empirical model for the toxicity of 16 PAHs to the duckweed *Lemna gibba* L. G-3. *Environmental Toxicology and Chemistry*, **16**, 2296–2303.

Huang X-D, Glick BR and Greenberg BM (2001). Combining remediation techniques for removal of persistent organic contaminants from soil. In Greenberg BM, Hull RN, Roberts MH Jr and Gensemer RW (eds), *Environmental Toxicology and Risk Assessment*, vol. 10. ASTM STP 1403. American Society for Testing Materials, West Conshohocken, PA, pp. 271–282.

Hulzebos EM, Adema DMM, Dirven-van Breemen EM, Henzen L and van Gestel CAM (1991) QSARs in phytotoxicity. *Science of the Total Environment*, **109**, 493–497.

Jones KC, Stratford JA, Tidridge P, Waterhouse KS and Johnston AE (1989) Polynuclear aromatic hydrocarbons in an agricultural soil: long-term changes in profile distribution. *Journal of Environmental Pollution*, **56**, 337–351.

Kapustka LA (1997) Selection of phytotoxicity tests for use in ecological risk assessments. In Wang W, Gorsuch JW and Hughes JS (eds), *Plants for Environmental Studies*. Lewis, CRC Press, Boca Raton, FL, pp. 515–548.

Kelly LD, Mcguiness LR, Hughes JE and Wainright SC (1999) Effects of phenanthrene on primary production of phytoplankton in two New Jersey estuaries. *Bulletin of Environmental Contamination and Toxicology*, **63**, 646–653.

Krylov SN, Huang XD, Zeiler LF, Dixon DG and Greenberg BM (1997) Mechanistic QSAR model for the photoinduced toxicity of polycyclic aromatic hydrocarbons. I. Physical model based on chemical kinetics in a two compartment system. *Environmental Toxicology and Chemistry*, **16**, 2283–2295.

Lewis MA (1992) Periphyton photosynthesis as an indicator of effluent toxicity: relationship to effects on animal test species. *Aquatic Toxicology*, **23**, 279–288.

Liste H-H and Alexander M (2000) Plant promoted pyrene degradation in soil. *Chemosphere*, **40**, 7–10.

Maliszewska-Kordybach B and Smreczak B (2000) Ecotoxicological activity of soils polluted with polycyclic aromatic hydrocarbons (PAHs) — effect on plants. *Environmental Technology*, **21**, 1099–1110.

Mallakin A, McConkey BJ, Miao G, McKibben B, Sneikus V, Dixon DG and Greenberg BM (1999) Impacts of structural photomodification on the toxicity of environmental contaminants: anthracene photooxidation products. *Ecotoxicology and Environmental Safety*, **43**, 204–212.

Marwood CA, Smith REH, Solomon KR, Charlton MN and Greenberg BM (1999) Intact and photomodified aromatic hydrocarbons inhibit photosynthesis in natural assemblages of Lake Erie phytoplankton exposed to solar radiation. *Ecotoxicology and Environmental Safety*, **44**, 322–327.

Marwood CA, Solomon KR and Greenberg BM (2001) Chlorophyll fluorescence as a predictive bioindicator of effects on growth in aquatic macrophytes from mixtures of polycyclic aromatic hydrocarbons. *Environmental Toxicology and Chemistry*, **20**, 890–898.

Mayer P, Halling-Sorensen B, Sijm DTHM and Nyholm N (1998) Toxic cell concentrations of three polychlorinated biphenyl congeners in the green alga *Selenastrum capricornutum*. *Environmental Toxicology and Chemistry*, **17**, 1848–1851.

McCann JH and Solomon KR (2000) The effect of creosote on membrane ion leakage in *Myriophyllum spicatum* L. *Aquatic Toxicology*, **50**, 275–284.

McIntire T and Lewis GM (1997) The advancement of phytoremediation as innovative environmental technology for stabilization, remediation and restoration of contaminated sites. *Journal of Soil Contamination*, **6**, 227–232.

Owczarek M, Guidotti M, Blasi G, De Simone C, De Marco A and Spadoni M (2001) Traffic pollution monitoring using lichens and bioaccumulators of heavy metals and polycyclic aromatic hydrocarbons. *Fresenius Environmental Bulletin*, **10**, 42–45.

Plewa MJ (1991) The biochemical basis of the activation of promutagens by plant cell systems. In Gorsuch JW, Lower WR, Wang W and Lewis MA (eds), *Plants for Toxicity Assessment*, vol. 2. American Society for Testing and Materials, ASTM STP 1115, West Conshohocken, PA, pp. 287–296.

Pradham SP, Conrad JR, Paterek JR and Srivastave VJ (1998) Potential of phytoremediation for treatment of PAHs in soil at MGP sites. *Journal of Soil Contamination*, **7**, 467–480.

Ren L, Zeiler LF, Dixon DG and Greenberg BM (1996) Photoinduced effects of polycyclic aromatic hydrocarbons on *Brassica napus* (Canola) during germination and early seedling development. *Ecotoxicology and Environmental Safety*, **33**, 73–80.

Salanitro JP, Dorn PB, Huesemann MH, Moore KO, Rhodes IA, Rice-Jackson LM, Vipond TE, Western MM and Wisniewski HL (1997) Crude oil hydrocarbon bioremediation and soil ecotoxicity assessment. *Environmental Science and Technology*, **31**, 1769–1776.

Salt DE, Smith RD and Raskin I (1998) Phytoremediation. *Annual Review of Plant Physiology and Plant Molecular Biology*, **49**, 643–668.

Schoffl F, Prandl R and Reindl A (1998) Regulation of the heat shock response. *Plant Physiology*, **117**, 1135–1141.

Schwab AP, Al-Assi AA and Banks MK (1998) Adsorption of naphthalene onto plant roots. *Journal of Environmental Quality*, **27**, 220–224.

Siciliano SD and Germida JJ (1997) Bacteria inoculants of forage grasses enhance degradation of 2-chlorobenzoic acid in soil. *Environmental Toxicology and Chemistry*, **16**, 1098–1104.

Stephenson GL, Solomon KR, Hale B, Greenberg BM and Scroggins RP (1997) Development of a suitable test method for evaluating the toxicity of contaminated soils to a battery of plant species relevant to soil environments in Canada. In Dwyer FJ, Doane TR and Hinman ML (eds), *Environmental Toxicology and Risk Assessment,* vol. 6. ASTM STP 1317. American Society for Testing and Materials, West Conchohocken, PA, pp. 474–489.

Thomas G, Sweetman AJ, Ockenden WA, Mackay D and Jones KC (1998) Air–pasture transfer of PCBs. *Environmental Science and Technology*, **32**, 936–942.

Tung G, McIlveen WD and Jones RD (1995) Synergistic effect of flyash and SO_2 on development of cucumber (*Cucumis sativus* L.) leaf injury. *Environmental Toxicology and Chemistry*, **14**, 1701–1710.

Wang W (1985) Use of millet root elongation for toxicity tests of phenolic compounds. *Environment International*, **11**, 95–98.

Wang W (1991) Literature review on higher plants for toxicity testing. *Water Air and Soil Pollution*, **59**, 381–400.

Wang W (1994) Rice seed toxicity tests for organic and inorganic substances. *Environmental Monitoring*, **29**, 101–107.

Wetzel A and Werner D (1995) Ecotoxicological evaluation of contaminated soil using the legume root nodule symbiosis as effect parameter. *Environmental Toxicology and Water Quality*, **10**, 127–134.

15

Assessing Risks from Photoactivated Toxicity of PAHs to Aquatic Organisms*

GERALD T. ANKLEY, LAWRENCE P. BURKHARD, PHILIP M. COOK,
STEPHEN A. DIAMOND, RUSSELL J. ERICKSON AND DAVID R. MOUNT
*Mid-Continent Ecology Division, US Environmental Protection Agency,
Duluth, MN, USA*

15.1 INTRODUCTION

Polycyclic aromatic hydrocarbons (PAHs) are one of the most ubiquitous classes of environmental contaminants (LaFlamme and Hites 1978; Neff 1979, 1985). Although most PAHs are toxic only at concentrations large enough to cause narcosis (Neff 1979; Veith *et al.* 1983; Landrum *et al.* 1986), the toxicity of some can be greatly enhanced through mechanisms that involve molecular activation or excitation. The fact that certain PAHs can act as photosensitizing agents, and produce detrimental effects in mammalian models in the presence of solar ultraviolet (UV) radiation, has been recognized for some time (Doniach and Mottram 1937; Mottram and Doniach 1938). Research in subsequent years utilized a variety of *in vitro* and *in vivo* models to assess the occurrence of, and mechanisms responsible for, UV-enhanced toxicity of PAHs. Some of this work used aquatic species such as paramecia and brine shrimps (Epstein *et al.* 1964; Morgan and Warshawsky 1977); however, it was not until the early 1980s that the potential importance of toxicity associated with UV/PAH interactions in aquatic environments began to be appreciated more broadly. This coincided initially with a publication by Bowling *et al.* (1983), which

* The information in this chapter has been funded wholly (or in part) by the US Environmental Protection Agency. It has been subjected to review by the National Health and Environmental Effects Research Laboratory and approved for publication. Approval does not signify that the contents reflect the views of the Agency, nor does mention of trade names or commercial products constitute endorsement or recommendation for use.

described the UV-enhanced toxicity of anthracene to bluegill sunfish held in outdoor mesocosms. Since that initial work, there have been a number of studies documenting the fact that UV radiation can greatly increase the toxicity of PAHs in a broad phylogenetic spectrum of aquatic species, including bacteria (Tuveson *et al*. 1987; McConkey *et al*. 1997), algae (Cody *et al*. 1984; Gala and Giesy 1993; 1994), higher plants (Huang *et al*. 1993; 1996; Ren *et al*. 1994; Duxbury *et al*. 1997; McConkey *et al*. 1997), protozoans (Joshi and Misra 1986), annelids (Ankley *et al*. 1995, 1997; Spehar *et al*. 1999), molluscs (Steinert *et al*. 1998; Spehar *et al*. 1999; Weinstein 2001), crustaceans (Allred and Giesy 1985; Newsted and Giesy 1987; Holst and Giesy 1989; Oris *et al*. 1990; Foran *et al*. 1991; Wernersson and Dave 1997, 1998; Boese *et al*. 1997, 1998; Pelletier *et al*. 1997; Spehar *et al*. 1999; Diamond *et al*. 2000), insects (Borovsky *et al*. 1987; Bleeker *et al*. 1998; Spehar *et al*. 1999), fish (Kagan *et al*. 1985, 1987; Oris and Giesy 1985, 1986, 1987; Oris *et al*. 1990; McClosky and Oris 1991, 1993; Weinstein *et al*. 1997; Spehar *et al*. 1999), and amphibians (Kagan *et al*. 1984; Fernandez and L'Haridon 1992; Walker *et al*. 1998; Hatch and Burton 1998; Monson *et al*. 1999).

15.1.1 MECHANISMS

Although there is evidence that UV radiation can cause enhanced toxicity through chemical modifications of PAHs (i.e. production of metabolites; e.g. Huang *et al*. 1997; Krylov *et al*. 1997), most aquatic toxicology research concerning the effects of UV on PAH toxicity has focused on the process of 'activation' of the compounds, through alterations in their electronic state. Hereafter, we refer specifically to this process as photoactivated toxicity (PAT). The primary toxic mode of action for PAHs that cause PAT, particularly in shorter-term exposures, involves the occurrence of oxidative stress. The following synopsis of the mechanistic basis of PAT is not intended to be comprehensive, but to present major concepts relative to exposure/effects approaches and models appropriate for risk characterization. For those interested in further detail concerning the mechanism(s) of action of PAT of PAHs, and how these mechanisms were elucidated, there are several reviews available (e.g. Foote 1968, 1987; Kearns and Khan 1969; Larson and Berenbaum 1988).

The interaction of UV radiation with PAHs initially involves elevation of the ground state molecule to excited singlet states, which are short-lived species. The energy of these excited states can be lost through one of three pathways: return to the ground state with concomitant loss of thermal energy; return to the ground state with energy loss through fluorescence (photon emission); or conversion to a lower-energy excited triplet state. The triplet state molecules, which in contrast to singlet molecules have unpaired electrons, are much more stable species than the singlet states. As a consequence, they can serve as efficient donors of energy to a variety of molecules. Interactions of these triplet state molecules in biological systems traditionally have been described as Type I and Type II mechanisms. Type I mechanisms involve transfer of energy via hydrogen atoms or electrons to produce radicals or radical ions, which in turn react with oxygen to produce a variety of potentially toxic oxygenated products. The Type II mechanism, generally believed to be primarily responsible for PAT of PAHs, involves transfer of energy from the triplet state molecule to molecular oxygen to form reactive singlet oxygen. Singlet oxygen interacts with a variety of biological substrates, including proteins and saturated fatty acids; interactions with the latter class of molecules produce lipid hydroperoxides, which are the basis of much of the cellular membrane/organelle damage associated with PAT. Organisms possess a variety of mechanisms for dealing with oxidative stress of the type produced by interactions of UV radiation with PAHs; these include enzymatic systems that transfer, via different routes, reducing 'equivalents' from NADPH to oxygen radicals, and small molecules (e.g. some vitamins) that serve as direct quenchers of radical species (Kappus 1986). When these defenses are overwhelmed, toxicity can occur.

15.1.2 PHOTOACTIVATED PAHs: EXPOSURE AND TOXICITY OVERVIEW

To date much of the research concerning PAT of PAHs has been done in the laboratory with organisms exposed to single chemicals in the water, a scenario not typical of most PAH-contaminated sites, which are dominated by complex mixtures. Some researchers have assessed the potential for PAT associated with complex mixtures of PAHs in environmental samples (such as aquatic sediments) in both the laboratory and the field; e.g. Davenport and Spacie (1991) reported that sediment elutriates from a PAH-contaminated site exhibited PAT to cladocerans. Ankley *et al.* (1994) found that field-collected sediments from PAH-contaminated sites were more toxic to epibenthic and benthic invertebrates (oligochaetes, chironomids, amphipods) under UV radiation than under background laboratory fluorescent light. Monson *et al.* (1995) corroborated and expanded upon this observation in a series of *in situ* experiments with sediments from the same site. Additional studies with the sediments revealed the presence of a variety of alkyl-substituted PAHs, as well as nitrogen heterocycles, that potentially could cause PAT (Ankley *et al.* 1996; Kosian *et al.* 1998; West *et al.* 2001). Ireland *et al.* (1996) described a field study similar to that of Monson *et al.* (1995), in which it was shown that cladocerans exposed *in situ* to creosote-contaminated sediments were more sensitive in the presence of full sunlight than in shaded containers. Winger *et al.* (2000) also reported that whole sediments and sediment extracts from PAH-contaminated locations from a river in the south-eastern USA caused PAT to amphipods. In other studies with complex environmental samples, it has been found that weathered crude oil exhibits PAT to a variety of aquatic species, presumably due to the presence of PAHs (Calfee *et al.* 1999; Little *et al.* 2000; Cleveland *et al.* 2000). In related work, Oris *et al.* (1998) reported the occurrence of PAT of PAHs associated with boat motor exhaust to cladocerans and fish in a relatively pristine field setting (Lake Tahoe, NV, USA).

Laboratory studies have clearly demonstrated that PAT of PAHs can be induced at PAH concentrations and UV levels consistent with values measured in the field. The limited number of field studies conducted to date appear to confirm that PAT can occur in natural ecosystems. The implications of these findings for widespread ecological risk become more serious given the ubiquity of PAHs in the environment, and the fact that most environmental guidelines (or criteria) for PAHs do not explicitly consider PAT, because the laboratory toxicity studies on which they are based use ordinary fluorescent light, which does not induce PAT. For example, draft sediment quality guidelines developed for PAHs by Hansen *et al.* (2001) are derived assuming a narcosis mode of action. From an analysis of the US Environmental Protection Agency Environmental Monitoring and Assessment Program (EMAP) data for PAHs in sediments, it is estimated that about 6% of the sediments sampled have PAH concentrations above a concentration estimated to induce chronic toxicity in sensitive organisms via a narcosis mode of action (Hansen *et al.* 2001). From laboratory studies, it is clear that PAT can increase the potency of PAHs by one or two orders of magnitude above that expected from narcosis. Assuming that PAT would actually cause effects in the field at 10% or 1% of the narcosis threshold derived by Hansen *et al.* (2001), the EMAP data suggest that about 40% of the samples would exceed a concentration of 10% of the narcosis threshold, and almost 90% of the samples would be above 1% of the narcosis threshold. We are not suggesting that the actual risk of PAT is in fact this large, but the potential shown by these comparisons makes it clear that a framework to assess risk in aquatic ecosystems is needed.

15.2 PREDICTION OF PAH PHOTOACTIVATED TOXICITY

15.2.1 CONCEPTUAL MODELING FRAMEWORK

Figure 15.1 outlines a conceptual model for predicting effects from PAT of PAHs. As described above, the PAT response is a function of exposure to both UV and PAHs. Basic dose–response models are thus required to integrate and translate UV and PAH dosimetry measurements into projected toxic effects (Figure 15.1A). Because large numbers of individual PAHs occur as mixtures, dose–response predictions require methods to estimate potency for many compounds, including those that have not been directly tested, and approaches to account for the interactive toxicity of PAH mixtures. These dose–response models should also account for differences in species sensitivity.

Prediction of UV exposure (Figure 15.1B) also has many complexities. Considerable progress has been made in developing models to predict incoming solar radiation as a function of astronomical and meteorological conditions.

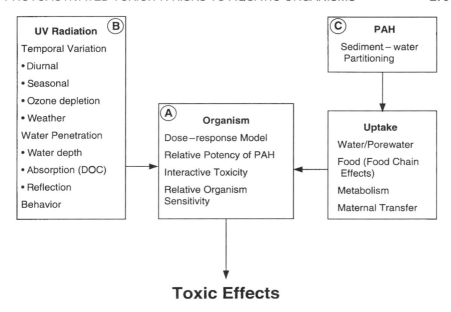

Figure 15.1   Conceptual model of components of prediction of the ecological risk of PAT of PAHs. See text for details

However, estimating exposure for aquatic organisms subject to PAT involves consideration of UV radiation attenuation in the water column, which is itself a function of several variables. Perhaps the most daunting uncertainty relates to organism behavior; since the aquatic environment is highly heterogeneous with respect to UV, the selection of microenvironments by an particular organism can greatly influence the UV radiation exposure it will receive.

Prediction of PAH exposure is also required or, more specifically, tissue residues of PAHs. Assessment of PAH bioaccumulation (Figure 15.1C) benefits from the extensive research concerning bioaccumulation of non-polar organic compounds. This includes consideration of partitioning of chemical between sediment and water column, transfer of chemical through the food chain, and potential maternal transfer of chemical through gametes. One issue of particular concern for PAH bioaccumulation is the effect of metabolism, since many organisms are known to metabolize PAHs (e.g. see Chapters 12 and 13), thus influencing the internal dose available for PAT.

In the following sections we describe in detail aspects of the conceptual framework depicted in Figure 15.1.

15.2.2   IDENTIFICATION OF PHOTOTOXIC PAHs

There are literally thousands of unsubstituted and substituted PAHs that could simultaneously be present in aquatic environments (Neff 1979), but only a subset of these would be expected to exhibit PAT. There has been some comparative testing of different PAH structures for PAT (e.g. Newsted and Giesy 1987), but these empirical data sets are necessarily limited with respect to representing the universe of PAHs that could potentially cause PAT (e.g. Kosian *et al*. 1998). One approach to addressing this challenge is via predictive quantitative structure–activity relationship (QSAR) models.

There have been various attempts to develop QSARs capable of predicting the degree to which the toxicity of PAHs, as well as other chemicals, might be enhanced by UV radiation. Early studies indicated that the ability of a PAH to absorb UV light, as well as the number of fused rings present, were critical chemical characteristics contributing to the occurrence of UV-enhanced toxicity (Epstein *et al*. 1964; Morgan and Warshawsky 1977). Specifically, those studies found that PAHs with four or five fused rings tended to be more toxic in the presence of UV radiation than molecules with fewer or more rings. In more refined work, Newsted and Giesy (1987) reported that the toxic potency of a series of photoactivated PAHs could be predicted accurately using empirically-derived physicochemical characteristics related to characteristics of the excited-state molecule, such as triplet energy and phosphorescence lifetime. An extremely important aspect of their study was that toxicity determinations were based upon fixed tissue concentrations of the PAHs, i.e. potency was a function solely of differences in photochemical properties of the molecules, not inherent differences in their bioconcentration related to different hydrophobicities of the test chemicals.

In recent years even more mechanistic models have been developed to predict the UV-enhanced toxicity of PAHs (Mekenyan *et al*. 1994a, 1994b; Krylov *et al*. 1997; Huang *et al*. 1997). With respect specifically to PAT, Mekenyan *et al*. (1994a, 1994b) described a first principle QSAR model based upon calculated values of the gap between the highest occupied and lowest unoccupied molecular orbitals (HOMO–LUMO) of PAHs (Figure 15.2). The model presumes that the PAT of PAHs is dictated by the energy of light absorbed and stability of the molecules, which can be described by the HOMO–LUMO gap. This relationship results in a HOMO–LUMO gap 'window', that reflects not only these relevant molecular properties of the (excited-state) PAHs, but the occurrence of environmentally realistic UV spectra (Figure 15.2). Mekenyan *et al*. (1994a) utilized data from Newsted and Giesy (1987) to initially quantify a HOMO–LUMO gap window of 6.8–7.6 eV for PAHs that cause PAT. In subsequent modeling work, Veith *et al*. (1995) assessed the likely effects of a variety of substituents on the HOMO–LUMO gap and predicting PAT of several model PAHs. They found that most substituents did not greatly alter the HOMO–LUMO gap calculated for the parent molecule, suggesting that knowledge of the parent structure often would be sufficient to predict potential for PAT. Recently, published toxicity data by Sinks *et al*. (1997) and Boese *et al*. (1998) for a variety of substituted PAHs support the modeling predictions made by Veith *et al*. (1995). It should be noted that PAH accumulation is an important uncertainty

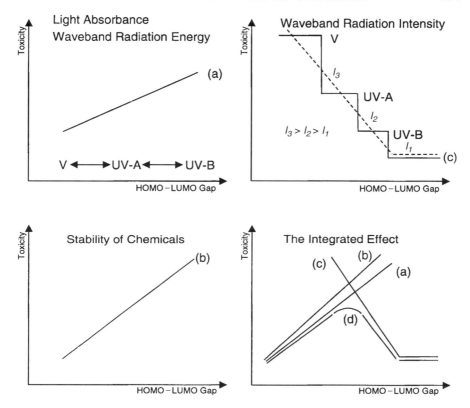

Figure 15.2 Conceptual QSAR model based on the HOMO–LUMO (highest occupied molecular orbital–lowest unoccupied molecular orbital) gap incorporating determinants of PAT of PAHs: (a) light absorbance by PAHs; (b) chemical stability of PAHs; and (c) ambient UV radiation intensity; into (d) integration of factors (a), (b), and (c) in causing PAT. Reproduced from Mekenyan *et al.* 1994a, with permission from Elsevier Science

associated with quantitative use of the HOMO–LUMO gap model; if a chemical is not bioaccumulated/retained it does not matter that it absorbs light at a certain wavelength/intensity. Thus, the model is most appropriately applied to those situations where tissue residues are known or can be estimated, as described below.

15.2.3 PREDICTION OF PAH EXPOSURE/ACCUMULATION

Because PAT of PAHs occurs within the organism, it follows that the most direct index of PAH exposure would be tissue residues of the parent compound. Studies with a variety of species and PAHs support the contention that the fraction of chemical responsible for PAT is more accurately predicted based on the PAH concentration present in the organism, rather than the PAH concentration in the aquatic environment (e.g. Bowling *et al.* 1983; Newsted and Giesy 1987; Ankley *et al.* 1995; Erickson *et al.* 1999; Weinstein 2001). The most direct approach to determining PAH dose is to measure residues in species of concern, but this is often impractical, if not impossible (e.g. if PAT occurs in sensitive species, they would not be sampled). In addition, most existing PAH concentration data for aquatic environments are for water and/or sediments. Thus, it is necessary to have appropriate factors to convert PAH concentrations in water and/or sediment to residues in organisms of concern.

The factors for conversion of chemical concentrations in water and sediment to aquatic organisms are the bioaccumulation factor (BAF) and the biota–sediment accumulation factor (BSAF), respectively. The BAF is the ratio of lipid normalized chemical concentration in the organism to the bioavailable (freely dissolved) chemical concentration in water, and the BSAF is the ratio of lipid normalized chemical in the organism to the organic carbon normalized chemical concentration in the sediment. With these ratios, PAH residues can be readily estimated from sediment and water chemical concentrations.

In developing BAFs and BSAFs for PAHs, metabolism of the PAHs by aquatic organisms themselves is an important consideration affecting the chemical residue in the animal. Little is known about specific metabolism rates for PAHs in aquatic species. However, fish in early life stages appear to have slower rates of PAH metabolism than juvenile/adult stages (Petersen and Kristensen 1998), and benthic invertebrates, mussels, and plankton also appear to have slower rates of PAH metabolism than adult fish. Consequently, PAH residues will, on average, be larger in invertebrates and early life stages of fishes in comparison to adult fishes. For chemicals of similar hydrophobicities, the BAFs and BSAFs for PAHs in fish will be smaller than those for non-metabolized chemicals, such as PCBs and DDE.

In the case of benthic invertebrates, BSAFs can be estimated fairly readily using equilibrium partitioning theory (Di Toro *et al.* 1991), which assumes a dynamic three-phase equilibrium between chemical present in sediment organic carbon, that freely dissolved in interstitial water (porewater), and chemical present in organism lipid. Depending upon the assumption of the preference of a chemical for the lipid and sediment organic carbon phases, BSAFs ranging from 1 (equality) to ~ 2.4 would be estimated using this theory, if no metabolism occurs. Field measured BSAFs for benthically-coupled organisms, compiled and analyzed by Tracey and Hanson (1996), are generally consistent with equilibrium partitioning theory, e.g. mean BSAFs (coefficient of variation) for PAHs, PCBs, and pesticides were 0.29 (57%), 2.1 (74%), and 2.69 (42%), respectively. The lower BSAF for PAHs is consistent with the effects of metabolism. This lower BSAF could also be caused by the presence of coal and soot phases in the sediments. PAHs associated with coal and soot phases appear to be less bioavailable, compared to sorbed PCBs and pesticides (Gustafsson *et al.* 1997).

Even with these uncertainties, it seems as if a reasonable first approximation for PAH BSAFs for benthic invertebrates would be 1.0.

Bioaccumulation factors for fishes can be estimated by assuming equality with the bioconcentration factor (BCF) of a chemical; the BCF is typically determined in the laboratory using exposures via water only. Bioconcentration factors have been measured for a few PAHs for a small number of aquatic species. For larval and embryonic fishes where *in vivo* metabolism was not significant, BCFs can be derived from the following equation of Petersen and Kristensen (1998):

$$\log BCF_l = 0.64 + 0.92 * \log K_{ow} \qquad (15.1)$$

where BCF_l is the ratio of the lipid normalized chemical concentration in the fish to the chemical concentration in the water and K_{ow} is the N-octanol–water partition coefficient of the chemical. Given the chemical concentration in the ambient water and K_{ow} of the chemical, the resulting chemical residue in larval and embryonic fishes can be estimated.

For juvenile and adult fishes, reported PAH BCFs are limited and vary in quality. If one assumes equality of the BCF_l and K_{ow} (see Mackay 1982 for theoretical justification), measured BCF_l values for PCBs and other non-metabolizable chemicals are consistent with this assumption. In contrast, measured BCF_l and BAF_l values for PAHs diverge downwards from equality with K_{ow} at increasing K_{ow} values (Figure 15.3). The BCF_l data suggests that for chemicals with log K_{ow} values less than ~ 4.6, chemical residues in the juvenile and adult fishes can be approximated by multiplying the chemical concentration in the ambient water by the K_{ow} of the chemical. For chemicals with log K_{ow} values exceeding ~ 4.6, much greater uncertainty exists in estimating the chemical residue in fish, and the illustrative line (Figure 15.3) provides a suggested initial estimate

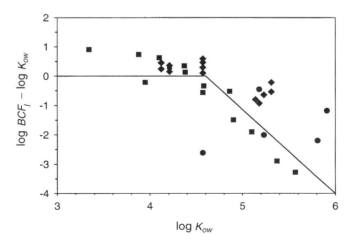

Figure 15.3   Measured BCF$_{ls}$ and BAF$_{ls}$ from Carlson *et al*. (1979), Burkhard and Lukasewycz (2000), and Baussant *et al*. (2001) for PAHs. The line is not fitted, but illustrative

for the BCF_l as a function of K_{ow}. To more accurately predict BCF values for individual PAHs, it is necessary to consider chemical-specific differences in rates of metabolism in the appropriate food chain(s).

15.2.4   PREDICTION OF UV RADIATION EXPOSURE

Prediction of exposure to relevant sunlight wavelengths is also critical to assessing the potential risk associated with PAT (Gilbert and Baggott 1991). The UV wavelengths of relevance in aquatic systems are 280–400 nm. Of this wavelength range, the UVA portion (320–400 nm) is of greatest concern because this is the absorbance range for most PAHs. Also, shorter wavelengths (UVB; 280–320 nm), while biologically very harmful, comprise only about 8% of the total UV radiation present, and are filtered from the water column much more effectively than longer UV wavelengths. An additional component of PAH photoactivation is duration of exposure; the UV dose received is a product of intensity and duration. In most systems, the law of reciprocity suggests that equal damage will be caused by equivalent photon doses (Dworkin 1958), regardless of the rate at which they enter the system.

In natural aquatic systems, dissolved organic carbon (DOC) is the primary determinant of UVA penetration (Williamson *et al.* 1996). Log-transformed UV intensity values plotted against depth typically produce linear relationships, with the slope representing the rate at which radiation is attenuated or reduced with depth. These extinction slopes have been shown to correlate strongly with DOC concentration, which can differ among water bodies by orders of magnitude (Smith *et al.* 1999; Arts *et al.* 2000; Peterson *et al.* 2002). The effect on subsurface UVA can be dramatic, e.g. the extinction coefficient for a near-shore area in Lake Superior (near Duluth, MN, USA) was estimated to be $-0.15/m$, whereas in a St. Louis Harbor (Duluth, MN, USA) site, the extinction slope was estimated to be $-41.3/m$ (Figure 15.4a). The depth at which 90% of the above-surface UVA intensity would be absorbed was estimated to be 15 m for Lake Superior, and 11 cm for the St. Louis Harbor location, a 136-fold difference. Given equivalent tissue PAH concentrations, and assuming that 10% of surface irradiance is sufficient to photoactivate PAHs, organisms would be at risk in 15 m of the Lake Superior water column vs. 11 cm in the St. Louis location water column.

These estimates of UV penetration are expressed as broad-spectrum values and do not reflect potential differences in the spectrum of radiation present, both between sites and at different depths. These differences arise because DOC attenuates shorter UV wavelengths more efficiently than longer wavelengths, and also because DOC composition can differ dramatically, based on its sources in the landscape. The importance of such spectral variability has been demonstrated by Diamond *et al.* (2000) in exposures of brine shrimp (*Artemia salina*) nauplii to three PAHs in combination with different UV spectra. In these assays,

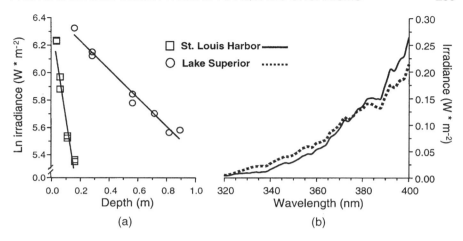

Figure 15.4 Extinction plots for UVA radiation (a) and spectra at 50% UV penetration depth (b) for PAH-contaminated site in St. Louis Bay, Duluth, MN, and near-shore location in Lake Superior. The 50% penetration depths shown are 10 cm and 80 cm for the St. Louis Harbor and Lake Superior sites, respectively

the overlap of UVA radiation with absorbance spectra of pyrene, fluoranthene, and anthracene was manipulated using various filters. Where the radiation spectrum overlapped significant portions of the PAH absorbance spectrum, toxicity did not differ. Where spectra differed in the extent of their overlap, toxicity differed significantly. Most importantly, the variation in the spectra of UVA used was consistent with variation possible in natural aquatic systems (see comparison of sites in Figure 15.4b). The interaction of UVA spectra and PAH absorbance spectra can be summarized by the phototoxicity weighting function (PWF):

$$PWF = \int_{\lambda=320}^{\lambda=400} \epsilon_{\lambda}^{*}\, I_{\lambda}^{*} d\lambda \qquad (15.2)$$

where: PWF = photoactivated toxicity weighting function, ϵ_{λ} = wavelength-specific molar absorptivity and I_{λ} = wavelength-specific irradiance.

Aquatic life risk assessments for typical chemical stressors are simplified by the assumption that exposure is uniform throughout the water column. Accordingly, accumulation of PAHs by aquatic organisms can be predicted with reasonable accuracy without accounting for variation among specific microhabitats within the system. However, despite the ability to estimate incoming solar radiation to aquatic systems, actual exposure will be greatly influenced by the life history and behavior of different species. Attenuation through the water column, as well as physical shading, creates a highly heterogeneous environment with respect to UV radiation. Species that reside in sediments, vegetation, or deep water during daylight will have much reduced UV radiation exposure. Even more complicated is calculation of UV exposure for organisms that move actively among shaded and unshaded habitats. Except in cases where behavior is well understood, risk assessment for PAT may be limited to using representative exposure scenarios as bounding conditions for possible effects, rather than as specific estimates of expected effects.

With these caveats in mind, it is possible to utilize basic modeling approaches linked to site-specific water chemistry characteristics (e.g. DOC concentrations) to derive first-order approximations of the UV dose experienced by aquatic organisms. Diamond *et al.* (2002) have estimated UV radiation doses in wetlands using this approach. Typical UV radiation doses were estimated by first generating maximal solar radiation doses for each day, using a radiative transfer model, SBDART (Richiazzi *et al.* 1998). The model produced values for the full spectrum of sunlight (280–700 nm) for cloudless conditions. These maximal values were then modified based on cloud cover effect estimates from 30 years of historical solar radiation data (National Renewable Energy Laboratory, Department of Energy; http://rredc.nrel.gov/solar/). The values calculated from this procedure were daily, terrestrial solar radiation doses at 2 nm increments (280–700 nm). Water column doses were then derived from site-specific extinction coefficients and spectral attenuation data described by Peterson *et al.* (2002). Although the focus of the study of Diamond *et al.* (2002) was to characterize risk of UVB radiation to amphibians, the procedure is directly applicable to determination of PAT risk, and the resulting UVA radiation, spectral doses can be directly incorporated into the PWF calculation described above.

15.2.5 INTERACTION BETWEEN PAH AND UV RADIATION

Because both PAH and UV exposures vary independently in the environment, dose–response models are needed to integrate combinations of UV and PAH exposure into a single expression of effect. Based on the Bunsen–Roscoe law of photochemical reciprocity (Dworkin 1958), PAT potency should be proportional to the product of UV light and PAH exposures; in other words, reduced PAH in the tissue of an organism can be made toxic by increasing exposure to UV radiation, and vice versa. Early PAT studies with aquatic organisms did, in fact, demonstrate that the UV radiation intensity/duration and the PAH water concentration associated with effects were to some degree inversely related to each other (Oris and Giesy 1985, 1986). This concept can be enhanced through a model that directly considers internal PAH dose (Ankley *et al.* 1995, 1997).

This model involves three assumptions. First, PAHs are assumed to accumulate in an organism by simple first-order kinetics:

$$\frac{dR}{dt} = k_1 \cdot C - k_2 \cdot R \tag{15.3}$$

where C is the PAH exposure concentration, R is the PAH concentration in tissue, k_1 is an uptake rate constant, and k_2 is an elimination rate constant. The model does not attempt to describe multicompartment kinetics or identify specific sites of action, but rather uses the whole-body accumulation as a surrogate for the chemical activity at the site of action.

Second, damage to the organism is assumed to accumulate in proportion to both the tissue PAH concentration and radiation intensity:

$$\frac{dD}{dt} = k_3 \cdot R \cdot I - k_4 \cdot D \qquad (15.4)$$

where D is the cumulative damage, I is the UV light intensity incident on the organism, k_3 is a rate constant for accrual of damage, and k_4 is a rate constant for damage repair. This damage involves a variety of biochemical and physiological steps which are not fully understood and cannot be quantitatively described, so damage is approximated to be proportional to the photochemical reaction rate. Internal light intensity is not addressed in the model because, for a given organism type and size, the average internal intensity would be proportional to the incident intensity by a constant which can be considered as part of k_3.

The third and final assumption is that a given effect is assumed to occur when damage accrues to a critical level, D_C. Such a threshold model is most applicable to quantal endpoints such as death, but might approximately apply to a specified magnitude of a continuous endpoint. However, such continuous endpoints would probably be better served by a model which integrates the degree of damage over time.

This model can be applied to any time series of PAH and UV exposures. For example, if death is the endpoint, PAH accumulation (R) and the light intensity (I) are constant during the period in question, and damage repair is negligible, the model can be expressed as:

$$\frac{R \cdot I \cdot t_D}{D_C / k_3} = 1 \qquad (15.5)$$

where time to death (t_D), accumulation, and light intensity are reciprocally related to each other, their product equaling the constant D_C / k_3, which is a measure of the chemical potency. This would result in a log–log plot of t_D vs. $R \cdot I$ that is linear with a slope of -1 and an intercept at $t = 1$ of D_C / k_3. This relationship would be expected to deviate from linearity in a fashion predictable from the model if either accumulation or light intensity varied during the exposure.

This model was tested by exposing oligochaetes (*Lumbricus variegatus*) to anthracene, fluoranthene, or pyrene in the dark, and then transferring the animals to clean water and exposing them to a range of UV radiation intensities (Ankley *et al.* 1995, 1997). Median lethal times were generally inversely related to the product of light intensity and initial PAH accumulation (Figure 15.5), with some deviation from linearity attributable to PAH depuration. With the

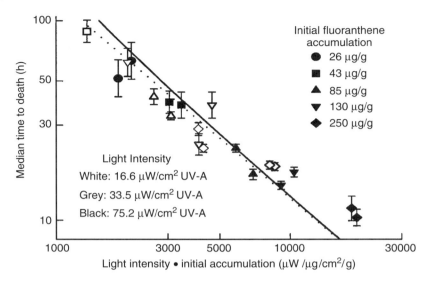

Figure 15.5 Median time to death vs. product of UV radiation intensity and initial fluoranthene concentration in the oligochaete *Lumbriculus variegatus*. Solid line denotes expected relationship based on measured fluoranthene elimination rates and fitted intercept. Dotted line denotes expected relationship assuming no elimination (data from Ankley *et al.* 1995)

model fit based only on the intercept, there was close agreement of the data and model, except that death was slightly delayed relative to model predictions at very high exposures and short times to death. Over the time frame of this data, there was no indication of damage repair, which would have been manifested in deviations from linearity at the longer exposure times. Weinstein (2001) has shown comparable adherence to this model for the PAT of fluoranthene to the glochidium larvae of *Utterbackia imbecillis*.

15.2.6 PREDICTION OF MIXTURE TOXICITY

Because PAHs occur as complex mixtures in the environment, it is necessary to know not only which are likely to cause PAT, but how mixtures of the different photoactivated PAHs interact to produce effects. It is generally accepted that chemicals which operate via the same modes/mechanisms of action should exhibit additive toxicity describable through concentration–addition models (Broderius 1991; Broderius *et al.* 1995). Concentration–addition models designed to predict the effects of mixtures of PAHs exhibiting toxicity via narcosis have been developed and validated (e.g. Swartz *et al.* 1995), but until recently comparable models have not been described for activated PAHs. Results of experiments by Swartz *et al.* (1997) and Boese *et al.* (1999) were not inconsistent with the concept of concentration–addition as a basis for predicting the PAT of PAH mixtures, but these studies were not conducted in a manner suitable for a quantitative assessment of the nature of this interaction.

Erickson *et al.* (1999) demonstrated that acute lethality associated with mixtures of photoactivated PAHs can be predicted via a concentration–addition model. Oligochaetes (*L. variegatus*) were exposed to various levels of individual PAHs and binary mixtures of PAHs and then exposed to UV radiation of various intensities for 96 h. Using the model described above, median times to death from exposures using individual PAHs were used to estimate D_C/k_3 for each chemical. The toxicity of the mixtures was then assessed by evaluating the following additive toxicity index:

$$\sum_i \frac{R_i \cdot I \cdot f_i(t_D)}{D_C/k_{3i}} \tag{15.6}$$

where i denotes different chemicals and $f_i(t_D)$ represents the functional solution of the model for the particular time dependence of chemical accumulation. If additivity exists, this index should equal 1. Across the various levels of light and chemical concentration, this index averaged 0.98 (95% CI 0.88–1.08) for the

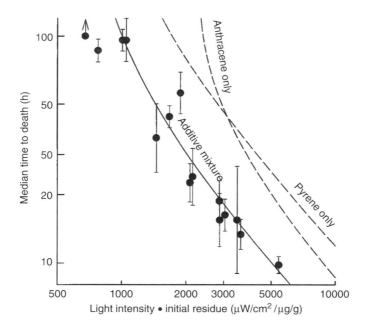

Figure 15.6 Median time to death vs. product of UV radiation intensity and initial combined anthracene and pyrene concentrations in the oligochaete *Lumbriculus variegatus*. Dashed lines denote expected relationship based on models fitted to single chemical exposure to anthracene and pyrene. Solid line denotes predicted relationship based on additive combination of single chemical models (data from Erickson *et al.* 1999)

anthracene/fluoranthene mixture, 0.95 (0.85 – 1.05) for the anthracene/pyrene mixture, and 1.07 (0.95 – 1.19) for the fluoranthene/pyrene mixture, indicating close adherence to additivity. The additivity of these mixtures is also evident in plots of observed median times to death compared to model predictions (Figure 15.6).

15.3 PROSPECTUS

PAHs known to be photoactivated are present at concentrations in aquatic systems such that animals can achieve tissue concentrations of the PAHs sufficient to cause PAT. Further, although UV penetration can vary dramatically among PAH-contaminated sites, it is likely that at least some portion of the aquatic community will be exposed to UV radiation at levels sufficient to initiate PAT.

Based on this, there is a strong suggestion that PAT of PAHs could add significantly to ecological risk in PAH-contaminated sites. At the present time, however, the ability to conduct PAH – PAT risk assessments of acceptable uncertainty is limited by comprehensive information on species exposure (specifically, tissue concentration) to both PAH and UV radiation during all life stages. The information gaps in these areas are somewhat daunting, and have been discussed by Diamond and Mount (1998). Briefly, PAH exposure and uptake, as well as UV exposure, are likely to vary considerably among species and life stage as they migrate into and out of contaminated locations and areas of high and low UV penetration. For all but sessile species, these patterns of movements are the greatest determinant of risk for PAT. Unfortunately, the species of greatest concern, or highest probability of PAT effects, are generally absent from contaminated sites, making clear documentation of those effects very difficult.

The risk of PAT in PAH-contaminated systems is unlikely to diminish in the near future. PAHs generally reside deep in contaminated-site sediments, are slow to degrade, and continue to be added to the environment via a number of pathways. Also, the environmental conditions that mediate both PAH and UV exposure, particularly water quality and weather factors, are presently undergoing significant alteration due to global warming and associated climate changes. The need for information sufficient to conduct reliable risk assessments is thus likely to increase rather than decrease in coming decades. The framework and predictive tools presented in this manuscript provide a conceptually rigorous approach via which to conduct this type of assessment.

ACKNOWLEDGMENTS

Review comments on an earlier draft were provided by R. Spehar and B. Boese. R. LePage and D. Spehar assisted in manuscript preparation.

REFERENCES

Allred PM and Giesy JP (1985) Solar radiation-induced toxicity of anthracene to *Daphnia pulex*. *Environmental Toxicology and Chemistry*, **4**, 219–226.

Ankley GT, Collyard SA, Monson PD and Kosian PA (1994) Influence of ultraviolet light on the toxicity of sediments contaminated with polycyclic aromatic hydrocarbons. *Environmental Toxicology and Chemistry*, **13**, 1791–1796.

Ankley GT, Erickson RJ, Phipps GL, Mattson VR, Kosian PA, Sheedy BR and Cox JS (1995) Effects of light intensity on the phototoxicity of fluoranthene to a benthic macroinvertebrate. *Environmental Science and Technology*, **29**, 2828–2833.

Ankley GT, Mekenyan OG, Kosian PA, Makynen EA, Mount DR, Monson PD and Call DJ (1996) Identification of phototoxic polycyclic aromatic hydrocarbons in sediments through sample fractionation and QSAR analysis. *SAR QSAR Environmental Research*, **5**, 177–183.

Ankley GT, Erickson RJ, Sheedy BR, Kosian PA, Mattson VR and Cox JS (1997) Evaluation of models for predicting the phototoxic potency of polycyclic aromatic hydrocarbons. *Aquatic Toxicology*, **37**, 37–50.

Arts MT, Robarts RD, Kasai F, Waiser MJ, Tumber VP, Plante AJ, Rai H and De Lange HJ (2000) The attenuation of ultraviolet radiation in high dissolved organic carbon waters of wetlands and lakes in the northern Great Plains. *Limnology and Oceanography*, **45**, 292–299.

Baussant T, Sanni S, Jonsson G, Skadsheim A and Borseth JF (2001) Bioaccumulation of polycyclic aromatic compounds: 1. Bioconcentration into marine species and in semipermeable membrane devices during chronic exposure to dispersed crude oil. *Environmental Toxicology and Chemistry*, **20**, 1175–1184.

Bleeker EAJ, van der Geest HG, Kraak MHS, de Voogt P and Admiraal W (1998) Comparative ecotoxicity of NPAHs to larvae of the midge *Chironomus raparius*. *Aquatic Toxicology*, **41**, 51–62.

Boese BL, Lamberson JO, Swartz RC and Ozretich RJ (1997) Photo-induced toxicity of fluoranthene to seven marine benthic crustaceans. *Archives of Environmental Contamination and Toxicology*, **32**, 389–393.

Boese BL, Lamberson JO, Swartz RC, Ozretich R and Cole F (1998) Photoinduced toxicity of PAHs and alkylated PAHs to a marine infaunal amphipod (*Rhepoxynius abronius*). *Archives of Environmental Contamination and Toxicology*, **34**, 235–240.

Boese BL, Ozretich RJ, Lamberson JO, Swartz RD, Cole FA, Pelletier J and Jones J (1999) Toxicity and phototoxicity of mixtures of highly lipophilic PAH compounds in marine sediment: can the summed PAH model be extrapolated? *Archives of Environmental Contamination and Toxicology*, **36**, 270–280.

Borovsky D, Linley JR and Kagan J (1987) Polycyclic aromatic compounds as phototoxic mosquito larvicides. *Journal of the American Mosquito Control Association*, **3**, 246–250.

Bowling JW, Leversee GJ, Landrum PF and Giesy JP (1983) Acute mortality of anthracene-contaminated fish exposed to sunlight. *Aquatic Toxicology*, **3**, 79–90.

Broderius SJ (1991) Modeling the joint toxicity of xenobiotics to aquatic organisms: basic concepts and approaches. In Mayes MA and Barron MG (eds), *Aquatic Toxicology and Risk Assessment*, Vol. 14. ASTM STP 1124. American Society for Testing and Materials, Philadelphia, PA, pp. 107–127.

Broderius SJ, Kahl MD and Hoglund MD (1995) Use of joint toxic response to define the primary mode of toxic action for diverse industrial organic chemicals. *Environmental Toxicology and Chemistry*, **9**, 1591–1605.

Burkhard LP and Lukasewycz MT (2000) Some bioaccumulation factors and biota-sediment accumulation factors for polycyclic aromatic hydrocarbons in lake trout. *Environmental Toxicology and Chemistry*, **19**, 1427–1429.

Calfee RD, Little EE, Cleveland L and Barron MG (1999) Photoenhanced toxicity of a weathered oil on *Ceriodaphnia dubia* reproduction. *Environmental Science and Pollution Research*, **6**, 207–212.

Carlson RM, Oyler AR, Gerhart EH, Caple R, Welch KJ, Kopperman HL, Bodenner D and Swanson D (1979) Implications to the aquatic environment of polynuclear aromatic hydrocarbons liberated from Northern Great Plains coal. EPA- 600/3-79/093. US Environmental Protection Agency, Duluth, MN.

Cleveland L, Little EE, Calfee RD and Barron MG (2000) Photoenhanced toxicity of weathered oil to *Mysidopsis bahia*. *Aquatic Toxicology*, **49**, 63–76.

Cody TE, Radike MJ and Warshawsky D (1984) The phototoxicity of benzo[α]pyrene in the green alga *Selenastrum capricornicum*. *Environmental Research*, **35**, 122–132.

Davenport R and Spacie A (1991) Acute phototoxicity of harbor and tributary sediments from lower Lake Michigan. *Journal of Great Lakes Research*, **17**, 51–56.

Diamond SA and Mount DR (1998) Evaluating the role of photo-activated PAH toxicity in risk assessment. *SETAC News, Learned Discourses*, **18**, 17–18.

Diamond SA, Mount DR, Burkhard LP, Ankley GT, Makynen EA and Leonard EN (2000) Effect of irradiance spectra on the photoinduced toxicity of three polycyclic aromatic hydrocarbons. *Environmental Toxicology and Chemistry*, **19**, 1389–1398.

Diamond SA, Peterson GS, Tietge JE, Ankley GT (2002) Assessment of the risk of solar ultraviolet radiation effects on amphibians. III. Prediction of impacts in selected northern midwestern wetlands. *Environmental Science and Technology*, **36**, 2866–2874.

Di Toro DM, Zarba CS, Hansen DJ, Berry WJ, Swartz RC, Cowan CE, Pavlou SP, Allen AE, Thomas NA and Paquin PR (1991) Technical basis for establishing sediment quality criteria for nonionic organic chemicals using equilibrium partitioning. *Environmental Toxicology and Chemistry*, **10**, 1541–1583.

Doniach I and Mottram JC (1937) Sensitization of the skin of mice to light by carcinogenic agents. *Nature*, **140**, 588.

Duxbury CL, Dixon DG and Greenberg BM (1997) Effects of simulated solar radiation on the bioaccumulation of polycyclic aromatic hydrocarbons by the duckweed *Lemna gibba*. *Environmental Toxicology and Chemistry*, **16**, 1739–1748.

Dworkin M (1958) Endogenous photosensitization in a carotenoidless mutant of *Rhodopseudomonus speroides*. *Journal of General Physiology*, **41**, 1099–1112.

Epstein SS, Small M, Falk HL and Mantel N (1964) On the association between photodynamic and carcinogenic activities in polycyclic compounds. *Cancer Research*, **24**, 855–862.

Erickson RJ, Ankley GT, DeFoe DL, Kosian PA and Makynen EA (1999) Additive toxicity of binary mixtures of phototoxic polycyclic aromatic hydrocarbons to the oligochaete *Lumbriculus variegatus*. *Toxicology and Applied Pharmacology*, **154**, 97–105.

Fernandez M and L'Haridon J (1992) Influence of lighting conditions on toxicity and genotoxicity of various PAH in the newt *in vivo*. *Mutation Research*, **298**, 31–41.

Foote CS (1968) Mechanisms of photosensitized oxidation. *Science*, **162**, 963–970.

Foote CS (1987) Type I and type II mechanisms of photodynamic action. In Heitz JR and Downum KR (eds), *Light-activated Pesticides*. American Chemical Society, Washington, DC, pp. 22–38.

Foran JA, Holst LL and Giesy JP (1991) Effects of photoenhanced toxicity of anthracene on ecological and genetic fitness of *Daphnia magna*. *Environmental Toxicology and Chemistry*, **10**, 425–427.

Gala WR and Giesy JP (1993) Using the carotenoid biosynthesis inhibiting herbicide, fluridone, to investigate the ability of carotenoid pigments to protect algae from the photo-induced toxicity of anthracene. *Aquatic Toxicology*, **27**, 61–70.

Gala WR and Giesy JP (1994) Flow cytometric determination of the photoinduced toxicity of anthracene to the green algae *Selenastrum capricornutum*. *Environmental Toxicology and Chemistry*, **13**, 831–840.

Gilbert A and Baggott J (1991) *Essentials of Molecular Photochemistry*. Blackwell Science, Paris.

Gustafsson O, Haghseta F, Chan C, MacFarlane J and Gschwend PM (1997) Quantification of the dilute sedimentary soot-phase: implications for PAH speciation and bioavailability. *Environmental Science and Technology*, **31**, 203–209.

Hansen DJ, DiToro DM, McGrath JA, Swartz RC, Mount DR, Burgess RM, Ozretich RJ, Bell HE, Reiley MC and Linton TK (2001) Equilibrium partitioning sediment guidelines (ESGs) for the protection of benthic organisms: PAH mixtures (draft). US EPA Office of Water, Office of Science and Technology, Washington, DC.

Hatch AC and Burton GA Jr (1998) Effects of photoinduced toxicity of fluoranthene on amphibian embryos and larvae. *Environmental Toxicology and Chemistry*, **17**, 1777–1785.

Holst L and Giesy J (1989) Chronic effects of the photoenhanced toxicity of anthracene on *Daphnia magna* reproduction. *Environmental Toxicology and Chemistry*, **8**, 933–942.

Huang XD, Dixon DG and Greenberg BM (1993) Impacts of UV radiation and photomodification on the toxicity of PAHs to the higher plant *Lemna gibba* (duckweed). *Environmental Toxicology and Chemistry*, **12**, 1067–1077.

Huang XD, Zeiler LF, Dixon DG and Greenberg BM (1996) Photoinduced toxicity of PAHs to the foliar regions of *Brassica napus* (canola) and *Cucumbis sativus* (cucumber) in simulated solar radiation. *Ecotoxicology and Environmental Safety*, **35**, 190–197.

Huang XD, Krylov SN, Ren L, McConkey BJ, Dixon DG and Greenberg BM (1997) Mechanistic quantitative structure-activity relationship model for the photoinduced toxicity of polycyclic aromatic hydrocarbons: II. An empirical model for the toxicity of 16 polycyclic aromatic hydrocarbons to the duckweed *Lemna gibba* L. G-3. *Environmental Toxicology and Chemistry*, **16**, 2296–2303.

Ireland DS, Burton GA Jr and Hess GG (1996) *In situ* toxicity evaluations of turbidity and photoinduction of polycyclic aromatic hydrocarbons. *Environmental Toxicology and Chemistry*, **15**, 574–581.

Joshi PC and Misra RB (1986) Evaluation of chemically-induced phototoxicity to aquatic organism using *Paramecium* as a model. *Biochemical and Biophysical Research Communications*, **139**, 79–84.

Kagan J, Kagan PA and Buhse HE Jr (1984) Light-dependent toxicity of α-terthienyl and anthracene toward late embryonic stages of *Rana pipiens*. *Journal of Chemical Ecology*, **10**, 1115–1122.

Kagan J, Kagan ED, Kagan IA, Kagan PA and Quigley S (1985) The phototoxicity of non-carcinogenic polycyclic aromatic hydrocarbons in aquatic organisms. *Chemosphere*, **14**, 1829–1834.

Kagan J, Sinnott D and Kagan ED (1987) The toxicity of pyrene in the fish *Pimephales promelas*: synergism by piperonyl butoxide and by ultraviolet light. *Chemosphere*, **16**, 2291–2298.

Kappus H (1986) Overview of enzyme systems involved in bioreduction of drugs and in redox cycling. *Biochemical Pharmacology*, **35**, 1–6.

Kearns DR and Khan AU (1969) Sensitized photooxygenation reactions and the role of singlet oxygen. *Photochemistry and Photobiology*, **10**, 193–210.

Kosian PA, Makynen EA, Monson PD, Mount DR, Spacie A, Mekenyan OG and Ankley GT (1998) Application of toxicity-based fractionation techniques and structure–activity relationship models for the identification of phototoxic polycyclic aromatic hydrocarbons in sediment pore water. *Environmental Toxicology and Chemistry*, **17**, 1021–1033.

Krylov SN, Huang XD, Zeiler LF, Dixon DG and Greenberg BM (1997) Mechanistic quantitative structure–activity relationship model for the photoinduced toxicity of polycyclic aromatic hydrocarbons: I. Physical model based on chemical kinetics in a two-compartment system. *Environmental Toxicology and Chemistry*, **16**, 2283–2295.

LaFlamme RE and Hites RA (1978) The global distribution of polycyclic aromatic hydrocarbons in recent sediments. *Geochimica et Cosmochimica Acta*, **42**, 289–303.

Landrum PF, Giesy JP, Oris JT and Allred PM (1986) The photoinduced toxicity of polycyclic aromatic hydrocarbons to aquatic organisms. In Vandermeulin JH and Hrudley S (eds), *Oil in Fresh Water: Chemistry, Biology, Technology*. Pergamon, New York, pp. 304–318.

Larson RA and Berenbaum MR (1988) Environmental phototoxicity. *Environmental Science and Technology*, **22**, 354–360.

Little EE, Cleveland L, Calfee R and Barron MG (2000) Assessment of the photoenhanced toxicity of a weathered oil to the tidewater silverside. *Environmental Toxicology and Chemistry*, **19**, 926–932.

Mackay D (1982) Correlation of bioconcentration factors. *Environmental Science and Technology*, **16**, 274–278.

McCloskey JT and Oris JT (1991) Effect of water temperature and dissolved oxygen concentration on the photoinduced toxicity of anthracene to juvenile bluegill sunfish (*Lepomis macrochirus*). *Aquatic Toxicology*, **21**, 145–156.

McCloskey JT and Oris JT (1993) Effect of anthracene and solar ultraviolet radiation exposure on gill ATPase and selected hematological measurements in the bluegill sunfish (*Lepomis macrochirus*). *Aquatic Toxicology*, **24**, 207–218.

McConkey BJ, Duxbury CL, Dixon DG and Greenberg BM (1997) Toxicity of a PAH photooxidation product to the bacteria *Photobacterium phosphoreum* and the duckweed *Lemna gibba*: effects of phenanthrene and its primary photoproduct, phenanthrenequinone. *Environmental Toxicology and Chemistry*, **16**, 892–899.

Mekenyan OG, Ankley GT, Veith GD and Call DJ (1994a) QSARs for photoinduced toxicity: I. Acute lethality of PAHs to *Daphnia magna*. *Chemosphere*, **28**, 567–582, Elsevier.

Mekenyan OG, Ankley GT, Veith GD and Call DJ (1994b) QSAR estimates of excited states and photoinduced acute toxicity of polycyclic aromatic hydrocarbons. *SAR QSAR Environmental Research*, **2**, 237–247.

Monson PD, Ankley GT and Kosian PA (1995) Phototoxic response of *Lumbriculus variegatus* to sediments contaminated by polycyclic aromatic hydrocarbons. *Environmental Toxicology and Chemistry*, **14**, 891–894.

Monson PD, Call DJ, Cox DA, Liber K and Ankley GT (1999) Photoinduced toxicity of fluoranthene to northern leopard frogs (*Rana pipiens*). *Environmental Toxicology and Chemistry*, **18**, 208–312.

Morgan DD and Warshawsky D (1977) The photodynamic immobilization of *Artemia salina* nauplii by polycyclic aromatic hydrocarbons and its relationship to carcinogenic activity. *Photochemistry and Photobiology*, **25**, 39–46.

Mottram JC and Doniach I (1938) The photodynamic action of carcinogenic agents. *Lancet*, **1**, 1156–1159.

Neff JM (1979) *Polycyclic Aromatic Hydrocarbons in the Aquatic Environment*. Applied Science, London.

Neff JM (1985) Polycyclic aromatic hydrocarbons. In Rand GM and Petrocelli SR (eds), *Fundamentals of Aquatic Toxicology*. Hemisphere, New York, p. 666.

Newsted JL and Giesy JP (1987) Predictive models for photoinduced acute toxicity of polycyclic aromatic hydrocarbons to *Daphnia magna*, Strauss (Cladocera, Crustacea). *Environmental Toxicology and Chemistry*, **6**, 445–461.

Oris JT and Giesy JP Jr (1985) The photoenhanced toxicity of anthracene to juvenile sunfish (*Lepomis* spp.). *Aquatic Toxicology*, **6**, 133–146.

Oris JT and Giesy JP Jr (1986) Photoinduced toxicity of anthracene to juvenile bluegill sunfish (*Lepomis macrochirus* Rafinesque): photoperiod effects and predictive hazard evaluation. *Environmental Toxicology and Chemistry*, **5**, 761–768.

Oris JT and Giesy JP Jr (1987) The photo-induced toxicity of polycyclic aromatic hydrocarbons to larvae of the fathead minnow (*Pimephales promelas*). *Chemosphere*, **16**, 1395–1404.

Oris JT, Hall AT and Tylka JD (1990) Humic acids reduce the photo-induced toxicity of anthracene to fish and *Daphnia*. *Environmental Toxicology and Chemistry*, **9**, 575–583.

Oris JT, Hatch AC, Weinstein JE, Findlay RH, Diamond SA, Burton GA and Allen B (1998) Modeling site specific phototoxicity of PAH in natural waters: case study of Lake Tahoe, California/Nevada, USA. In *SETAC 19th Annual Meeting Abstract Book*. Society of Environmental Toxicology and Chemistry, Charlotte, NC, November 15–19, p. 45.

Pelletier MC, Burgess RM, Ho KT, Kuhn A, McKinney RA and Ryba SA (1997) Phototoxicity of individual polycyclic aromatic hydrocarbons and petroleum to marine invertebrate larvae and juveniles. *Environmental Toxicology and Chemistry*, **16**, 2190–2199.

Petersen GI and Kristensen P (1998) Bioaccumulation of lipophilic substances in fish early life stages. *Environmental Toxicology and Chemistry*, **17**, 1385–1395.

Peterson GS, Johnson LB, Axler RP and Diamond SA (2002) *In situ* characterization of solar ultraviolet radiation in amphibian habitats. *Environmental Science and Technology*, **36**, 2859–2865.

Ren L, Huang XD, McConkey BJ, Dixon DG and Greenberg BM (1994) Photoinduced toxicity of three polycyclic aromatic hydrocarbons (fluoranthene, pyrene, and naphthalene) to the duckweed (*Lemna gibba* L.). *Ecotoxicology and Environmental Safety*, **28**, 160–171.

Ricchiazzi P, Yang S, Gautier C, Sowle D (1998) SBDART: a research and teaching software tool for plane-parallel radiative transfer in the earth's atmosphere. *Bulletin of the American Meteorological Society*, **79**, 2101–2114.

Sinks GD, Schultz TW and Hunter RS (1997) UVB-induced toxicity of PAHs: effects of substituents and heteroatom substitution. *Bulletin of Environmental Contamination and Toxicology*, **59**, 1–8.

Smith REH, Furgal JA, Charlton DM, Greenberg BM, Hiriart V and Marwood C (1999) Attenuation of ultraviolet radiation in a large lake with low dissolved organic matter concentrations. *Canadian Journal of Fisheries and Aquatic Science*, **56**, 1351–1361.

Spehar RL, Poucher S, Brooke LT, Hansen DJ, Champlin D and Cox DA (1999) Comparative toxicity of fluoranthene to freshwater and saltwater species under fluorescent and ultraviolet light. *Archives of Environmental Contamination and Toxicology*, **37**, 496–502.

Steinert SA, Streib Montee R and Sastre MP (1998) Influence of sunlight on DNA damage in mussels exposed to polycyclic aromatic hydrocarbons. *Marine Environmental Research*, **46**, 355–358.

Swartz RC, Schults DW, Ozretich RJ, Lamberson JO, Cole FA, DeWitt TH, Redmond MS and Ferraro SP (1995) PAH: a model to predict the toxicity of field-collected marine sediment contaminated by polynuclear aromatic hydrocarbons. *Environmental Toxicology and Chemistry*, **14**, 1977–1987.

Swartz RC, Ferraro SP, Lamberson JO, Cole FA, Ozretich RJ, Boese BL, Schults DW, Behrenfeld M and Ankley GT (1997) Photoactivation and toxicity of mixtures of polycyclic aromatic hydrocarbon compounds in marine sediment. *Environmental Toxicology and Chemistry*, **16**, 2151–2157.

Tracey GA and Hansen DJ (1996) Use of biota-sediment accumulation factors to assess similarity of nonionic organic chemical exposure to benthically-coupled organisms of differing trophic mode. *Archives of Environmental Contamination and Toxicology*, **30**, 467–475.

Tuveson RW, Kagan J, Shaw MA, Moresco GM, Behne EMV, Pu H, Bazin M and Santus R (1987) Phototoxic effects of fluoranthene, a polycyclic aromatic hydrocarbon, on bacterial species. *Environmental and Molecular Mutagenesis*, **10**, 245–261.

Veith, GD, Call DJ and Brooke LT (1983) Structure–toxicity relationships for the fathead minnow, *Pimephales promelas*: narcotic industrial chemicals. *Canadian Journal of Fisheries and Aquatic Sciences*, **40**, 743–748.

Veith GD, Mekenyan OG, Ankley GT and Call DJ (1995) A QSAR analysis of substituent effects on the photoinduced acute toxicity of PAHs. *Chemosphere*, **30**, 2129–2142.

Walker SE, Taylor DH and Oris JT (1998) Behavioral and histopathological effects of fluoranthene on bullfrog larvae (*Rana catesbiana*). *Environmental Toxicology and Chemistry*, **17**, 734–739.

Weinstein JE (2001) Characterization of the acute toxicity of photoactivated fluoranthene to glochidia of the freshwater mussel, *Utterbackia imbecillis*. *Environmental Toxicology and Chemistry*, **20**, 412–419.

Weinstein JE, Oris JT and Taylor DH (1997) An ultrastructural examination of the mode of UV-induced toxic action of fluoranthene in fathead minnow, *Pimephales promelas*. *Aquatic Toxicology*, **39**, 1–22.

Wernersson AS and Dave G (1997) Phototoxicity identification by solid phase extraction and photoinduced toxicity to *Daphnia magna*. *Archives of Environmental Contamination and Toxicology*, **32**, 268–273.

Wernersson AS and Dave G (1998) Effects of different protective agents on the phototoxicity of fluoranthene to *Daphnia magna*. *Comparative Biochemistry and Physiology*, **120C**, 373–381.

West CW, Kosian PA, Mount DR, Makynen EA, Pasha MS, Sibley PK and Ankley GT (2001) Amendment of sediments with a carbonaceous resin reduces bioavailability of polycyclic aromatic hydrocarbons. *Environmental Toxicology and Chemistry*, **20**, 1104–1111.

Williamson CE, Stemberger RS, Morris DP, Frost TM and Paulsen SG (1996) Ultraviolet radiation in North American lakes: attenuation estimates from DOC measurements and implications for plankton communities. *Limnology Oceanography*, **41**, 1024–1034.

Winger PV, Lasier PJ, White DH and Seginak JT (2000) Effects of contaminants in dredge material from the Lower Savannah River. *Archives of Environmental Contamination and Toxicology*, **38**, 128–136.

16

Biomarkers and PAHs — Prospects for the Assessment of Exposure and Effects in Aquatic Systems

ROLF ALTENBURGER[1], HELMUT SEGNER[2] AND
RON VAN DER OOST[3]

[1]*UFZ Centre for Environmental Research, Leipzig, Germany*
[2]*Centre for Fish and Wildlife Health, University of Berne, Berne, Switzerland*
[3]*OMEGAM Environmental Research Institute, Amsterdam, The Netherlands*

16.1 INTRODUCTION

This chapter aims to provide an overview of biomarker studies for the assessment of exposure of aquatic organisms to PAHs and their molecular, biochemical, and cellular actions, as well as their organismic and ecological effects. A discussion on efforts to validate the indicative potential of specific biomarkers is also provided. The central focus of the paper is to demonstrate that, in the case of PAHs, (eco)toxicological assessment approaches focusing on a single biomarker response or single substances are very restricted in scope, due to the many confounding factors, such as compound- and species-dependent biotransformation of compounds with different effect qualities, mixture exposure to various PAHs with multiple-response profiles, photoactivation of compounds, etc. Instead, mode-of-action-oriented integrative assessment approaches are favored to gain rational approaches for assessment of exposure and effects of PAHs in aquatic systems.

The term 'biomarker' is generally used in a broad sense to include almost any measurement reflecting an interaction between a biological system and a potential hazard, which may be chemical, physical, or biological (WHO 1993). A biomarker is defined as a change in a biological response (ranging from molecular through cellular and physiological responses to behavioral changes)

PAHs: An Ecotoxicological Perspective. Edited by Peter E.T. Douben.
© 2003 John Wiley & Sons Ltd

that can be related to exposure to or toxic effects of environmental chemicals (Peakall 1994). Van Gestel and Van Brummelen (1996) redefined the terms 'biomarker', 'bioindicator', and 'ecological indicator', linking them to different levels of biological organization:

- *Biomarker*: any biological response to an environmental chemical at the sub-individual level, measured inside an organism or in its products (urine, feces, hair, feathers, etc.), indicating a deviation from the normal status that cannot be detected in the intact organism.
- *Bioindicator*: an organism giving information on the environmental conditions of its habitat by its presence or absence or by its behavior.
- *Ecological indicator*: ecosystem parameter, describing the structure and functioning of ecosystems.

Good biomarkers are sensitive indices of both bioavailability and early biological responses. Biomarkers may be used after exposure to dietary, environmental, or occupational sources, to elucidate cause–effect and dose–effect relationships in health risk assessment, in clinical diagnoses, and for monitoring purposes. Generally, biomarker responses are considered to be intermediates between sources and higher-level effects (Suter 1993). The most compelling reason for using biomarkers is that they can give information on the biological effects of pollutants, rather than a mere quantification of their environmental levels. Biomarkers may provide insight into the potential mechanisms of contaminant effects. By screening multiple biomarker responses, important information will be obtained about organism toxicant exposure and stress. A pollutant stress situation normally triggers a cascade of biological responses, each of which may, in theory, serve as a biomarker (McCarthy *et al.* 1991). Above a certain threshold (in pollutant dose or exposure time), the pollutant-responsive biomarker signals deviate from the normal range in an unstressed situation, finally leading to the manifestation of a multiple effect situation at higher hierarchical levels of biological organization. Improper application or interpretation of biomarker responses, however, may lead to false conclusions as to pollutant stress or environmental quality, e.g. certain responses established for one species are not necessarily valid for other species. Moreover, ecotoxicological data obtained by laboratory studies can be difficult to translate into accurate predictions of effects that may occur in the field (ECETOC 1993). Since both overestimation and underestimation of effects may occur, laboratory observations on biomarkers must always be validated when used for site-specific assessments. Since biomarkers can be applied in both the laboratory and the field, they may provide an important linkage between laboratory toxicity and field assessment. For field samples, biomarkers provide an important index of the total external load that is biologically available in the 'real-world' exposure.

According to the NRC (1987) and the WHO (1993), biomarkers can be subdivided into three classes:

- *Biomarkers of exposure*: covering the detection and measurement of an exogenous substance or its metabolite or the product of an interaction between a xenobiotic agent and some target molecule or cell that is measured in a compartment within an organism.
- *Biomarkers of effect*: including measurable biochemical, physiological, or other alterations within tissues or body fluids of an organism that can be recognized as associated with an established or possible health impairment or disease.
- *Biomarkers of susceptibility*: indicating the inherent or acquired ability of an organism to respond to the challenge of exposure to a specific xenobiotic substance, including genetic factors and changes in receptors which alter the susceptibility of an organism to that exposure.

In this chapter, biomarkers are understood as detecting a biological response to a chemical exposure or prevalence of a stress factor and comparing this with a reference. Commonly, suborganismic effects detected in individuals are used for biomarker studies. Biomarkers do not provide toxicological information *eo ipso* but detect alterations of a structural or functional feature in comparison to a reference situation — not necessarily a control. Effect propagation from molecular to organismic and from organismic to population or community responses, as well as the toxicological significance, therefore have to be validated in separate investigations, which will be considered in Section 16.4. The most promising features of biomarker investigations are the early and adjustable indications of potentially toxic effects. The early indication can be seen in the perspective of both time and concentration.

Regarding our current understanding of the different mechanisms of action of PAHs, there are several established biomarkers reflecting different aspects of molecular interaction, such as CYP450 induction, DNA adduct formation, or PAH metabolites in fish bile. Within the framework of the Oslo and Paris Conventions for the Prevention of Marine Pollution (OSPAR) and in collaboration with the International Council for the Exploration of the Sea (ICES), a PAH-specific suite of biomarkers has been suggested, consisting of CYP1A induction, DNA adduct formation and PAH metabolites in bile (Stagg 1998). Further biochemical and physiological responses, such as Phase II metabolic transformation activities, or modulation of gap junction communication, have promise as biomarkers for PAH exposure and effect, while parameters such as the prevalence of tumors or scope for growth may serve as indicators of effect propagation to higher levels of biological response. General information on these biomarkers can be obtained from an extensive review by Van der Oost *et al.* 2003.

When considering PAHs and their ecotoxicity, we have to be aware that PAHs do not constitute a homogenous group of chemicals, but that the various PAHs can vary considerably in their physicochemical characteristics, their environmental fate and metabolism, and their toxicokinetic and toxicodynamic properties (Van Brummelen *et al.* 1998). While kinetic considerations focus on the fate of chemicals in an organism and estimation of internal dosage, toxicodynamics is concerned with elucidating modes and mechanisms of toxic action. To date, focus in (eco)toxicological studies has mainly been given to PAHs that can form carcinogenic metabolites, while the environmental impact of non-metabolized, non-carcinogenic PAHs is less well characterized.

With the intention of identifying prospects for the assessment of exposure and effects of PAHs in aquatic systems using biomarkers, a consideration of our current understanding of the toxicology of PAHs in organisms and its relation to biomarker responses will be the starting point. Subsequently, the possibilities of biomarkers to identify, detect, and assess occurrence, primary actions, and effects in aquatic systems will be discussed in separate aspects and reviewed for site-specific case studies.

16.2   BIOAVAILABILITY AND TOXICOKINETICS OF PAHs

The bioavailability of PAHs depends on: (a) physicochemical parameters of the compounds themselves; (b) their abiotic partitioning in the aquatic environment; (c) their environmental transformation, e.g. by photooxidation or microbial degradation; and (d) biological parameters, be it biological mixing of sediments by benthic organisms (bioturbation), or physiological and ecological properties of the exposed organisms (Forbes and Forbes 1994; Van Brummelen *et al.* 1998). Whilst some aspects are covered in Chapters 6 and 8, here we evaluate the interface between these aspects and biomarkers.

16.2.1   PAH UPTAKE AND DISTRIBUTION

The uptake routes of PAHs by aquatic animals occur via gills, skin, and the digestive tract. The bioaccumulation of PAHs by various marine organisms has been extensively reviewed by Meador *et al.* (1995). Lifestyle and microhabitat of the exposed species clearly influence PAH uptake (Van Brummelen *et al.* 1998). Following uptake, PAHs are subject to internal distribution between tissues and organs, to biotransformation and to elimination. The balance between these processes determines the net bioaccumulation. The uptake route may strongly influence PAH absorption and tissue disposition; e.g. in a study on rainbow trout exposed to either dietary or waterborne benzo[a]pyrene, Sandvik *et al.* (1998) observed that uptake of benzo[a]pyrene from the water led to high concentrations in the gills, liver, bile, intestine, olfactory organ, kidney and skin, while in dietary-exposed trout, high concentrations occurred only in the intestine and bile.

16.2.2   PAH BIOTRANSFORMATION IN VERTEBRATES

It is particularly the ability for biotransformation that is of prime importance to understand PAH bioaccumulation and to explain variability in tissue residues among different species in a specific habitat (Van Brummelen *et al.* 1998). A major catalyst for the oxidative metabolism of tumorigenic PAHs, such as 7,12-dimethylbenzo[a]anthracene or benzo[a]pyrene, is
CYP1A (Stegeman and Lech 1991; Miranda *et al.* 1997). Biotransformation
of benzo[a]pyrene in English sole, *Parophrys vetulus*, produces as primary
metabolites 7,8-dihydrodiol, 9,10-dihydrodiol, 1-hydroxy-benzo[a]pyrene and
3-hydroxy-benzo[a]pyrene, which are subsequently conjugated by phase II
enzymes and finally end up in the bile (Nishimoto *et al.* 1992). Furthermore,
DNA adducts derived from both the *trans*-7β,8α-dihydroxy-9α,10α-epoxy-
7,8,9,10-tetrahydro-benzo[a]pyrene (*anti*-BPDE), an ultimate carcinogen, and
trans-7β,8α-dihydroxy-9β,10β-epoxy-7,8,9,10-tetrahydro-benzo[a]pyrene (*syn*-
BPDE) have been detected in the liver of sole exposed to benzo[a]pyrene
(Varanasi 1989), indicating the presence of a regio- and stereoselective PAH
metabolism. Also, in carp, *Cyprinus carpio*, and in tilapia, *Oreochromis* sp.,
regio- and stereoselective metabolism of benzo[al]pyrene into hydroxylated
and *trans*-dihydrodiol metabolites was observed (Ueng *et al.* 1994). In a
study on the liver metabolism of dibenzo[al]pyrene in rat and rainbow trout,
Yuan *et al.* (1999) found comparable metabolic rates and qualitatively similar
metabolites, but the relative proportion of the individual metabolites varied significantly between the two species. The proportion of the ultimate carcinogen
dibenzo[al]pyrene-11,12-diol was over two-fold higher in the trout than in the
rat, whereas the non-carcinogen 8,9-diol metabolite was produced at higher
relative quantities in the rat. Similarly, brown bullhead, *Ictalurus nebulosus*, produced preferentially carcinogenic metabolites from the PAHs, benzo[a]pyrene,
chrysene, and phenanthrene, while non-carcinogen metabolites were generated
at a low rate only (Pangrekar *et al.* 1995).

As a consequence of the biotransformation of PAHs in exposed vertebrate
organisms, exposure to PAHs generally cannot be assessed by simply measuring
their tissue levels (e.g. Van der Oost *et al.* 1994). The biota–sediment accumulation factors (BSAF values), i.e. the ratio between the lipid weight (LW)-based
biota concentration and the concentration in organic matter (OM) of sediments,
range from 0.01 for total PAH in eel (Van der Oost *et al.* 1994) to 10 for phenanthrene in brown bullhead (Baumann and Harshbarger 1995). All LW:OM-based
BSAF values were much lower than the values predicted by the model of Van
der Kooij *et al.* (1991), based upon the equilibrium partitioning theory (EPT),
indicating a reduced uptake or an increased clearance of these compounds.
BSAF values of PAHs in sunfish declined with increasing K_{ow}, probably due to
low gut assimilation efficiency and increased metabolism (Thomann and Komlos 1999). PAH congeners in the aquatic environment can be transformed by
chemical (photo)oxidation or biological transformations (Varanasi 1989). PAH

biotransformation occurs in many aquatic organisms, but it is most effective in the liver of fish. Reported half-lives of parental PAHs in rainbow trout range from 1 day for acenaphthylene to 9 days for phenanthrene (Meador *et al.* 1995). Since the metabolism and elimination of PAHs appears to be rather efficient in fish, no strong bioaccumulation of these compounds has generally been demonstrated here. In field studies, highest PAH tissue levels are frequently observed in fish from the sites with the lowest PAH sediment levels, most probably because a low induction level in fish from sites with low contamination will limit PAH metabolism, while increased metabolic clearance will occur in fish from the polluted sites due to induced phase I enzyme activity (Van der Oost *et al.* 1991). PAH fish tissue levels are, therefore, not indicative of the levels to which the animals were exposed and cannot be used as bioaccumulation markers for exposure assessment (see also subsequent discussion on multiple biomarker approaches, Section 16.4.4).

PAH toxicokinetics and metabolism in fish has been shown to be modified by exogenous factors such as temperature or salinity, but also by endogenous factors such as hormones, organ disposition, or enzymatic characteristics of the target tissue. The major organ for xenobiotic biotransformation in fish is the liver; however, a number of other organs also possess metabolic capabilities (Saraquette and Segner 2000). Even though specific metabolic enzyme activities of these organs are lower than those of the liver, the local generation of reactive metabolites may be of high toxicological relevance. In considering the uptake routes, PAHs can be metabolized during their passage both through the gills — which have been repeatedly shown to possess significant metabolic capabilities (e.g. Stegeman and Lech 1991) — and through the gastrointestinal tract (Van Veld 1990). The uptake pathway is relevant with respect to first-pass metabolism of the PAHs and the response of the biotransformation enzymes in those organs. Biotransformation in the uptake organs may be particularly important when low concentrations of xenobiotics are present. In catfish, the K_m values for intestinal biotransformation enzymes were found to be in the micromolar range, so that low PAH levels should be metabolized readily in the intestine (James *et al.* 1993). Further, the presence of tissue-specific isoforms with different substrate selectivity cannot be excluded, although it has been not unequivocally demonstrated for fish.

16.2.3   PAH BIOTRANSFORMATION IN INVERTEBRATES

In invertebrates, the occurrence of most phase I and phase II biotransformation enzymes has been demonstrated, although relatively few of them have been characterized in detail (Livingstone 1998). Also for algae, the capability to metabolize PAHs has been demonstrated (Kirso and Irha 1998). *In vivo* studies have demonstrated that PAHs accumulated in aquatic invertebrates or algae become oxygenated, which suggests the presence of cytochrome P450 systems in these species (James and Boyle 1998; Kirso and Irha 1998).

Possibly, however, it is not CYP1A that is the major catalyst for PAH monooxygenation in these species, but other P450 iso-enzymes may be involved. This is indicated from the metabolite profiles generated from PAHs by invertebrates (James and Boyle 1998), as well as from the observation that the enzyme activity of 7-ethoxyresorufin-O-deethylase (EROD), which is catalyzed by the CYP1A subfamily in fish and mammals, generally appears to be low or absent in invertebrates, whereas benzo[a]pyrene hydroxylase activity, which is catalyzed by a variety of P450 enzymes, is widespread in invertebrates (Livingstone 1998).

It is often assumed that invertebrates possess less active and/or less inducible biotransformation enzymes than vertebrates, and therefore they should be less efficient in metabolizing PAHs. In fact, Livingstone (1998) found that the biotransformation of benzo[a]pyrene in fish is faster than in certain aquatic invertebrates. As a result, PAH body burdens in invertebrates may be more indicative of PAH exposure than levels quantified in teleost fish. However, the rates of metabolism may also vary widely among invertebrate species (Van Brummelen *et al.* 1998); e.g. the spiny lobster, *Panulirus argus*, rapidly metabolized benzo[a]pyrene after intrapericardial injection, while in the American lobster, *Homarus americanus*, a poor biotransformation of benzo[a]pyrene was demonstrated under comparable conditions (James *et al.* 1995). Care should thus be taken in the interpretation of PAH body residues of invertebrates in monitoring programs.

16.3 TOXICODYNAMICS: THE MECHANISTIC BASIS OF PAH BIOMARKERS

Looking at the mechanisms of PAH toxicity in organisms one finds that, as there is no homogeneous group of PAHs, there are also different interactions with biological structures at the cellular level that may contribute to observable adverse effects. Figure 16.1 provides a schematic summary of known interactions.

PAHs as lipophilic substances are easily abstracted from a water phase and partitioned into cellular membranes. Presence in membranes in concentrations in the millimolar range will subsequently lead to acute lethality via non-polar narcosis, i.e. disturbance of vital membrane functions; e.g. exposure of the aquatic macrophyte *Myriophyllum spicatum* to PAH-rich creosote leads to membrane ion leakage and gives rise to creosote toxic effects on *Myriophyllum* growth (McCann *et al.* 2000). Partitioning, of course, is not restricted to the outer cell membrane, but PAHs may protrude into all intracellular membrane systems (Van Brummelen *et al.* 1998). The narcotic effect can be expressed as a linear function of the octanol–water partition coefficient, K_{ow}.

In addition to narcosis or baseline toxicity, PAHs display multiple specific toxicity mechanisms, mostly involving biochemical activation and subsequent induction of oxidative stress or mutagenic and carcinogenic effects. Biochemical activation occurs during PAH metabolism, which is executed by phase I and II

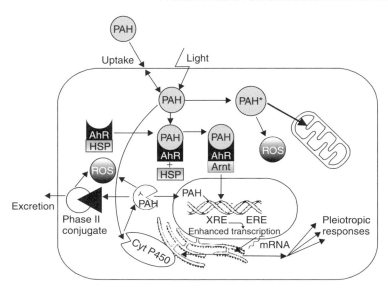

Figure 16.1 Schematic illustration of different modes of cellular interaction of PAHs (according to Ma 2001; Nie *et al*. 2001; Safe 2001; Klinge *et al*. 1999; Van Brummelen *et al*. 1998; Stegeman and Hahn 1994; Winston and Di Giulio 1991). AhR, aryl hydrocarbon receptor; Arnt, arylhydrocarbon nuclear translocator; HSP, heat shock protein; ROS, reactive oxygen species

metabolism, as described in Section 16.2. An important feature, particularly with respect to the derivation of biomarkers for PAHs, is the inducibility of various enzymes involved in PAH metabolism, especially of CYP1A. Enzyme induction means that PAHs are able to stimulate the synthesis of enzymes involved in their metabolism via gene activation. In this way, the chemicals accelerate their own metabolic conversion and elimination. The competence of vertebrate organisms to express biotransformation enzymes and actively metabolize PAHs leads to quickly reduced PAH body burdens, thus possibly generating bioactivated compounds with increased reactivity and toxic potency.

The mechanisms and consequences of enzyme induction have been particularly well studied for CYP1A. The induction of CYP1A occurs through ligand binding of PAHs to a cytoplasmic receptor, the so-called arylhydrocarbon receptor (AhR). The resulting ligand–receptor complex then leads to gene activation. A PAH has to fulfill certain structural requirements, particularly planarity and a size coming close to 3×10 Å (Mekenyan *et al*. 1996) in order to serve as a ligand of the AhR. The AhR binding and effect chain has been described in detail for 2,3,7,8-tetrachloro dibenzodioxin (TCDD) (Safe 2001). Binding of the ligand activates the receptor, which leads to dissociation of the heat shock (Hsp) and immunophilin-related (Irp) proteins. Upon heterodimerization with the arylhydrocarbon nuclear translocator (Arnt), the nuclear translocation of the complex proceeds (Safe 2001; Nie *et al*. 2001). This complex, together with further co-activators, is thought to specifically bind to dioxin- (or xenobiotic-) responsive elements (DRE or XRE) at the DNA upstream of promoters, e.g. of CYP1A genes (Safe 2001). Transcription factors now have ready access to the promotor region of the CYP1A gene. This results in an upregulation of gene transcription and subsequent rises in CYP1A mRNA, CYP1A protein and CYP1A catalytic activity (Stegeman and Hahn, 1994). The AhR-mediated transcription is tightly regulated, so that it is controlled at a physiologically adequate level. In addition to CYP1A, AhR regulates a number of other enzymes, e.g. specific forms of the glutathione-S-transferase (phase II) family (George 1994). AhR-mediated gene expression provides a mechanistic model for induction of metabolic enzymes by xenobiotics. At the DNA transcription level, crosstalk with estrogen-responsive gene promoter regions via transcription of adjacent downstream sequences (Safe 2001; Klinge *et al*. 1999) and with hypoxia-responsive enhancer (HRE) via competition for Arnt is assumed (Nie *et al*. 2001), and may partly explain observable multiple cellular responses.

PAHs such as benzo[a]pyrene or 3-methylcholanthrene are activated by CYP1A to reactive electrophiles, since reactive groups introduced by CYP1A into the 'bay region' of these PAHs cannot be further metabolized and detoxified by conjugation. These electrophiles in turn bind covalently to the DNA to form DNA adducts, possibly inducing mutations that may ultimately result in neoplastic changes and tumor formation (Guengerich 2000). Since PAH metabolism is essential for genotoxicity and carcinogenicity, it is obvious that those processes involved in the induction of cytochrome P450 enzymes, such as binding to the AhR, are important risk determinants of carcinogenicity. Genotoxicity and carcinogenicity are hallmarks of the toxicity of many PAHs and a considerable amount of research has been devoted to predicting the tumor-initiating potential of PAHs, based on chemical structure. Furthermore, PAHs are carcinogens that are involved not only in tumor initiation but also in the epigenetic events of tumor promotion (Upham *et al*. 1998).

There are plenty of reports of multiple responses following a PAH exposure. PAH metabolism can result in the generation of reactive oxygen species (ROS), which may lead to membrane damage, uncoupling of electrochemical gradients in membranes, cytotoxicity, and other pathological changes (Winston and DiGiulio 1991; Segner and Braunbeck 1998). Endocrine, mainly anti-estrogenic, effects have been described repeatedly for PAHs, and there is evidence that this effect may be mediated via the AhR pathway (Navas and Segner 2000). A further, well-documented effect quality of PAHs is their immunotoxic potency, e.g. by suppression of phagocytosis activity or by altered oxidative burst response (Davila *et al*. 1997; Hart *et al*. 1998).

In addition to endogenous biochemical activation, PAHs with their well-developed π-electron system are easily accessible to photochemical activation by light in the environment. Certain compounds can be photoactivated (Mekenyan *et al*. 1994) and by triplet state energy transfer may form oxygen radicals, which in turn show unspecific reactivity with various biological macromolecules. Alternatively, they may transform into oxygenated PAHs, of which not many examples are known regarding their structures and biological properties. Some oxy-PAHs and azaarenes have been shown *in vitro* to also elucidate AhR-mediated responses (Machala *et al.* 2001). However, 1,2-dihydroxyanthraquinone, which is an oxy-PAH of anthracene, is known to additionally interfere with energy membranes in the chloroplast and in the mitochondrium. It basically inhibits the linear electron transfer chains in both organelles, being a good electron acceptor and a bad donor. In this context, mechanistic models of joint action have recently been proposed that will be considered in further detail below. Photoactivation can enhance PAH toxicity and is of particular relevance for the aquatic organisms living in shallow littoral areas or close to the water surface (Arfsten *et al.* 1996).

PAHs occur in complex mixtures of different components, rather than as single pure compounds which may only be used in laboratory testing. Joint action from chemical mixtures is of course no specificity of PAHs. Whenever we intend to assess the combined effect of a chemical mixture, we need to use references on which we base an assessment, e.g. the response of the mixture is more than what we expected, i.e. acted synergistically (Altenburger *et al.* 2003 (in press). The crucial question for the suitability of an assessment reference, however, is related to mode of action information as far as we know (Faust *et al.* 2000). While different modes of action of components of a mixture containing PAHs would lead to the expectation of independent joint effects, similarly acting components would lead to the expectation of concentration additive behavior. This, however, may lead to the ambiguous situation that the same combined effect data may be assessed by different experts as synergistic or antagonistic joint action, depending on the reference model used.

The above-outlined current understanding of toxicokinetics and toxic modes of action form the basis on which we have to interpret PAH biomarker responses. It has to be emphasized that knowledge of PAH toxicity is mainly available for a rather small set of PAHs, in particular for those that are readily metabolized and have recognized tumorigenic potencies. In addition, toxicity information is available for some of the more abundant PAHs, such as naphthalene or phenanthrene, which appear not to induce the specific toxicity mechanisms discussed above and which apparently show narcotic action only. It would be important to have more thorough and detailed toxicity data available for the non-carcinogenic PAHs.

16.4 LINKING BIOMARKERS TO PAH EXPOSURE AND EFFECT ASSESSMENT

The primary value of using biomarkers is not the investigation of mechanisms of compound–biosystem interaction, but the tool character for studying environmental probes in order to:

- Detect possible hazards at an early stage of manifestation.
- Characterize possible relevant modes of action in contaminated environments.
- Identify components of toxicological relevance in complex contaminated samples.
- Analyze the importance of mixture effects.
- Study the habitat-dependency of adverse biological responses.

According to their tool character, biomarker measurements are not restricted to the analysis of biological samples collected from contaminated sites, but may also be utilized as biosystems to allow high-throughput of sample testing (bioassays) (Wells *et al.* 1998). Thus, specific fish models may be considered as surrogates for studying environmental carcinogenesis of specific PAHs, multi-endpoint assays may be used to identify relevant biological target structures, and animal and plant cell cultures may serve as surrogate tissues to measure biomarker responses. Substitute bioassays systems, such as *in vitro* cell cultures or fish embryo tests, have been used repeatedly to identify PAH pollution of environmental samples (Segner *et al.* 2001). Future developments will strive to tap the potentials of gene arrays ('genomics') and utilize protein profiling techniques ('proteomics') in order to obtain patterns of specific and multiple responses to PAH exposure.

In the following paragraphs, the potential role of biomarkers in environmental risk assessment of PAHs, reflecting the above-outlined understanding of toxicokinetics and toxicodynamics of PAHs, as well as selected examples and case studies illustrating the various uses of biomarkers, will be presented.

16.4.1 POTENTIAL USE OF FISH BIOMARKERS IN ENVIRONMENTAL RISK ASSESSMENT (ERA)

Site-specific assessment of environmental hazards and risks using biomarker information requires coherent signaling for selected parameters. In the following, an attempt is made to provide a synoptic view of published results for different biomarkers used in environmental hazard characterization of PAHs and other organic pollutants. For a solid interpretation of biomarker responses, it is of paramount importance that aspects of the habitat dependency of adverse biological responses, as well as the toxicological significance of biomarkers, are considered. These aspects will be briefly discussed in Section 16.4.4.

When summarizing literature data on the responses of *phase I parameters* to PAH exposure, four parameters (total cytochrome P450, cytochrome b_5, CYP1A and EROD) were mainly found to be responsive. The relative frequency of pollution-induced responses is illustrated for these parameters in Figure 16.2. In these graphs, the percentages of negative or positive pollutant-induced biomarker responses for all fish species, as reported in the literature considered for a review paper on environmental risk assessment (ERA) using

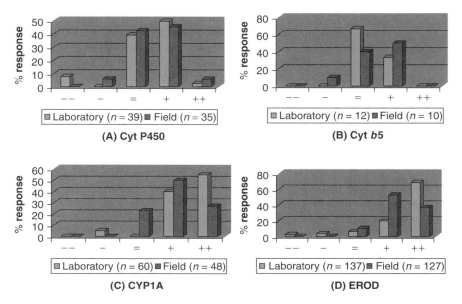

Figure 16.2 Frequencies of pollutant-induced responses of phase I-related enzymes in fish: (A) cytochrome P450 (cyt P450); (B) cytochrome b_5 (cyt b_5); (C) cytochrome P450 1A (CYP1A); (D) ethoxyresorufin O-deethylase (EROD); --, strong decrease (< 20% of control); -, decrease; =, no (significant) response; +, increase; ++, strong increase (> 500% of control). Adapted from Van der Oost et al. (2003)

fish biomarkers (Van der Oost *et al.* 2003), are visualized for both laboratory and field studies. The response frequencies are divided into five categories: –, strong inhibition (less then 20% of control value); -, inhibition (significantly lower than control); =, no response (not significantly different from control); +, induction (significantly higher than control); and ++, strong induction (more than 500% of control value).

Figure 16.2 indicates that phase I enzymes are quite responsive to environmental pollutants. Although a positive response in cyt P450 levels was observed in more than 50% of the reported laboratory and field studies (Figure 16.2A), its value as a biomarker for ERA in aquatic ecosystems is limited, since the responses of individual isoenzymes (notably CYP1A) are more specific and sensitive. Although cyt b_5 levels may be elevated in some fish species after exposure to organochlorine compounds (Figure 16.2B), its value as a biomarker in ERA procedures remains questionable. Further research is required to elucidate the mechanism of cyt b_5 induction. In all fish species considered, hepatic CYP1A protein levels seem to be a very sensitive biomarker of exposure to PAHs and HAHs (Figure 16.2C), which will certainly be feasible in ERA procedures. The CYP1A response has been validated for use in ERA monitoring programs (Bucheli and Fent 1995), assuming that all potential variables that may affect this parameter are considered in the experimental design. EROD activities in fish liver are very sensitive biomarkers (Figure 16.2D), and may thus be of great value in ERA processes. Although certain chemicals may inhibit EROD induction or activity, this interference is generally not a drawback to the use of EROD induction as a biomarker (Whyte *et al.* 2000). Together with levels of CYP1A protein and mRNA, the induction of CYP1A catalytic activities may be used, both for the assessment of exposure and as early-warning signs for potentially harmful effects of many organic trace pollutants. Our current understanding of mechanisms of CYP1A-induced toxicity suggests that EROD activity may not only indicate chemical exposure, but may also precede effects at various levels of biological organization (Whyte *et al.* 2000) (see Sections 16.3 and 16.4.4). CYP1A and EROD determinations may be used in various steps of the ERA process, such as quantification of impact and exposure of various organic trace pollutants, environmental monitoring of organism and ecosystem 'health', identifying subtle early toxic effects, triggering of regulatory action, and identification of exposure to specific compounds (Stegeman *et al.* 1992). It should be emphasized that PAH exposure is only partly responsible for the induction of the phase I system in fish liver, since the response is also due to the exposure to other xenobiotic substances, such as chlorinated dibenzodioxins and furans and planar PCBs (Safe 1998). It is of paramount importance that certain 'confounding variables', i.e. non-pollution-related parameters that may affect the enzyme activities (see Section 16.4.4.1), are being considered when interpreting the responses of the phase I parameters.

The relative frequencies of the responses of four *phase II parameters* [reduced glutathione (GSH), oxidized glutathione (GSSG), glutathione-*S*-transferases (GSTs) and UDP-glucuronyl-transferases (UDPGTs)] are illustrated in the graphs of Figure 16.3. In these graphs, the percentages of negative or positive pollutant-induced biomarker responses for all fish species, as above (Van der Oost *et al.* 2003), are visualized for both laboratory and field studies. Because of the limited number of observations on the responses on GSH and GSSG levels (Figures 16.3A and 16.3B), these parameters cannot yet be considered as valid biomarkers for ERA purposes. The key role played by glutathione in detoxifications and the responsiveness of this system to xenobiotics, however, motivates continued research on its feasibility as a biomarker. Although it seems that pollutant-induced effects on GSH levels are restored by feedback mechanisms, the hepatic GSH:GSSG ratio may be a potential biomarker for oxidative stress. Hepatic total GST activity in fish does not seem to be feasible as a biomarker for ERA, since increased activities are only observed in a limited number of fish species (Figure 16.3C). In addition, the exposure to pollutants such as PCDDs and PAHs may cause both induction and inhibition of the enzyme activity. However, more research on this parameter, which is of paramount importance for major detoxification processes, may elucidate specific isoenzymes that have a more sensitive and selective response to pollutants. Although not as sensitive as phase I enzymes, the UDPGT activity appears to be the phase II parameter that is most responsive

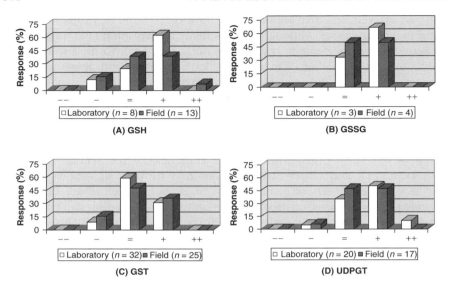

Figure 16.3 Frequencies of pollutant-induced responses of phase II enzymes and co-factors in fish: (A) reduced glutathione (GSH); (B) oxidized glutathione (GSSG); (C) glutathione *S*-transferase (GST); (D) UDP-glucuronyl transferase (UDPGT). --, strong decrease (<20% of control); -, decrease; =, no (significant) response; +, increase; ++, strong increase (>500% of control). Adapted from Van der Oost *et al.* (2003)

to pollutant exposure (Figure 16.3D). More work will be required to investigate the potential pollution-induced responses of the various UDPGT isoenzymes. As for the phase I enzymes, the 'confounding variables' that may affect the phase II enzyme activities must be considered when interpreting the responses reflected in these parameters.

A strong increase (>500% of control) in the *biliary levels of PAH metabolites* has been observed in most of the laboratory studies in which various fish species were exposed to PAHs (Van der Oost *et al.* 2002). These results are confirmed by field studies, in which a significant increase of fluorescent aromatic compounds (FAC) levels was observed in the bile of fish from polluted environments (Van der Oost *et al.* 2003). A significant decrease in FAC levels was not observed in any of the laboratory or field studies. The FAC response frequencies for all fish species from 15 laboratory studies and 24 field studies are summarized in Figure 16.4A. A strong significant increase in biliary FAC levels was observed in the majority of laboratory and field studies. Levels of biliary PAH metabolites are certainly sensitive biomarkers to assess recent exposure to PAHs. Since PAH exposure cannot be reliably determined by measuring fish tissue levels, this parameter is a promising biomarker for fish exposure in ERA processes concerning PAH-contaminated sites.

Both laboratory and field studies on *DNA adduct* formation in PAH-exposed fish have been reviewed by Pfau (1997). A strong increase (>500% of control)

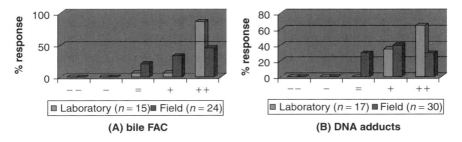

Figure 16.4 Frequencies of pollutant-induced responses of (A) fluorescent PAH metabolites in bile (bile FAC) and (B) DNA adducts; --, strong decrease (<20% of control); -, decrease; =, no (significant) response; +, increase; ++, strong increase (>500% of control). Adapted from Van der Oost *et al.* (2003)

in the hepatic levels of DNA adducts was observed in most laboratory studies in which various species of fish were exposed to PAHs (Van der Oost *et al.* 2002). These results are confirmed by several field studies, in which significant increases of DNA adduct levels were observed in the livers of fish from polluted environments (Van der Oost *et al.* 2003). The DNA adduct responses for all fish species from 17 laboratory studies and 30 field studies are summarized in Figure 16.4B. A significant increase in hepatic DNA adduct levels was observed in 100% of the laboratory studies and 70% of the field studies. Due to the strong and consistent responses of hepatic DNA adduct levels to PAH exposure, this parameter is considered to be an excellent biomarker for the assessment of PAH exposure, as well as a sensitive biomarker for the assessment of potentially genotoxic effects. Neoplasms in fish, particularly of the liver, are regarded as a major chronic adverse effect due to sublethal exposure concentrations against genotoxic compounds. Evidence has been accumulated that PAH–DNA adducts indeed provide a good predictive biomarker for the potential of neoplasia to occur (Reichert *et al.* 1998; Vethaak *et al.* 1996) and these will be considered in Section 16.4.4. It is important to combine the measurements of DNA adduct formation as a molecular dosimeter with analysis of carcinogen metabolism and determination of tumor formation to provide insights into the mechanisms involved in chemical carcinogenesis (Maccubbin 1994). This type of integrated study may be used in ERA to provide information about potential exposure and risk of environmental carcinogenesis from contamination that may be found in waterways. In addition to their use as a biomarker for exposure and effects of genotoxins, DNA adducts may provide information on the biological effect and potential risk of a chemical, since it has been suggested that any chemical that forms DNA adducts, even at very low levels, should be considered to have carcinogenic and mutagenic potential (Maccubbin 1994). An important criterion for a valid biomarker, the relative ease with which it can be measured, is not met when DNA adducts are determined using the ^{32}P-postlabeling assay, which is expensive and time-consuming. In this respect it might be interesting to improve immunological methods, which are less sensitive but much easier to perform, such as the ELISA assay (Van Schooten 1991), which uses polyclonal antibodies against benzo[a]pyrene-7,8-diol-9,10-epoxide – DNA, or the immuno-slot-blot technique (Law *et al.* 1998), using various diethylnitrosamine adducts.

16.4.2   IDENTIFICATION OF TOXIC ACTION OF CONTAMINANTS IN THE FIELD

The intention to use biomarkers in order to gain information on the type and biological quality of a given impacted site is often addressed. Adams (2001) provided a review of studies linking specific anthropogenic activities and subsequent emissions to specific exposure and effect responses on the biomarker level. The hypothesis is that the more specific a biomarker is for a biological process or level of biological organization, the higher its identification value will be for specific pollutants, but the more difficult its interpretation in terms of adverse health effects will be. Examples of using quantification of CYP1A levels or activity in fish to indicate exposure to inducing compounds are numerous (see review by Whyte *et al.* 2000), but only a few studies could link the biomarker responses to sources of ecotoxic concern.

Investigations in Canada found that a waterborne factor from pulp mill effluents induced CYP1A in white suckers, *Catostomus commersoni*. CYP1A induction was associated with disturbances of fish physiology and reproduction (Munkittrick *et al.* 1992; Hodson *et al.* 1992). It was originally assumed that halogenated aromatic hydrocarbons in the pulp mill effluents could be responsible for CYP1A induction, since among them are well-known inducers of EROD and of adverse reproductive changes. However, subsequent toxicity identification studies provided evidence that the compound responsible was not a compound of anthropogenic origin, but a substance belonging to the pinosylvin family, a group of naturally occurring substances in coniferous trees (Burnison *et al.* 1998).

Brack *et al.* (1999) undertook an identification of toxicants present in sediment probes of the Spittelwasser, a creek tributing to the river Mulde and finally to the Elbe river (Germany). The Spittelwasser drains the former chemical production site of Bitterfeld and Wolfen and, using a whole biotest battery for bioassay-directed fractionation, it was demonstrated that the Spittelwasser carried various toxicants of high potency in its sediments. Part of the contamination pattern that proved phyto- and cytotoxic could be attributed to fractions containing aromatic hydrocarbons. Elegant fractionation procedures allowed separation of halogenated and non-halogenated fractions, both of which showed fractions containing EROD-inducing activities, as determined with the RTL-W1 EROD assay with rainbow trout liver cell culture. Figure 16.5 presents two PAH compounds that were identified in this work, viz. dinaphthofuran and chrysene,

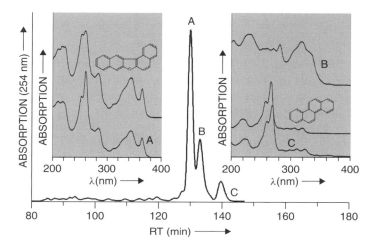

Figure 16.5 PAH compounds identified in a bioassay directed fractionation procedures applied to a contaminated sediment from a creek in the Bitterfeld area, Germany. The structures are identified by comparison of HPLC retention times for peaks (A) and (C) and by comparison of their UV-spectra with pure reference compounds, shown as lane A (dinapthofuran) and C (chrysene) in the inset DAD-spectra (adapted from Brack *et al*. 2002)

of which the former is a PAH that is not commonly analyzed in surveillance efforts, and thus show the potential of this methodological approach.

A major limitation when using CYP1A and other AhR-regulated responses as markers for PAH exposure is the fact that the AhR pathway is not only activated by certain PAHs but also by various dioxins, furans, and PCBs. PCBs and PAHs appear to act additively on CYP1A (Behrens, Altenburger and Segner, unpublished); however, the halogenated compounds have much higher induction potencies than PAHs. Therefore, in a mixture CYP1A induction may primarily indicate dioxin or PCB but not PAH exposure, unless PAHs are by far the dominating contaminants. This was well illustrated in a study by Engwall *et al*. (1997), who injected polyaromatic fractions of sediment extracts into chicken eggs, measured the resulting EROD response and performed chemical analyses on the fractions. They could estimate the relative contributions of PAHs and PCBs to the EROD response and demonstrated that the contents of PAHs could account for only a minor part of the EROD induction response in the chicken eggs.

16.4.3 ASSESSMENT OF PAH MIXTURE EFFECTS USING BIOMARKERS

Regarding the goal to study combined effects on biomarkers after exposure to mixtures, different scopes may be distinguished. First, there are mechanistically based ideas on the joint action of PAHs and other pollutants that might improve our understanding of observable combined effects. Second, as it is widely recognized that PAHs occur universally as mixtures, procedures have been developed to account for these in effect assessments. Third, one might be confronted with the task to predict mixture effects from information on the activity of single compounds. The different scopes have their specific instrumentation and some examples will be highlighted.

As PAHs do not only occur in mixtures but also show different mechanisms of toxicity and above being prone to transformation reactions, it is no surprise that *joint action of PAHs* in mixtures and together with other contaminants have been anticipated for the different pharmacological phases, such as co-solubility, uptake, distribution, metabolism, excretion and effect propagation.

Various combination effects of PAHs and other chemicals have been described in the literature. It has been demonstrated that heavy metals are able to decrease CYP1A induction due to PAH exposure. The EROD induction in human HEPG2 cells after B[a]P exposure was decreased in the presence of 5 μmol/L amounts of metals (Vakharia *et al.* 2001): e.g. arsenic (57%), cadmium (82%), mercury (4%), and lead (20%). Jett *et al.* (1999) demonstrated that some PAHs (e.g. B[a]P) have anti-cholinesterase activity, and that they contribute in an additive manner to the inhibitory effect of organophosphorous insecticides (e.g. chlorpyrifos) on acetylcholinesterase (AChE) *in vitro*. The immunotoxic responses of combinations of B[a]P and TCDD were investigated in different mouse strains. The immunosuppression was consistent with additivity, regardless of the composition arrangement or the phenotype of the mouse strain, although some antagonism could not be excluded with certainty (Silkworth *et al.*, 1995). A binary mixture of B[a]P (an indirect-acting mutagen) and 2,4,6-trinitrotoluene (TNT, a direct-acting mutagen) failed to induce the positive mutagenic response in *Salmonella typhimurium* that was observed when the bacteria were exposed to the individual compounds (Washburn *et al.*, 2001). Mass spectroscopy revealed a number of products that may account for the antagonistic action of TNT on B[a]P-induced mutagenicity. Further, kinetic studies indicated that TNT inhibited the incorporation of B[a]P into cells. Interaction of nucleophilic substances, which may derive from oxidative stress or photoactivated PAH compounds with benzo[a]pyrene, have been reported to show enhanced DNA–B[a]P adduct formation in lung cell tissue cultures (Borm *et al.*, 1997).

When addressing the question of *hazard assessment of combined effects from PAH mixtures* using biomarker responses, the concept of (toxic) equivalency factors (TEF) or, more specifically, induction equivalency factors (IEF) is of widespread use, as proposed and advocated by Safe (1990). It reduces the problem of combined effects to the quantitative aspect of calculating concentrations of mixtures evoking a predefined effect level. The rationale assumes that substances that act via a common mode of action, i.e. induce CYP1A by AhR binding, should be different only in their different potency, i.e. behave like dilutions of one another. For determining equivalency factors, the highly toxic 2,3,7,8-tetrachlorodibenzo-*p*-dioxin (2,3,7,8-TCDD) has been taken as a reference compound in various assays (Safe 1990). TEF values are assay-, effect-level- and compound-specific but have been published for several groups of halogenated and non-halogenated compounds such as PCBs, PCDDs, PCDFs, PAHs, and various bioassays. An assessment of a known mixture of PAHs or similar compounds for a defined response may then be based on the sum of all component concentrations multiplied by their individual TEF for that response (Equation 16.1), and comparing the expected response with an observed effect for that mixture.

$$\text{TEQ} = \sum (C_i \bullet \text{TEF}_i) \qquad (16.1)$$

If the TEQ (toxic equivalent) equals 1, the effect of the reference compound (2,3,7,8-TCDD), e.g. 50% of maximal CYP450 inducibility, is expected for the mixture. If the value is greater than 1, a higher effect is expected, and vice versa.

The TEF/TEQ concept is a special form of the reference concept of concentration addition for the prediction of combined effects from information on the components (see below), specifically adapted for response parameters with open scaling, such as induction potency or enzyme activities. Willet *et al.* (1997) provide an example of how this concept is commonly applied to assess mixtures of PAHs present in environmental samples. In their study they characterized the EROD-inducing potency for 23 different PAH compounds individually, using a H4IIE rat hepatoma cell bioassay. An induction equivalency factor was calculated by dividing the half-maximum induction activity (in mol/L) for the reference compound 2,3,7,8-TCDD (1.1×10^{-10} mol/L) by the half-maximum induction activity for the individual PAH. The mixture assessment was based on the calculated TEQ for a mixture of the 23 components that had been analytically quantified in oyster exposure studies and compared with results of an induction experiment for a dilution series of the mixture. As calculated and observed responses were comparable, the joint action of the components was considered as additive (Willet *et al.* 1997). Similarly, Engwall and co-workers (1997) investigated contamination of settling particulate matter, blue mussel (*Mytilus edulis*) tissue, and sediment samples from different sites of the Baltic Sea. After separating organic sample extracts into three fractions containing monoaromatic/aliphatic, diaromatic/polychlorinated PCDDs/Fs, and polyaromatic compounds, they employed EROD induction measurements in chicken embryo liver culture to assess bioactivity. Analytical identification of compounds and subsequent testing for their EROD-inducing potential and calculation of TEQs, and comparison with mixture activities, were used to assess the contributions of the different fractions to the observed toxic effects. From the higher EROD activities of the extracted PAH-containing fraction as compared to the TEQ calculation, it was concluded that more than the analytically identified components might be present as contaminants (Engwall *et al.*, 1997). Basu *et al.* (2001) evaluated whether EROD-inducing activity of PAH mixtures in rainbow trout *in vivo* can be predicted from the IEF of the individual compounds. They found that the mixtures in general showed additivity. This result is interesting, since due to factors such as different toxicokinetics of the individual PAHs, mixtures may be expected to deviate from additivity, and this has been observed indeed in an *in vitro* study with rat hepatocytes (Till *et al.* 1999). However, the influence of toxicokinetics on the EROD response to mixtures may depend on the treatment time. With too short an exposure time, compound-specific differences in metabolic turnover may not become effective. Alternatively, an extended half-life of the EROD induction cascade may obscure toxicokinetic differences in the metabolism of individual PAHs.

There are, however, several drawbacks inherent to the IEF/TEF/TEQ approach. For one, the effect-level considered is often set at a half-maximum level, which leads to biased assessments when different maxima occur, as is common in CYP1A induction by PAHs. This may be overcome by defining fixed response levels (Engwall *et al.* 1997). Furthermore, TEQs allow only point assessments, i.e. due to individual concentration response relationships, assessments may vary for different effect levels and there is no information as to how far away a TEQ value of say 1.2 or 0.8 is from the effect of concern, as long as the steepness of concentration–response relationships is not considered. Thus, many comparative assessments presented are based purely on presumed plausibility. Indeed, an improvement would be to use whole concentration response functions for assessments, if adequate information for the individual components of a mixture is available, as is demonstrated in Figure 16.6. Moreover, care has to be taken when transferring the TEF concept from dioxins and PCBs to PAHs for toxicological reasons: Safe (1990), when introducing the TEF concept, did this on the basis of comparing toxicity endpoints, e.g. lethality or immune dysfunction. It was then observed that EROD induction potency of the various dioxins and PCBs parallels their toxic potency; therefore, EROD became important as a kind of 'indicator' value for the toxic potency of dioxins and PCBs, although EROD induction itself is not a toxicity parameter.

Other indices in assessing activities of mixtures of PAHs that are numerically equivalent to the TEF approach, such as the summation of toxic units (Altenburger *et al.* 2003), have been employed. The mixture contributions of yet untested PAHs to lethality (LC_{50}) in amphipods after 10 day spiked-sediment exposure were estimated using a log K_{ow}-based QSAR and fed into a toxic unit summation model (ΣPAH model) (Swartz *et al.* 1995). In a multistep procedure, starting from field analytical data on PAH concentrations in sediments, interstitial water concentrations were calculated as exposure concentrations, followed by toxic unit estimation for the identified components. These were summed and translated to a graded response. This last step, a concentration–response model relating the sum of toxic units to observed joint effects, implicitly assumes constant efficacy ratios of the individual components or parallel concentration–response functions of the constituents. Although this is not demonstrated, and is probably difficult to prove, it seems to suffice for assessing the combined

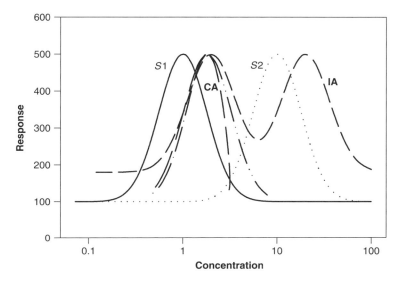

Figure 16.6 Prediction of combined effects of a binary mixture of substances (*S1 and S2*), using the reference concepts concentration addition (CA) and independent action (IA) for simulated effects of the individual components modeled with Gaussian concentration response functions of the type $y(d) = Yb + (Ym - Yb)\,exp\{-C[\ln(d) - \ln(dm)]^2\}$

effects of multiple mixtures of similar-acting compounds. The capability of this approach was satisfactory for sediment samples with PAHs constituting the contaminants of concern, while in cases where PAHs were not the principal contaminants, the predicted mortality underestimated the observed responses.

When aiming at *predicting mixture toxicities*, concentration addition, which is the inherent predictive concept to the TEF approach, has been shown to be a valid reference for mixtures whose components have the same sites and modes of action; independent action, in contrast, has been demonstrated to work better for mixtures of dissimilarly acting compounds (Faust *et al.* 2000). Independent action implies different sites and modes of action of the mixture components and calculates the expected combined effects from multiplying fractional responses. Nesnow and colleagues (1998), performing response analysis on tumorigenic responses of mice after co-exposure to quintary PAH mixtures, found independent action a model well capable of describing the observed responses.

A specific problem that has been largely ignored in the assessment and prediction of PAH mixtures arises for responses that are not steadily increasing with dosage, as illustrated in Figure 16.6. If cases occur where e.g. an EROD induction follows a bell-shaped curve with increasing concentration and may be best described by an appropriate function such as a Gauss function (Kennedy *et al.* 1996), the prediction, even for a binary mixture using concentration addition or derived indices like TEQs in contrast to IA predictions, is no longer unambiguous. Several solutions are possible, which means that predictions are not straightforward. We may use conventions to overcome this for well-defined mixtures, but we do not know what is appropriate for an environmental sample.

Finally, when considering the combined effects of multiple mixtures and the relevance of components at very low individual concentrations, it may be helpful to acknowledge the current techniques and evidence for the relevance of contributions from low-level contamination to joint action (Walter *et al.* 2002).

16.4.4   PERSPECTIVES OF BIOMARKER APPROACHES FOR PAH EFFECT ASSESSMENT

Biomarkers may provide insight into the potential risks and hazards of a given environmental contamination. By screening multiple biomarker responses, important information will be obtained about the type of toxicant organisms are exposed to and the stress qualities involved. For field samples, biomarkers provide an important index of the total external load that is biologically available under 'real-world' exposure. Individual biomarkers may be used as an early warning for potential (environmental) hazards. If causative relationships are required, or if the group of pollutants inducing a certain response needs to be identified, then a multiple biomarker approach is needed. The advantages of using a multiple biomarker approach were clearly demonstrated in a study by Van der Oost *et al.* (1996b). By using a suite of biomarkers in feral eel (*Anguilla anguilla*), it was demonstrated that the exposure to organic pollutants was highest at a site polluted with chemical waste, whereas opposite conclusions would have been drawn if only PAH metabolite levels in fish bile were considered. Since low 1-OH pyrene levels indicated that recent PAH exposure was relatively low at that site, the high CYP1A induction was most probably due to elevated levels of dioxins and PCBs that have been detected in the sediments (Van der Oost *et al.*, 1996a). High hepatic DNA adduct levels in eel at the polluted site might be due to an increased bioactivation of low PAH amounts for a longer period of time, as a consequence of the highly induced CYP1A activity. For more susceptible species of fish, the pollution levels at this site may be a serious health risk. If various biomarkers are used simultaneously, false interpretations due to confounding factors will be reduced. Another important topic for the interpretation of multiple biomarker data is the toxicological significance of the biomarker responses.

16.4.4.1   Confounding factors

Many non-pollution-related variables may have an additional impact on the various enzyme systems, and may thus interfere with biomarker responses. Examples of such 'confounding' or 'modifying' factors are the organisms' health, condition, sex, age, nutritional status, metabolic activity, migratory behavior, reproductive and developmental status, and population density, as well as factors such as season, ambient temperature, heterogeneity of the environmental pollution, etc. Modulation of biomarker responses due to a varying habitat has been studied for different factors likely to affect biomarker responses, such as CYP450 induction by temperature, medium in cell culture-based assays, or presence of natural estrogens. This type of investigation leads to clearer understanding of the appropriate use and interpretation of specific biomarker tools and the role of endogenous and confounding factors for observable responses. Biomarker studies may also be used to understand conditions that can be crucial for elucidating adverse PAH effects, e.g. timing, dosing, ways of translation of primary reactions to effects on the organismic level, or species specificity.

Unfortunately, most available toxicity data rarely quantify the potency that confounding factors are likely to exhibit in natural environments. Moreover, estimates of confounding factor interactions are scarce, as evidenced by the extensive use of uncertainty factors in risk assessment to address unknowns. Due to confounding factors, a certain over- or underestimation of the potential environmental risks may occur. The chances of making these misinterpretations in evaluating the environmental quality (both false-negatives and false-positives) will decrease if a multi-biomarker approach is used.

16.4.4.2   Toxicological significance of biomarkers

While there is a tremendous body of literature available on PAH exposure and biomarker and other responses, it is still difficult to establish consistent or even quantitative relationships between primary interactions and lesions on an organism level. Most evidence is rather of an epidemiological, i.e. statistical, nature and focuses on a few PAH model compounds, such as benzo[a]pyrene. This is even more so for the evaluation of environmental samples. The reasons for this derive mainly from two unresolved questions: (a) what is to be considered as an adverse effect relevant at least for the organism level?; and (b) how does the effect propagate specifically?

CYP1A induction in fish is not only an effective indicator of exposure to specific classes of chemicals but, in association with AhR-mediated effects, may eventually allow it to serve as an accurate predictor of toxicity (Whyte et al., 2000). Many of the toxic effects of EROD-inducing compounds are believed to be mediated through the binding of xenobiotic ligands to the AhR. Although CYP1A is only one of the protein products of this induction pathway, it may play a major role in toxicity through the generation of reactive intermediate compounds and reactive oxygen species (ROS) (Stegeman and Hahn, 1994). EROD activity in fish may also serve to predict biochemical changes caused by other members of the Ah gene battery, as they may be induced by the same mechanism (Whyte et al., 2000). Another member of the Ah gene battery that may be related to oxidative damage is GST. Glutathione (GSH) is an important cellular antioxidant, serving as a scavenger for oxygen-free radicals (Pichorner *et al.*, 1995). After induction, GST may disrupt the normal cycle of GSH utilization in the cell, leading to elevated levels of ROS. Additional detrimental effects, not related to ROS, that may result from the induction of the Ah gene battery have been reviewed by Nebert *et al.* (1993). The use of CYP1A induction as a biomarker for endocrine effects (e.g. steroid hormone modulation, anti-estrogenic effects) is currently not warranted, especially in the light of the variability caused by hormonal changes in fish during reproductive cycles (Whyte *et al.*, 2000). The immune system is a potential target for the toxic effects of AhR agonists such as PCBs, PCDDs, and PAHs (Safe, 1994). The role of the AhR in immunosuppression by these compounds, however, is not well understood.

Regarding the question of adverse effects, acute lethality caused by exposure to PAHs is commonly rather low and related to their unspecific actions (Van Brummelen *et al.* 1998), except for exposure situations, where phototoxicity may become important. Since CYP1A enzymes may play an important role in carcinogen activation in natural fish populations, their levels and activities can be used as biomarkers for the presence of potential carcinogens (Kantoniemi *et al.* 1996). Neoplasms in fish, particularly of the liver, are regarded as a major chronic adverse effect due to sub-lethal exposure concentrations. In this specific case, extensive evidence has been accumulated that PAH–DNA adducts indeed provide a good predictive biomarker for the potential of neoplasia to occur (Reichert *et al.* 1998). However, even here the effect chain, i.e. the study of epigenetic events, is still in its beginning (Upham *et al.* 1998), leaving room for many surprises regarding individual compounds' activities or confounding effects.

Approaching effect translation, i.e. the chain of reactions from primary interactions between PAHs and subcellular targets to effects visible at organismic levels, several issues have to be considered, such as polymorphism, tissue sensitivities, or stress adaptation strategies, most of which can account for observed intra- and interspecific differences in sensitivities to PAH exposure. A major point in the assessment of the adverse consequences of an interaction between a PAH and biological macromolecules is the consideration of locations of interference within an organism. Again, most experimental evidence comes from investigations of CYP protein distribution within different tissues of fish, recently reviewed by Saraquette and Segner (2000).

Although not all effects are directly related to CYP1A, the induction of EROD in conjunction with other members of the Ah gene battery allows it to serve as a warning for potential effects. More information is necessary to assess the types of toxicity that may occur at specific levels of EROD induction. Biliary PAH metabolites may serve as indicators of recent exposure to biodegradable PAHs, which can be used indirectly to assess health risks. The levels of DNA adducts have a high ecotoxicological significance, since these adducts may be precursors to tumor promotion.

16.5 CONCLUSIONS

Prospects and suitability of biomarkers in the assessment arena crucially depend on specific objectives, and therefore may only be summarized in general terms. The type of information that may be gained from a biomarker study includes the estimation of the amount of a specific PAH compound or environmental probe that is bioavailable, the quantification of a primary organismal reaction, the description of a biological response to exposure, or the tapping of biological variability in response.

With respect to PAH contaminations in the environment, biomarkers have been established for fish with sufficient understanding regarding their physiological meaning. For invertebrates and other aquatic organisms, the picture is more or less vague.

Levels of biliary PAH metabolites are specific and sensitive biomarkers to assess recent exposure to PAHs. Since PAH exposure cannot be reliably determined by measuring fish tissue levels, this parameter is a valid fish biomarker for ERA processes concerning PAH-contaminated sites. Levels of PAHs in invertebrates (such as crustaceans and molluscs) may also be a good indicator for prolonged PAH exposure, since these organisms are generally less capable of biotransforming PAHs, due to the lack of an efficient MO enzyme system.

Hepatic CYP1A induction in fish as an indicator of contaminant exposure has been studied for more than three decades. During this time it has been well characterized in the laboratory and extensively validated in field investigations. EROD induction has been used successfully as a biomarker in many contaminated aquatic systems. One of the main strengths, and at the same time a drawback, of this biomarker is its inducibility by a wide range of toxic chemicals, among others PAHs. CYP1A induction by different classes and mixtures of contaminants is still a highly active area of research. Most of the current research is focused on the linkages between EROD induction and toxic impacts on fish. Regardless of indication of toxic impact, EROD induction in fish has proven utility for revealing the biological uptake of contaminants with known toxic properties in aquatic systems. The 'confounding variables' affecting the phase I enzyme levels and activities must be considered when interpreting the responses to these parameters.

Hepatic phase II enzymes and co-factors in fish (possibly with the exception of UDPGT) do not seem to be sensitive biomarkers for PAH assessment, since increased activities are only observed in a limited number of fish species. In addition, exposure to pollutants such as PCDDs and PAHs may cause both induction and inhibition of the enzyme activity. More research is needed to evaluate the value of specific phase II isoenzymes as biomarkers.

Due to the strong and consistent responses of hepatic DNA adduct levels to PAH exposure, this parameter is considered to be an appropriate biomarker for the indication of genotoxic events. As DNA adducts like the CYP induction are not exclusively due to PAH exposure, although perhaps, to a great extent, DNA adducts may be considered to provide integral information about the potential risk of a site-specific xenobiotic mixture. It has been suggested that any chemical that forms DNA adducts, even at very low levels, should be considered to have carcinogenic and mutagenic potential.

Biomarkers, such as hepatic CYP1A, EROD, GST, UDPGT and DNA adducts, and biliary PAH metabolites, may be useful tools in field studies to identify and monitor sites of relevant contamination, provided that appropriate controls are being used and that the potential impact of confounding factors (sex, age, etc.) is recognized. An important aspect with respect to environmental monitoring is that, at least for EROD activity, PAH mixtures seem to show concentration additive responses that provide scope for combined effect assessment.

We advocate the use of biomarker batteries whenever PAH exposure is to be identified unequivocally. Multiple biomarker approaches will improve the signal/noise ratio and are generally recommended to improve the quality of environmental risk assessment for aquatic systems. It is, for instance, important to combine the measurements of DNA adduct formation as a molecular dosimeter with analysis of carcinogen metabolism and determination of tumor formation to provide insights in the mechanisms involved in chemical carcinogenesis.

PAH biomarkers may therefore be utilized as:

- Early warning indicators of exposure and effect (i.e. field surveys on feral fish).
- Tools in active biomonitoring programs (e.g. caged fish at hot spots).
- Bioassays in diagnostic studies aimed to establish causal relationships between exposure and adverse effects on organisms.

While the use of biomarkers as bioanalytical tool to detect sites with significant PAH contamination is quite well established, with respect to both field monitoring programs using feral animals or laboratory screening tests on environmental samples, much less is known about the toxicological and ecological implications of the biomarker responses. Furthermore, most biomarker work is focused almost exclusively on tumorigenic PAHs, and we have almost no understanding of the effects of the other PAHs.

REFERENCES

Adams SM (2001) Biomarker/bioindicator response profiles of organisms can help differentiate between sources of anthropogenic stressors in aquatic ecosystems. *Biomarkers*, **6**, 33–44.
Altenburger R, Nendza M and Schüürmann G (2003) Mixture toxicity and its modeling by quantitative structure–activity relationships. *Environmental Toxicology and Chemistry*, (in press).
Arfsten DP, Schaeffer DJ and Mulveny DC (1996) The effects of near-ultraviolet radiation on the toxic effects of polycyclic aromatic hydrocarbons in animals and plants: a review. *Ecotoxicology and Environmental Safety*, **33**, 1–24.

Basu N, Billiard S, Fragosos N, Omoike A, Tabash S, Brown S and Hodson PV (2001) Ethoxyresorufin-*O*-deethylase induction in trout exposed to mixtures of polycyclic aromatic hydrocarbons. *Environmental Toxicology and Chemistry*, **20**, 1244–1251.

Baumann PC and Harshbarger JC (1995) Decline in liver neoplasms in wild brown bullhead catfish after cooking plant closes and environmental PAHs plummet. *Environmental Health Perspectives*, **103**, 168–170.

Borm PJA, Knaapen AM, Schins RPF, Godschalk RWL and van Schooten F-J (1997) Neutrophils amplify the formation of DNA adducts by benzo[a]pyrene in lung target cells. *Environmental Health Perspectives*, **105**, 1089–1093.

Brack W, Schirmer K, Kind T, Schrader S, Schüürmann G (2002). Effect-directed fractionation and identification of cytochrome P4501A-inducing halogenated aromatic hydrocarbons in a contaminated sediment. *Environmental Toxicology and Chemistry*, **21**, 2654–2662.

Brack W, Altenburger R, Ensenbach U, Möder M, Segner H and Schüürmann G (1999) Bioassay-directed identification of organic toxicants in river sediment in the industrial region of Bitterfeld (Germany) — a contribution to hazard assessment. *Archives of Environmental Contamination and Toxicology*, **37**, 164–174.

Bucheli TD and Fent K (1995) Induction of cytochrome P450 as a biomarker for environmental contamination in aquatic ecosystems. *Critical Reviews in Environmental Contamination and Toxicology*, **25**, 201–268.

Burnison BK, Comba ME, Carey JC, Parrot J and Sherry JP (1998) Isolation and tentative identification of compound in bleached-kraft mill effluent capable of causing mixed function oxygenase induction in fish. *Environmental Toxicology and Chemistry*, **18**, 2882–2887.

Davila DR, Mounho BJ and Burchiel SW (1997) Toxicity of polycyclic aromatic hydrocarbons to the human immune system: models and mechanisms. *Toxicology and Ecotoxicology News*, **4**, 5–9.

ECETOC (1993) *Environmental Hazard Assessment of Substances*. Technical Report No. 51. European Centre for Ecotoxicology and Toxicology of Chemicals, Brussels, Belgium.

Engwall M, Bromann B, Näf C, Zebür Y and Brunström B (1997) Dioxin-like compounds in HPLC-fractionated extracts of marine samples from the east and west coast of Sweden: bioassay- and instrumentally-derived TCDD equivalents. *Marine Pollution Bulletin*, **34**, 1032–1040.

Faust M, Altenburger R, Backhaus T, Bodeker W, Scholze M and Grimme LH (2000) Predictive assessment of the aquatic toxicity of multiple chemical mixtures. *Journal of Environmental Quality*, **29**, 1063–1068.

Forbes VE and Forbes TL (1994) *Ecotoxicology in Theory and Practice*. Chapman & Hall, London.

George SG (1994) Enzymology and molecular biology of phase II xenobiotic-conjugating enzymes in fish. In Malins DC and Ostrander GK (eds), *Aquatic Toxicology; Molecular, Biochemical and Cellular Perspectives*. Lewis, CRC Press, Boca Raton, FL, pp. 37–85.

Guengerich FP (2000) Metabolism of chemical carcinogens. *Carcinogenesis*, **21**, 345–351.

Hart LJ, Smith SA, Smith BJ, Robertson J, Besteman EG and Holladay SD (1998). Subacute immunotoxic effects of the polycyclic aromatic hydrocarbon, 7,12-dimethylbenzanthracene (DMBA) on spleen and pronephros leukocytic cell counts and phagocytic cell activity in tilapia (*Oreochromis niloticus*). *Aquatic Toxicology*, **41**, 17–29.

Hodson PV, McWhirter M, Ralph K, Gray B, Thivierge D, Carey JC, van der Kraak G, Whittle DM and Levesque MC (1992) Effects of bleached kraft mill effluent on fish in the St. Maurice River, Quebec. *Environmental Toxicology and Chemistry*, **11**, 1635–1651.

James MO, Altman AH, Feistner H and Gahne A (1993) Intestinal and hepatic glucuronida-
tion and sulfation of 4,7- and 9-hydroxybenzo[a]pyrenes in the catfish. *Toxicologist*,
13, 63–70.

James MO, Altman AH, Mugadi S, Li CJ and Schell JD (1995) Oral bioavailability, biotransformation and hepatopancreas DNA binding of benzo[a]pyrene in the American
lobster, *Homarus americanus*. *Chemical–Biological Interactions*, **95**, 141–160.

James MO and Boyle SM (1998) Cytochromes P450 in crustacea. *Comparative Biochemistry and Physiology*, **121C**, 157–172.

Jett DA, Navoa RV and Lyons jr MA (1999) Additive inhibitory action of chlorpyrifos and
polycyclic aromatic hydrocarbons on acetylcholinesterase activity *in vitro*. *Toxicology
Letters*, **105**, 223–229.

Kantoniemi A, Vahakangas K and Oikari A (1996) The capacity of liver microsomes to
form benzo[a]pyrene–diolepoxide–DNA adducts and induction of cytochrome P450
1A in feral fish exposed to pulp mill effluents. *Ecotoxicology and Environmental
Safety*, **35**, 136–141.

Kennedy SW, Lorenzen A and Norstrom RJ (1996) Chicken embryo hepatocyte bioassay
for measuring cytochrome P4501A-based 2,3,7,8-tetrachlorodibenzo-*p*-dioxin equivalent concentrations in environmental samples. *Environmental Science and Technology*, **30**, 706–715.

Kirso U and Irha N (1998) Role of algae in fate of carcinogenic polycyclic aromatic
hydrocarbons in the aquatic environment. *Ecotoxicology and Environmental Safety*,
41, 83–89.

Klinge CM, Bowers JL, Kulakosky PC, Kamboj KK and Swanson HI (1999) The aryl
hydrocarbon receptor (AHR/AHR nuclear translocator, (ARNT) heterodimer interacts with naturally occurring estrogen response elements. *Molecular and Cellular
Endocrinology*, **157**, 105–119.

Law JM, Bull M, Nakamura J and Swenberg JA (1998) Molecular dosimetry of DNA
adducts in the medaka small fish model. *Carcinogenesis*, **19**, 515–518.

Livingstone DR (1998) The fate of organic xenobiotics in aquatic ecosystems: quantitative
and qualitative differences in biotransformation by invertebrates and fish. *Comparative
Biochemistry and Physiology*, **120A**, 43–49.

Ma Q (2001) Induction of CYP1A1. The AhR/DRE paradigm: transcription, receptor
regulation, and expanding biological roles. *Current Drug Metabolism*, **2**, 149–164.

Maccubbin AE (1994) DNA adduct analysis in fish: laboratory and field studies. In
Malins DC and Ostrander GK (eds), *Aquatic Toxicology; Molecular, Biochemical and
Cellular Perspectives*. Lewis, CRC Press, Boca Raton, FL, pp. 267–294.

Machala M, Ciganek M, Blaha L, Minksova K and Vondracek J (2001) Aryl hydrocarbon receptor-mediated and estrogenic activities of oxygenated polycyclic aromatic
hydrocarbons and azaarenes originally identified in extracts of river sediments. *Environmental Toxicology and Chemistry*, **20**, 2736–2743.

McCann JH, Greenberg BM and Solomon KR (2000) The effect of creosote on the growth
of axenic culture of *Myriophyllum spicatum* L. *Aquatic Toxicology*, **50**, 265–274.

McCarthy JF, Halbrook RS and Shugart LR (1991) *Conceptual Strategy for Design, Imple-
mentation, and Validation of a Biomarker-based Biomonitoring Capability*. 3072,
ORNL/TM-11783. Environmental Sciences Division, Oak Ridge National Laboratory,
TN, USA.

Meador JP, Stein JE, Reichert WL and Varanasi U (1995). Bioaccumulation of polycyclic
aromatic hydrocarbons by marine organisms. *Reviews in Environmental Contamination and Toxicology*, **143**, 79–165.

Mekenyan OG, Veith GD, Call DJ and Ankley GT (1996) A QSAR evaluation of Ah receptor
binding of halogenated aromatic xenobiotics. *Environmental Health Perspectives*,
104, 1302–1310.

Mekenyan OG, Ankley GT, Veith GD and Call DJ (1994) QSAR estimates of excited states and photoinduced acute toxicity of polycyclic aromatic hydrocarbons. *SAR and QSAR in Environmental Research*, **2**, 237–247.

Miranda CL, Henderson MC, Williams DE and Buhler DR (1997) *In vitro* metabolism of 7,12-dimethylbenz[a]anthracene by rainbow trout liver microsomes and trout P450 isoforms. *Toxicology and Applied Pharmacology*, **142**, 123–132.

Munkittrick KR, van der Kraak GJ, McMaster ME and Portt CB (1992) Reproductive dysfunction and MFO activity in three species of fish exposed to bleached kraft mill effluent at Jackfish Bay, Lake Superior. *Water Pollution Research in Canada*, **27**, 439–446.

Navas JM and Segner H (2000) Antiestrogenicity of β-naphthoflavone and PAHs in cultured rainbow trout hepatocytes: evidence for a role of the arylhydrocarbon receptor. *Aquatic Toxicology*, **51**, 79–91.

Nebert, DW, Puga A and Vasiliou V (1993) Role of the Ah receptor and the dioxin-inducible [Ah] gene battery in toxicity, cancer, and signal transduction. *Annals of the New York Academy of Science*, **685**, 624–640.

Nesnow S, Mass MJ, Ross JA, Galati AJ, Lambert GR, Gennings C, Carter Jr WH and Stoner GD (1998) Lung tumorigenic interactions in strain A/J mice of five environmental polycyclic aromatic hydrocarbons. *Environmental Health Perspectives*, **106**, 1337–1346.

Nie M, Blankenship AL and Giesy JP (2001) Interactions between aryl hydrocarbon receptor (AhR) and hypoxia signaling pathways. *Enviromental Toxicology and Pharmacology*, **10**, 17–27.

Nishimoto M, Yanagida GK, Stein JE, Baird WM and Varanasi U (1992) The metabolism of benzo[a]pyrene by English sole (*Parophrys vetulus*): comparison between isolated hepatocytes *in vitro* and liver *in vivo*. *Xenobiotica*, **22**, 949–961.

NRC: Committee on Biological Markers of the National Research Council (1987) *Environmental Health Perspectives*, **74**, 3–9.

Pangrekar J, Kandaswami C, Kole P, Kumar S and Sikka HC (1995) Comparative metabolism of benzo[a]pyrene, chrysene and phenathrene by brown bullhead liver microsomes. *Marine Environmental Research*, **39**, 51–55.

Parkinson A (1996) Biotransformation of xenobiotics. In Klaassen CD (ed.). *Casarett's and Doull's Toxicology*, 5th edn. McGraw-Hill, New York. pp. 113–186.

Peakall DW (1994) Biomarkers: the way forward in environmental assessment. *Toxicology and Ecotoxicology News*, **1**, 55–60.

Pfau W (1997) DNA adducts in marine and freshwater fish as biomarkers of environmental contamination. *Biomarkers*, **2**, 145–151.

Pichorner H, Metodiewa D and Winterbourne CC (1995) Generation of superoxide and tyrosine peroxide as a result of tyrosyl radical scavenging by glutathione. *Archives of Biochemistry and Biophysics*, **323**, 429–437.

Reichert WL, Myers MS, Peck-Miller K, French B, Anulacion BF, Collier TK, Stein JE and Varanasi U (1998) Molecular epizootiology of genotoxic events in marine fish: linking contaminant exposure, DNA damage and tissue-level alterations. *Mutation Research*, **411**, 215–225.

Safe S (1990) Polychlorinated biphenyls (PCBs), dibenzo-*p*-dioxins (PCDDs), bibenzofurans (PCDFs), and related compounds: environmental and mechanistic considerations which support the development of toxic equivalency factors (TEFs). *Critical Reviews in Toxicology*, **21**, 51–88.

Safe SH (1994) Polychlorinated biphenyls (PCBs): environmental impact, biochemical and toxic responses, and implications for risk assessment. *Critical Reviews in Toxicology*, **24**, 87–149.

Safe SH (1998) Development validation and problems with the toxic equivalency factor approach for risk assessment of dioxins and related compounds. *Journal of Animal Science*, **76**, 134–141.

Safe S (2001) Molecular biology of the Ah receptor and its role in carcinogenesis. *Toxicology Letters*, **120**, 1–7.

Sandvik M, Horsberg TE, Skaare JU and Ingebrigtsen K (1998) Comparison of dietary and waterborne exposure to benzo[a]pyrene: bioavailability, tissue disposition and CYP1A induction in rainbow trout (*Oncorhynchus mykiss*). *Biomarkers*, **3**, 399–410.

Saraquette C and Segner H (2000) Cytochrome P4501A (CYP1A) in teleostean fishes. A review of immunohistochemical studies. *Science of the Total Environment*, **247**, 313–332.

Segner H and Braunbeck T (1998) Cellular response profile to chemical stress. In Schüürmann G, Markert B (eds), *Ecotoxicology*. Wiley, New York/Spektrum, Heidelberg, pp. 521–569.

Segner H, Fritayre P and Chesne C (2001) Cell bioassays in environmental diagnosis and monitoring. In Hofmann M (ed.), *Focus in Biotechnology* (in press).

Silkworth JB, Lipinskas T and Stoner CR (1995) Immunosuppressive potential of several polycyclic aromatic hydrocarbons (PAHs) found at a Superfund site: new model used to evaluate additive interactions between benzo[a]pyrene and TCDD. *Toxicology*, **105**, 375–386.

Stagg RM (1998) The development of an international programme for monitoring the biological effects of contaminants in the OSPAR convention area. *Marine Environmental Research*, **46**, 307–313.

Stegeman JJ and Lech JJ (1991) Cytochrome P450 monooxygenase systems in aquatic species: carcinogen metabolism and biomarkers for carcinogen and pollutant exposure. *Environmental Health Perspectives*, **90**, 101–109.

Stegeman JJ and Hahn ME (1994) Biochemistry and molecular biology of monooxygenase: current perspective on forms, functions, and regulation of cytochrome P450 in aquatic species. In Malins DC and Ostrander GK, (eds), *Aquatic Toxicology; Molecular, Biochemical and Cellular Perspectives*. Lewis, CRC Press, pp. 87–206.

Stegeman JJ, Brouwer M, Richard TDG, Förlin L, Fowler BA, Sanders BM and Van Veld PA (1992) Molecular responses to environmental contamination: enzyme and protein systems as indicators of chemical exposure and effect. In Huggett RJ, Kimerly RA, Mehrle Jr. PM and Bergman HL (eds), *Biomarkers: Biochemical, Physiological and Histological Markers of Anthropogenic Stress*. Lewis, Chelsea, MI, pp. 235–335.

Suter II GW (1993) *Ecological Risk Assessment*. Lewis, Boca Raton, FL, USA, 538 pp.

Swartz RC, Schults DW, Ozretich RJ, Lamberson JO, Cole FA, DeWitt TH, Redmond MS and Ferraro SO (1995) ΣPAH: a model to predict the toxicity of polynuclear aromatic hydrocarbon mixtures in field collected sediments. *Ecotoxicology and Environmental Safety*, **14**, 197–1987.

Thomann RV and Komlos J (1999) Model of biota-sediment accumulation factor for polycyclic aromatic hydrocarbons. *Environmental Toxicology and Chemistry*, **18**, 1060–1068.

Till M, Riebniger D, Schmitz HJ and Schrenk D (1999) Potency of various polycyclic aromatic hydrocarbons as inducers of CYP1A in rat hepatocyte cultures. *Chemical–Biological Interactions*, **117**, 135–150.

Ueng TH, Ueng YF and Chou MW (1994) Regioselective metabolism of benzo[a]pyrene and 7-chlorobenzo[a]anthracene by fish liver microsomes. *Toxicology Letters*, **70**, 89–99.

Upham BL, Weis LM and Trosko JE (1998) Modulated gap junctional intercellular communication as a biomarker of PAH epigenetic toxicity: structure–function relationship. *Environmental Health Perspectives*, **106**, 975–981.

Vakharia DD, Liu N, Pause R, Fasco M, Bessette E, Zhang QY and Kaminsky LS (2001) Polycyclic aromatic hydrocarbon/metal mixtures: effect on PAH induction of CYP1A1 in human HEPG2 cells. *Drug Metabolism and Disposition*, **29**, 999–1006.

Van Brummelen TC, van Hattum B, Crommentuijn T and Kalf DF (1998) Bioavailability and ecotoxicity of PAHs. In Nielson, AH (ed.), *The Handbook of Environmental Chemistry, vol. 3, Part J. PAHs and Related Compounds*. Springer Verlag, Berlin, Heidelberg, pp. 203–263.

Van der Kooij LA, van de Meent D, van Leeuwen CJ and Bruggeman WA (1991) Deriving quality criteria for water and sediment from aquatic toxicity tests and product standards: application of the equilibrium partitioning method. *Water Research*, **25**, 697–705.

Van der Oost R, Opperhuizen A, Satumalay K, Heida H and Vermeulen NPE (1996a) Biomonitoring aquatic pollution with feral eel (*Anguilla anguilla*): I. Bioaccumulation: biota-sediment ratios of PCBs, OCPs, PCDDs and PCDFs. *Aquatic Toxicology*, **35**, 21–46.

Van der Oost R, Goksøyr A, Celander M, Heida H and Vermeulen NPE (1996b) Biomonitoring aquatic pollution with feral eel (*Anguilla anguilla*): II. Biomarkers: pollution-induced biochemical responses. *Aquatic Toxicology*, **36**, 189–222.

Van der Oost R, Van Schooten FJ, Ariese F, Heida H and Vermeulen NPE (1994) Bioaccumulation, biotransformation and DNA binding of PAHs in feral eel (*Anguilla anguilla*) exposed to polluted sediments: a field survey. *Environmental Toxicology and Chemistry*, **13**, 859–870.

Van der Oost R, Heida H, Opperhuizen A and Vermeulen NPE (1991) Bioaccumulation of organic micropollutants in different aquatic organisms: sublethal toxic effects on fish. In Mayes MA and Baron MG (eds), *Aquatic Toxicology and Risk Assessment*, vol. 14. ASTM STP 1124. American Society for Testing and Materials, Philadelphia, PA, pp. 166–180.

Van der Oost R, Beyer J and Vermeulen NPE (2003) Fish biomarkers and environmental risk assessment: a review. *Enviromental Toxicology and Pharmacology*, **13**, 52–149.

Van Gestel CAM and van Brummelen TC (1996) Incorporation of the biomarker concept in ecotoxicology calls for a redefinition of terms. *Ecotoxicology*, **5**, 217–225.

Van Schooten FJ (1991) Polycyclic Aromatic Hydrocarbon–DNA Adducts in Mice and Humans. Academic Thesis, State University of Leiden, The Netherlands.

Van Veld PA (1990) Absorption and metabolism of dietary xenobiotics by the intestine of fish. *Reviews in Aquatic Science*, **2**, 185–203.

Varanasi U (ed.) (1989) *Metabolism of Polycyclic Aromatic Hydrocarbons in the Aquatic Environment*. CRC Press, Boca Raton, FL.

Vethaak AD, Jol JG, Meijboom A, Eggens ML, Reinallt T, Wester PW, van de Zande T, Bergman A, Dankers N, Ariese F, Baan RA, Everts JM, Opperhuizen A and Marquenie, JM (1996) Skin and liver diseases induced in flounder (*Platichthys flesus*) after long-term exposure to contaminated sediments in large-scale mesocosms. *Environmental Health Perspectives*, **104**, 1218–1229.

Walter H, Consolaro F, Gramatica P, Scholze M and Altenburger R (2002) Mixture toxicity of priority pollutants at no observed effect concentrations (NOECs). *Ecotoxicology* **11**, 299–310.

Washburn KS, Donnelly KC, Huebner HJ, Burghardt RC, Sewall TC and Claxton LD (2001) A study of 2,4,6-trinitrotoluene inhibition of benzo[a]pyrene uptake and activation in a microbial mutagenicity assay. *Chemosphere*, **44**, 1703–1709.

Wells PG, Lee K and Blaise C (eds) (1998) *Microscale Testing in Aquatic Toxicology*. CRC Press: Boca Raton, FL, pp. 679.

Weis LM, Rummel AM, Masten SJ, Trosko JE and Upham BL (1998) Bay or baylike regions of polycyclic aromatic hydrocarbons were potent inhibitors of gap junctional intercellular communication. *Environmental Health Perspectives*, **106**, 17–22.

WHO International programme on chemical safety (IPCS) (1993) *Biomarkers and Risk Assessment: Concepts and Principles*. Environmental Health Criteria 155, World Health Organization, Geneva.

Whyte JJ, Jung RE, Schmitt CJ and Tillitt DE (2000) Ethoxyresorufin-*O*-deethylase (EROD) activity in fish as a biomarker of chemical exposure. *Critical Reviews in Toxicology*, **30**, 347–570.

Willett KL, Gardinali PR, Sericano JL, Wade TL and Safe SH (1997) Characterization of the H4IIE rat hepatoma cell bioassay for evaluation of environmental samples containing polynuclear aromatic hydrocarbons (PAHs). *Archives of Environmental Contamination and Toxicology*, **32**, 442–228.

Winston GW and Di Giulio RT (1991) Pro-oxidant and anti-oxidant mechanisms in aquatic organism. *Aquatic Toxicology*, **19**, 137–161.

Yuan ZX, Honey SA, Kumar SB and Sikka HC (1999) Comparative metabolism of dibenzo(a,l)pyrene by liver microsomes from rainbow trout and rats. *Aquatic Toxicology*, **45**, 1–8.

PART IV
Integration of Information on PAHs

17

Approaches to Developing Sediment Quality Guidelines for PAHs*

DAVID R. MOUNT[1], CHRISTOPHER G. INGERSOLL[2]
AND JOY A. MCGRATH[3]
*[1]US Environmental Protection Agency, Office of Research and Development,
National Health and Ecological Effects Laboratory, Duluth, MN, USA
[2]US Geological Survey, Columbia, MO, USA
[3]HydroQual Inc., Mahwah, NJ, USA*

17.1 INTRODUCTION

As described in previous chapters, the properties of PAHs are such that they often accumulate in aquatic sediments, resulting in exposure of benthic organisms. To understand and manage the associated risks, it is generally desirable to quantitatively relate sediment concentrations of PAHs or other sediment contaminants to levels of biological effect that are expected as a result. This need has fueled an array of research approaches to developing chemistry–toxicity relationships for sediment, generically referred to as 'sediment quality guidelines' or SQGs. This chapter describes several common approaches to developing SQGs, and discusses their strengths and weaknesses as potential regulatory tools.

Although PAHs in sediment may lead to exposure of pelagic aquatic organisms as well as terrestrial and avian species, most of the SQGs developed to date focus on effects on the benthic community, that is organisms living directly in or on bedded sediment. In general terms, SQGs that have been proposed can be categorized into two groups, depending on their derivation: empirically-based guidelines (sometimes called 'co-occurrence guidelines'), and

---

* This manuscript was reviewed according to the policies of the US Environmental Protection Agency and the US Geological Survey, and does not necessarily reflect the views of either organization or the US Government. Mention of trade names does not imply endorsement.

PAHs: An Ecotoxicological Perspective. Edited by Peter E.T. Douben.
© 2003 John Wiley & Sons Ltd

mechanistically-based guidelines. Empirically-based guidelines are derived from databases of paired sediment chemistry and biological effect (e.g. toxicity) data, and use various analysis approaches to relate chemical concentrations to the frequency of biological effects, or lack thereof. Mechanistically-based guidelines are those that predict sediment toxicity based on an understanding of the factors that influence the bioavailability of sediment-associated chemicals. Subsequent sections describe the features of individual SQGs in greater detail.

17.2 EMPIRICALLY-BASED GUIDELINES

Empirically-derived sediment quality guidelines have been developed and/or adopted by federal and state environmental agencies in North America, Europe, and Australia. The approaches selected by different jurisdictions vary according to differences in sediment management objectives, the receptors to be protected, the intended degree of protection, the geographic area to which the values apply, and their intended uses (e.g. screening/prioritization tools, targets for remedial action). Notable examples of empirical SQGs are outlined below.

17.2.1 SCREENING LEVEL CONCENTRATION (SLC)

Screening level concentrations (SLCs), originally proposed by Neff *et al.* (1987), are based on the relationship between sediment chemistry and benthic community structure. For each benthic organism for which adequate data are available, a frequency distribution of chemical concentration is developed for all sites at which the species occurs. The 90th percentile of this distribution is taken as the species screening level concentration (SSLC) for that organism. Following calculation of SSLCs for all of the species having adequate data, the frequency distribution of the SSLCs as a function of chemical concentration is created. The SLC is then calculated as the 5th percentile of this second distribution. The intent is to define the chemical concentration that should be tolerated by 95% of all species evaluated. Subsequent studies have used this same approach to calculate SQGs for freshwater sediments (EC and MENVIQ 1992; Persaud *et al.* 1993).

17.2.2 EFFECTS RANGES (ERL AND ERM)

The effects range approach was originally developed to provide informal tools for assessing the association between contaminants tested in the National Status and Trends Program (NSTP) and adverse effects on marine or estuarine sediment-dwelling organisms (Long and Morgan 1991). Paired information on sediment chemistry and biological effects from toxicity tests on spiked or field-collected sediments, field studies conducted in the USA, and various SQGs were compiled and sorted in ascending order of concentration. The 10th percentile of the concentrations causing effects was labeled the 'effects range-low' (ERL); this was considered to represent a threshold value below which adverse effects occur only infrequently. The 'effects range-median' (ERM) was defined as the 50th percentile of the distribution, intended as a concentration above which adverse effects are frequently observed. The original ERL and ERM values were updated in 1995 (Long *et al.* 1995). The same or a similar approach has been used by other authors to calculate additional SQGs for freshwater and marine sediments (Ingersoll *et al.* 1996; MacDonald 1997).

17.2.3   EFFECTS LEVELS (TEL AND PEL)

The effects level method is closely related to the effects range approach described above, but is based on an expanded database of chemical and bio-logical data, and also uses a different algorithm to derive SQG concentrations (McDonald *et al.* 1996). In this approach, concentrations of a chemical associ-ated with either a biological effect or the lack of an effect are sorted separately in ascending order. The threshold effect level (TEL) is calculated as the geometric mean of the 15th percentile of the effect data and the 50th percentile of the no-effect data, and is intended to represent the chemical concentration below which adverse effects are expected to occur only infrequently. This differs from the ERL not only in the percentile of effects data used, but also in jointly using no-effect data. The probable effects level (PEL) is intended to be the concentra-tion above which adverse effects are expected to be observed frequently; this is calculated as the geometric mean of the 50th percentile of the effects data set and the 85th percentile of the no effects data set. The effects level approach was initially developed and used to derive SQGs for the coastal waters of Florida, USA (MacDonald *et al.* 1996), but has also been used to calculate SQGs for freshwater sediments (Ingersoll *et al.* 1996; Smith *et al.* 1996; CCME 1999).

17.2.4   APPARENT EFFECTS THRESHOLD (AET)

The apparent effects threshold (AET) approach was conceived for use in the Puget Sound area of Washington State, USA (PTI 1991). AETs are based on measured chemistry and biological effect data for field-collected sediments within the region to which they will be applied. Paired chemistry/biology data are arrayed in order of increasing chemical concentration, and the AET is defined as the concentration above which significant ($p < 0.05$) adverse effects are always observed. Rather than pooling data for different measures of effects, AET values are generally created for specific biological effects, such as toxicity to benthic and/or water column species measured using sediment toxicity tests, or changes in benthic community structure. In the original derivation for Puget Sound, AETs were developed for toxicity tests conducted on amphipods, oyster larvae, and bacteria (Microtox®), and for the abundance of benthic macrofauna. Beyond the endpoint-specific AETs, the 'low AET' (AET-L) represents the lowest AET value for any of the biological endpoints assessed, while the 'high AET' or AET-H is the highest of the values, above which significant effects are always observed for all endpoints. The State of Washington in the USA has used AET values to establish sediment quality standards and minimum clean-up levels for contaminants of concern (WDOE 1990). Freshwater SQGs have been calculated using these same procedures (Ingersoll *et al.* 1996; Cubbage *et al.* 1997).

17.2.5   LOGISTIC REGRESSION MODELING

Logistic regression modeling focuses on establishing the probability of adverse effect as a function of chemical concentration, and was developed by Field *et al.* (1999). In that study, data were assembled for field-collected sediments that had both measured concentrations of sediment contaminants and effects data from sediment toxicity tests with marine amphipods. Prior to creating a regression for a specific contaminant, data were screened by eliminating toxic samples in which the concentration of that contaminant was lower than the average concentration in non-toxic samples; this was done on the presumption that it was unlikely that the contaminant would have contributed substantially to toxicity in those sediments. The screened data were then used to develop logistic regression models which relate chemical concentration to a predicted probability of toxicity. Because the logistic regression establishes a continuous relationship between chemical concentration and probability of effect, this approach can be used to define SQGs associated with any desired probability. In their original study, Field *et al.* (1999) calculated values for 10%, 50%, and 90% probabilities of toxicity (T10, T50, and T90, respectively) for four metals, two PAHs (phenanthrene and fluoranthene), and total PCBs. In a later publication (Field *et al.* 2002), the index probabilities were changed to 20%, 50%, and 80%, and SQGs were calculated for a much wider range of chemicals, including 22 different PAHs.

17.3   MECHANISTIC GUIDELINES

While many different approaches have been used to develop empirically-based guidelines, fewer mechanistic guidelines have been proposed, and even those share a common basis in 'equilibrium partitioning' (EqP) theory (Di Toro *et al.* 1991).

17.3.1   EQUILIBRIUM PARTITIONING: BACKGROUND

The concept behind EqP is clearly displayed by the experiments of Adams *et al.* (1985), who showed that the concentration response for midge larvae exposed to kepone in sediment varied substantially across three different sediments when expressed on a dry weight basis (Figure 17.1a). This suggested that the composition of a sediment was important to determining the bioavailability of chemical in that sediment. If one expressed these same toxicity results on the basis of kepone concentration in interstitial water, the data were resolved into a single concentration response curve (Figure 17.1b). Similarly, when these toxicity data were normalized on the basis of kepone concentration per unit organic carbon, a similar dose–response pattern was generated (Figure 17.1c). Di Toro et $al.$ (1991) rationalized these observations by assuming that a thermodynamic equilibrium ('equilibrium partitioning') exists among the interstitial water and the organic carbon phase of sediment, and that the potency of the sediment could be expressed on the basis of either phase. Since chemical activity in water is generally approximated by concentration in water, it follows that sediments with comparable chemical concentrations in interstitial water would have comparable chemical activity, and therefore comparable toxicity.

For non-ionic organic chemicals such as PAHs, organic carbon represents the primary binding phase in natural sediments. The relationship between chemical concentration in sediment and that in interstitial water at equilibrium can be described by a simple partitioning equation between sediment organic carbon and interstitial water:

$$c_{soc} = c_{iw}{}^* K_{oc} \qquad (17.1)$$

where c_{soc} is the concentration in sediment per unit organic carbon, c_{iw} is the concentration in interstitial water, and K_{oc} is the organic carbon partition coefficient. Expressing concentration in sediment on the basis of bulk sediment, this expands to:

$$c_{sed}{}^* f_{oc}^{-1} = c_{iw}{}^* K_{oc} \qquad (17.2)$$

where c_{sed} is the concentration per unit dry mass in sediment and f_{oc} is organic carbon content as a fraction of total dry mass. Based on these relationships, equilibrium partitioning theory asserts that the toxicity of a non-ionic organic chemical in different sediments can be predicted by normalizing the chemical concentration to organic carbon. Additional work showed the applicability of this principle in a variety of experiments with different sediments and different non-ionic organic chemicals, including three individual PAHs, acenaphthene, phenanthrene, and fluoranthene (Di Toro et $al.$ 1991; USEPA 2000).

To establish a sediment guideline ($c_{guideline}$) on the basis of EqP, the interstitial water concentration in Equations 17.1 or 17.2 is set equal to a concentration (c_{effect}) that is associated with the desired degree of protection:

$$c_{guideline} = c_{sed}{}^* f_{oc}^{-1} = c_{effect}{}^* K_{oc} \qquad (17.3)$$

In their draft 'Sediment Quality Criteria', the US Environmental Protection Agency (USEPA 1993a) selected the 'final chronic value' from their ambient water quality criteria (AWQC) (Stephan et $al.$ 1985). These are water column

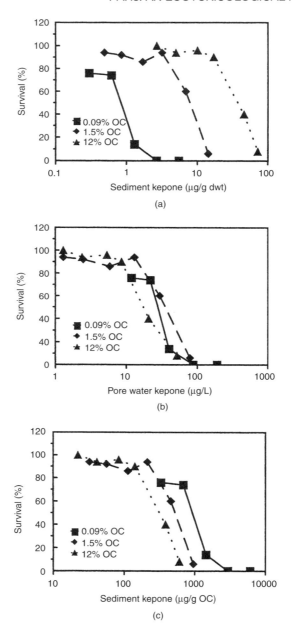

Figure 17.1 Response of midge larva of three kepone-spiked sediments with different organic carbon content, expressed as kepone in bulk sediment (a), kepone in interstitial water (b), and organic carbon-normalized kepone in sediment (c). Redrawn from Adams *et al.* (1985)

concentrations of chemical intended to protect most aquatic species from acute or chronic toxic effects. Supplementary analyses of toxicity data for scores of aquatic species suggested that the toxicant sensitivity of benthic/epibenthic organisms was generally comparable to that of pelagic organisms (USEPA 1993a), supporting the appropriateness of basing sediment guidelines on toxicity data from species of both types.

17.3.2 INCORPORATION OF MIXTURES INTO EqP GUIDELINES FOR PAH: THE ΣPAH MODEL

Based on these principles, USEPA originally published draft 'Sediment Quality Criteria' (SQC) for two pesticides (endrin and dieldrin) and three single PAHs, with values (in $\mu g \cdot g^{-1}$ organic carbon) of 230 for acenaphthene, 240 for phenanthrene, and 300 for fluoranthene (USEPA 1993b, 1993c, 1993d, 1993e, 1993f). Although these values were derived in a manner consistent with the demonstrated principles of EqP, they were derived for each PAH *as a single stressor* (i.e. absent concurrent exposure from other PAH or other chemicals). While this single-chemical approach had been taken previously in the development of regulatory criteria for many different chemicals, it is problematic for PAHs, because PAH exposure in the environment is essentially always to a complex mixture of PAH compounds rather than of individual PAHs. As a result, a sediment with a concentration of, for example, phenanthrene high enough to exceed the single PAH SQC would have toxicity far in excess of the desired effect threshold, because of the concurrent concentrations of PAHs other than phenanthrene. This created the potential for a sediment with toxic concentrations of PAHs to be judged acceptable because it met the three individual PAH SQCs; accordingly, these original single PAH SQCs were withdrawn by USEPA and were never published as final regulatory guidance.

Swartz *et al.* (1995) were the first to modify the EqP approach to explicitly address mixtures of PAHs in developing their 'ΣPAH' sediment toxicity model, which predicts the toxicity of PAH-contaminated sediments to estuarine amphipods based on the concentrations of 13 'parent' (i.e. unsubstituted) PAHs. The ΣPAH model uses EqP to relate sediment PAH concentrations to interstitial water PAH concentrations, then predicts the toxic units (TU; PAH concentration divided by the 10 day LC_{50}) for each individual PAH. The potencies of individual PAHs are predicted from a regression of K_{ow} vs. 10-day LC_{50}. Based on an assumption of additive toxicity, the TU predicted for each individual PAH are summed across the 13 PAHs to predict the TU for the mixture as a whole. A sample is predicted to cause 50% or more mortality of amphipods if the total TU ≥ 1. When the ΣPAH model was applied to sediment toxicity data for PAH-contaminated sediments from four geographically distinct regions, there was a strong relationship between predicted and observed toxicity.

While the ΣPAH model was a substantial advance over the single PAH SQC approach, Swartz *et al.* (1995) discussed several residual uncertainties in the approach that could benefit from additional development. For one, the ΣPAH model considers only 13 parent PAHs; other PAHs, such as alkylated PAHs, are not explicitly incorporated. Another uncertainty involved the slope of the toxicity–K_{ow} relationship used, which was based on a limited data set and extrapolated beyond the range of the data. A third uncertainty involved potential interspecific differences in sensitivity among the multiple amphipod species used; and lastly there was the uncertain relationship between chronic toxicity endpoints and the 10 day mortality endpoint used in the model. While the basic approach of the ΣPAH model appeared sound, it would be necessary to extend the model to provide a more robust approach.

17.3.3   THE EqP/NARCOSIS APPROACH

Just such an extension was proposed by Di Toro *et al.* (2000) and Di Toro and McGrath (2000). The keystone of this enhanced approach lies in the method used to assess the potency of individual PAHs. In the ΣPAH model, Swartz *et al.* (1995) had shown a relationship between the toxicity of individual PAHs and their K_{ow}, using a regression relationship between LC_{50} and K_{ow} to predict the toxicity of PAHs that were not actually tested as individual chemicals. In their analysis, Di Toro *et al.* (2000) considered this relationship as part of a larger phenomenon known as 'non-polar narcosis' (Veith *et al.* 1983).

Di Toro *et al.* (2000) analyzed toxicity data for 33 different species exposed to non-polar narcotic toxicants (both PAH and non-PAH compounds) and concluded that the relationship between LC_{50} and K_{ow} had a slope that was not statistically different across all species, a so-called 'universal narcosis slope' (Figure 17.2). At the *y* intercept, $K_{ow} = 1$ (10^0), which means that the concentration in the water is theoretically equal to the concentration in the lipid of the organism, assuming that octanol is a suitable surrogate for organismal lipid. This *y* intercept value therefore defines the critical body burden (body burden that causes 50% mortality) for that species. Differences in the *y* intercept of the log LC_{50} – log K_{ow} relationship correspond to relative species sensitivity to narcotic toxicants (lower intercept = higher sensitivity). By analyzing the distribution of critical body burdens across all species (Figure 17.3), the 95th percentile in acute sensitivity was determined, which is conceptually parallel to the final acute value used in deriving AWQC (Stephan *et al.* 1985).

Chronic effects were assessed by analyzing the ratio of the 48- or 96-h LC_{50} to concentrations associated with 'threshold' chronic toxicity to the same species, the so-called 'acute to chronic ratio' or ACR, which is also part of the AWQC derivation procedure. The 95th percentile in acute sensitivity was then divided by the geometric mean of ACRs for all species and chemicals (5.09) to arrive at an estimated 95th percentile in chronic sensitivity, parallel to the final chronic value used to derive AWQC. Applying the universal narcosis slope to

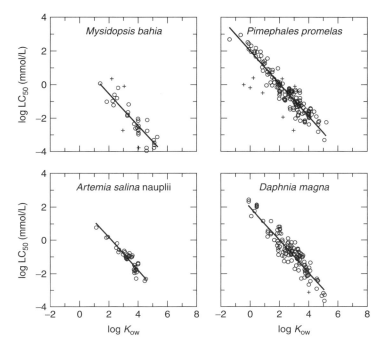

Figure 17.2 Relationship of log LC$_{50}$ to log K$_{ow}$ for four of the 33 species evaluated for response to narcotic chemicals. Slope for all relationships is the same; y intercept values represent the relative sensitivity of the species. Data marked with a + symbol are significant outliers discarded from the analysis. From Di Toro *et al.* (2000), reproduced by permission of SETAC Press

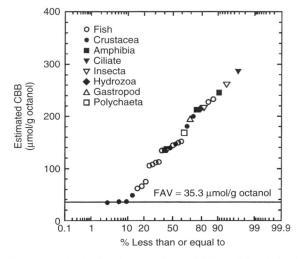

Figure 17.3 Frequency distribution for narcotic sensitivity and the derivation of the final acute value at the 95th percentile. From Di Toro *et al.* (2000), reproduced by permission of SETAC Press

this estimated final chronic value can then be used to estimate the final chronic value for any non-polar narcotic chemical, based solely on its K_{ow}.

Once PAH-specific final chronic values are derived, the remainder of the analysis is very similar in concept to the ΣPAH model. The final chronic values in water are converted to organic carbon-normalized sediment concentrations through an EqP calculation (Equation 17.3). Fractional potencies are calculated for each individual PAH by dividing measured PAH concentrations by the PAH-specific sediment guideline value. These fractional contributions are then summed across all PAHs; if the sum exceeds a value of 1, then the guideline is exceeded. Viewing PAH toxicity in the context of narcosis provided several important components: (a) it established a firmer basis for a K_{ow}-based prediction of toxicity for individual PAHs by incorporating a much larger toxicological data set; (b) it provided a mechanistic basis for the assumption of additive toxicity in PAH mixtures; (c) it incorporated a range of sensitivity across species; and (d) it provided a mechanism to consider chronic toxicity.

In parallel with the development of the EqP/narcosis approach, USEPA undertook work to develop a SQG for PAH mixtures. At the time of this writing, USEPA has not formally published this guideline, although the draft guidance is available from the Health and Ecological Criteria Division at the USEPA's Office of Water (USEPA 2000). The USEPA guideline derivation is very close to that described by Di Toro et al. (2000; Di Toro and McGrath 2000). The primary difference lies in the derivation of the final chronic value. Di Toro et al. used species-specific narcosis intercepts based on all non-polar narcotics, then applied a PAH-specific correction to account for the slightly greater potency of PAHs (see Di Toro et al. 2000 for more on this correction factor). In the draft USEPA (2000) guidance, only toxicity data specifically for PAHs are used; this approach increases the amount of PAH-specific data in the derivation, because Di Toro et al. used only data for species for which at least four narcotic chemicals had been tested. At the same time, using PAH-specific data eliminates the need to use a chemical class correction for PAHs. Individual test results were normalized by expressing them as the corresponding log LC_{50} – log K_{ow} intercept, calculated using the universal narcosis slope from Di Toro et al. (2000). From there, data were aggregated as described in the USEPA AWQC guidelines (Stephan et al. 1985) to calculate a final acute value. The ACR used to convert the final acute value to a final chronic value also differs; the USEPA ACR (4.16) is based only on PAH-specific data, while the ACR used by Di Toro et al. (5.09) uses data for all non-polar narcotics. The aggregate effect of these differences is relatively small, however, with the USEPA value being about 40% lower than the Di Toro and McGrath (2000) guideline.

The Di Toro and McGrath (2000) and USEPA (2000) derivations are designed around the level of protection afforded by AWQC, which theoretically protects all but the most sensitive species from acute or chronic toxicity. While this is conceptually consistent with the intent of many regulatory programs, it is worth noting that the approach in general is not restricted to this particular level of

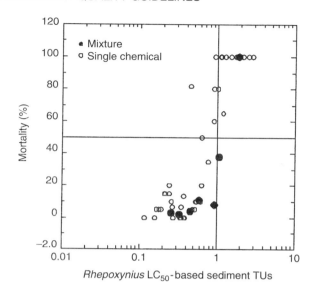

Figure 17.4 Use of EqP/narcosis approach to predict the response of the amphipod *Rhepoxynius abronius* to PAHs in single PAH and PAH mixture experiments. Data from Swartz *et al.* (1997). Figure from Di Toro and McGrath (2000), reproduced by permission of SETAC Press

protection. In cases where a different assessment endpoint is desired, all that need be done is to modify the water column-based effect concentration used to define PAH-specific guideline concentrations appropriately, and recalculate the PAH-specific guideline values. Alternatively, if one has sediment toxicity test data for contaminated sediments and wishes to evaluate the observed toxicity in light of measured PAH concentrations, a water-column toxicity value appropriate for the species and endpoint of the sediment toxicity test can be used to calculate guidelines specific to the interpretation of those data (Figure 17.4).

17.4 COMPARISON OF SQGs FOR PAHs

Table 17.1 lists both empirical and mechanistic guidelines for individual PAHs and 'total' PAHs. One of the difficulties in making comparisons across SQGs lies in whether the guidelines are expressed on a dry weight basis or on an organic carbon-normalized basis. Several common guidelines, such as ERL/ERM and TEL/PEL, are expressed on a dry weight basis, while others, such as the SLC and all of the EqP-based guidelines, are organic carbon-normalized (i.e. the guideline value for a specific sediment is dependent on the organic carbon content of the sediment). Following the suggestion of Swartz (1999), the dry

TABLE 17.1 Summary of selected empirical and mechanistic guidelines. Values in italics are not from the original publication but were estimated by Swartz (1999) or by parallel methods, by summing SQG values for the 13 individual PAHs listed. Values in parentheses are not guidelines themselves, but are intermediates in the calculation of the total PAH guideline value. Narrative intent of the guideline is interpreted according to Swartz (1999) and MacDonald et al. (2000). All guidelines are expressed as μg·g⁻¹OC. Total PAH values are based on unsubstituted 'parent' PAHs only

Approach	Screening level	Effects range		Effects level		Apparent effects threshold		Logistic regression			Screening level	Screening level	Effects level		Effects level		ΣPAH (EqP)		Narcosis EqP	Narcosis EqP
Source	Neff et al. (1987)	Long et al. (1995)		McDonald et al. (1996)		PTI (1991)		Field et al. (2002)			Persuad et al. (1993)	EC/MENVIQ (1992)	Smith et al. (1996)		Ingersoll et al. (1996)		Swartz (1999)		DiToro and McGrath (2000)	US EPA (2000)
Empirical/mechanistic	Empirical	Empirical		Empirical		Empirical		Empirical			Empirical	Empirical	Empirical		Empirical		Mechanistic		Mechanistic	Mechanistic
Marine or freshwater	Marine	Mixed		Marine		Marine (Puget Sound)		Marine			Freshwater	Freshwater	Freshwater		Freshwater		Marine		Both	Both
Data source/endpoint	Benthos	Mixed		Mixed		Toxicity/benthos		10 day amphipod mortality			Benthos	Benthos	Mixed		Amphipod 28 day toxicity		10 day amphipod mortality		AWQC-type final chronic value	AWQC-type final chronic value
SQG	SLC	ERL[a]	ERM[a]	TEL[a]	PEL[a]	AET-L[a]	AET-H[a]	T20[a]	T50[a]	T80[a]	LEL	MET	TEL[a]	PEL[a]	TEL[a]	PEL[a]	EPAH threshold	EPAH LC50	EqP SQG	ESG
Narrative intent	Threshold	Threshold	Median	Threshold	Median	Median	Extreme	Threshold	Median	Extreme	Threshold	Threshold	Threshold	Median	Threshold	Median	Threshold	Median	Threshold	Threshold
Naphthalene	41	16	210	3	39	210	270	3	22	157	19	40			2	14	13	71	(652)	(385)
Acenaphthylene	5	4	64	1	13		130	1	14	142							3	15	(766)	(452)
Acenaphthene	15	15	65	1	9	50	200	2	12	71							4	23	(852)	(491)
Fluorene	10	35	64	2	14	54	360	2	11	66					1	15	17	90	(911)	(538)
Phenanthrene	37	22	138	9	54	150	690	7	46	306	56	40	4	52	2	41	29	155	(1010)	(596)
Anthracene	16	9	110	5	24	96	1300	3	29	249	22				1	17	21	114	(1010)	(594)
Fluoranthene	64	60	510	11	149	170	3000	12	103	895	75	60	11	236	3	32	69	371	(1200)	(707)
Pyrene	66	67	260	15	140	260	1600	12	93	698	49	70	5	88	4	49	90	481	(1200)	(697)
Benz[a]anthracene	26	26	160	7	69	130	510	6	47	354	32	40	3	38	2	28	21	111	(1420)	(841)
Chrysene	38	38	280	11	85	140	920	8	65	519	34	60	6	86	3	41	31	169	(1425)	(844)
Benzo[b]fluoranthene						160	445	13	111	941							33	180	(1650)	(979)
Benzo[k]fluoranthene						160	445	7	54	412							29	155	(1650)	(981)
Benzo[a]pyrene	40	43	160	9	76	160	360	7	52	391	37	50	3	78	3	32	33	179	(1630)	(965)
Total PAHs	409	402	4480	87	409	1800	10300	83	659	5200	400				26	340	393	2110	194[b]	103[b]

[a] Converted from dry weight values assuming 1% organic carbon.

[b] Estimated values based on the average of individual PAH values listed, divided by 6.78, the 80th percentile value for TU_{34}/TU_{13} from Table 17.2.

weight-normalized guidelines were recalculated on an assumed 1% organic carbon content, which was the average organic carbon content in the sediment chemistry/effects database assembled by McDonald *et al.* (1996).

In an attempt to rationalize the many SQGs available, Swartz (1999) conducted an analysis in which SQGs for total PAHs were grouped according to their presumed intent, either as threshold effect concentrations (TEC; e.g. ERL, TEL, SLC, ΣPAH threshold response), median effect concentrations (MEC; e.g. ERM, PEL, AET-L, ΣPAH LC_{50}), or extreme effect concentrations (EEC; AET-H). The arithmetic mean of SQG in each group was calculated to create 'consensus' TEC, MEC, and EEC values, which were 290, 1800, and 10,000 $\mu g \cdot g/OC$, respectively (Figure 17.5). While this analysis did suggest a central tendency of the various guidelines, it should be remembered that part of this consistency was created by converting dry weight values to organic carbon-normalized

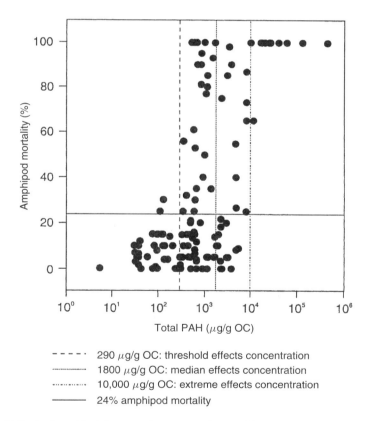

Figure 17.5 Sediment toxicity to marine or estuarine amphipods as a function of total PAH concentration (organic carbon-normalized). Vertical lines represent threshold, median, and extreme effect SQGs as derived by Swartz (1999). From Swartz (1999), reproduced by permission of SETAC Press

values through the assumed 1% organic carbon. Consistency will exist across sediments with differing organic carbon only if the organic carbon normalization of all guidelines is maintained. In other words, a dry weight-normalized SQG and an organic carbon-normalized SQG that are comparable at 1% OC will be 10-fold different at 10% OC. MacDonald *et al.* (2000) conducted a similar analysis of PAH SQGs for freshwater, and derived total PAH 'consensus' values parallel to Swartzs' TEC and MEC of 1.61 and 22.8 $\mu g \cdot g^{-1}$ dry weight (161 and 2280 $\mu g \cdot g^{-1}$ OC, assuming 1% organic carbon).

In addition to total PAH SQGs, Table 17.1 lists guideline values for individual PAHs derived by both empirical and mechanistic approaches. However, the interpretation of these individual PAH guidelines varies greatly between the two guideline types. Because the data for empirical guidelines come primarily from field sediments containing mixtures of PAHs, the values for individual PAHs reflect the concentration of that specific PAH, *when present in a mixture*, that has an aggregate effect as defined by the guideline. As such, it is fair to say that the concentration of that single PAH is *associated* with that level of effect, even though that single PAH concentration would represent only a small part of the overall potency of the mixture present. In contrast, the individual PAH guidelines for the mechanistic guidelines represent the concentration estimated to cause the index effect when acting alone, a condition that never applies in the environment. Accordingly, these values only have meaning when used in intermediate calculations for a mixture — they should never be applied individually to evaluate the effects of mixtures.

Swartz (1999) extends this argument even further, suggesting that even empirically-derived guidelines for individual PAHs should not be used, because they inappropriately focus attention on the individual PAHs rather than the aggregate mixture. Beyond appearances, focusing on total PAHs may have importance in reducing uncertainty in toxicity predictions. Data presented by Swartz (1999) show that the range of concentrations from the lowest concentration causing toxicity to the highest concentration not causing toxicity was about 300-fold for acenaphthene alone, but only about 30-fold for total PAHs; this suggests that total PAHs is a less variable indicator of PAH potency than is acenaphthene alone. This is a logical finding when one considers that the relative contribution of a single PAH to total PAHs will vary across sediments with different sources of contamination, and different degrees of environmental weathering. Based on these considerations, it does appear most appropriate to emphasize total PAHs in the development of SQGs for PAHs.

17.4.1   THE DEFINITION OF TOTAL PAHs: EFFECTS ON SQG INTERPRETATION

To this point, we have used the term 'total PAHs' in a conceptual sense, without specific definition. However, total PAHs is subject to many definitions, and this has great impact on the interpretation of PAH SQGs. A typical definition of 'total

PAHs' for empirical SQGs, as well as many other environmental assessments, is the summed concentration of 8–16 commonly measured, unsubstituted PAHs, oft-called 'priority pollutant' PAHs. However, as discussed elsewhere in this book, PAH mixtures found in the environment contain many more PAH structures — not only more ring configurations but, more importantly, many compounds with substitutions of alkyl groups or other substituents. Indeed, in some PAH sources such as petroleum, the aggregate concentration of alkyl-substituted PAH structures is far, far in excess of the sum of the unsubstituted structures, while in others the degree of alkylation is much lower. For example, data from weathered North Slope crude oil shows that > 95% of the total measured PAH mass was present as alkylated structures (Stubblefield *et al.* 1995), whereas sediments downstream of a coking facility contained about equal proportions of alkylated and unsubstituted PAHs (Burkhard *et al.* 1994). This difference alone could be expected to result in a roughly 10-fold difference in toxicity between these two sources when total PAHs is defined only in terms of unsubstituted PAHs.

The impact of this issue on SQGs for PAHs varies according to the derivation of the guideline. For empirical guidelines based on data for PAH-contaminated sediments from the field, the toxicological effects of unmeasured PAHs are captured by the aggregate measure of biological effect, whether it is a sediment toxicity test or a benthic community measure. However, the actual accuracy of the SQG across sites will depend in part on how consistent the composition of the PAH mixture is within the effects database. When the effects database contains data for sediments with differing ratios of measured:unmeasured PAHs, these differences can be expected to introduce variability into the observed relationship between the total PAHs measured and the effects observed. While it is difficult to show this effect quantitatively, one would intuitively expect that an empirical guideline derived from data with varying contamination sources would reflect some kind of 'average' PAH composition. For sites with more extreme PAH composition, one would expect poorer performance of the guideline relative to predicting biological effects (e.g. Hayward *et al.* 2002).

For an EqP/narcosis-based guideline, the impact of varying degrees of alkylation are more directly quantifiable. Because all PAHs, substituted or unsubstituted, will contribute to the overall narcotic potency, accounting for the full range of PAHs present is important. However, since there are hundreds of possible structural isomers of alkylated PAHs, the question becomes, 'how does one decide what constitutes a measurement of "total PAHs"?' In deriving the SQG for PAH mixtures, USEPA defined total PAHs as, at a minimum, the 34 PAHs measured in the Environmental Monitoring Assessment Program (EMAP) (USEPA 1996a,b; 1998). These 34 PAHs include 18 parent PAHs and 16 alkylated PAHs; or, more properly, 16 homolog groups of one-, two-, three-, or four-carbon alkylations of naphthalene, fluorene, phenanthrene/anthracene, fluoranthene/pyrene, and chrysene (see methodologies from Sauer and Boehm 1991 or Short *et al.*

1996). While not including all PAH structures, it is generally thought that the bulk of PAHs in most mixtures should be captured by this approach.

Although aggressive characterization of PAH mixtures will provide the most accurate estimate of sediment toxicity via the EqP/narcosis approach, there are often cases where SQGs must be applied to sediments with less rigorous chemical characterization. For such applications, some means of accounting for unmeasured PAHs must be incorporated, or the actual potency of the mixture will be underestimated. Commonly, environmental data sets include measurements of the 13–16 PAHs identified as priority pollutants (Keith and Telliard 1979) or the expanded list of 23 PAHs as provided by NOAA (1991). In their draft SQG for PAH mixtures, USEPA (2000) suggests adjustment factors to relate concentrations of parent PAHs to the expected potency of the overall mixture. These adjustment factors were derived using the EMAP data sets (USEPA 1996a,b; 1998) where the 34 PAH set was measured. For each sediment sample in the EMAP data, individual PAH concentrations were divided by their potencies to calculate relative TU. By summing these TUs for different sets of PAHs, aggregate potencies for the EMAP 34-PAH set (TU_{34}), the NOAA 23-PAH set (TU_{23}), and the basic 13-PAH parent set from the ΣPAH model (TU_{13}) were calculated. The ratios of TU_{34}/TU_{13} and TU_{34}/TU_{23} were computed and plotted on a log normal probability distribution (Figure 17.6). Using this approach, samples with data for only the 13 or 23 PAH sets could be assessed by multiplying the summed guideline units for the measured PAHs by the applicable correction factor to account for the likely contribution of unmeasured PAHs. Specific adjustment factors can be selected, depending on the desired uncertainty (Table 17.2). As an example, if 13 PAHs are measured and 95% certainty is desired, the summed guideline units for the 13 PAHs are multiplied by a factor of 11.5.

The concern with this approach lies in the appropriateness of applying factors derived from EMAP data to sites with specific types of contamination. Based on the distribution of specific PAHs, the PAHs in the EMAP data sets appear to be dominated by pyrogenic sources. For sediments where the source of PAH contamination is petrogenic, the adjustment factors may not be appropriate, since the composition of PAHs differs between petrogenic and pyrogenic sources. PAH mixtures from petrogenic sources tend to be more highly alkylated

TABLE 17.2 Adjustment factors for estimating potency of PAH mixtures from measured subsets (USEPA 2000)

Percentile of the distribution	TU_{34}/TU_{13}	TU_{23}/TU_{13}
50	2.75	1.64
80	6.78	2.8
90	8.45	3.37
95	11.5	4.14
99	16.9	6.57

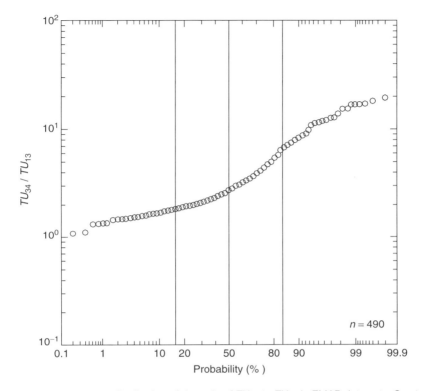

Figure 17.6 Frequency distribution of the ratio of TU_{34} to TU_{13} in EMAP data sets. See text for description of TU calculation

than those from pyrogenic sources. Because the primary difference in analytes between the 13 and 34 PAH measurements is alkylated compounds, TU_{34}/TU_{13} correction factors derived from largely pyrogenic sources (e.g. EMAP data) would generally underestimate the potency for sediments with petrogenic contamination.

Potency correction factors specific to petrogenic PAH sources have not been developed, but there is evidence that they would be substantially higher. Hellweger *et al.* (2002) compared the toxicity of water-accommodated fractions of unweathered and weathered *Exxon Valdez* crude oil to fathead minnows. The calculated potency ratios of TU_{34}/TU_{13} for the unweathered and weathered oils were 6 and 19, respectively, the latter of which is beyond the 99th percentile from the EMAP-based analysis (Table 17.2). The higher ratio for the weathered oil is directly attributable to the greater proportion of alkylated PAHs remaining after weathering, which is not captured when only parent PAHs are quantified. The appropriate corrections for a petrogenically contaminated site will depend on the characteristics of the source oil and the degree of weathering it has undergone. In any case, the many issues involved in predicting 'total' PAHs from smaller subsets of PAHs should make clear the desirability of having good analytical characterization of PAH contamination when making predictions about expected ecological effects. Even in cases where an assessment will be based largely on parent PAHs, it might be advisable to more extensively characterize a small subset of representative samples, so that a site-specific correction factor can be developed.

17.4.2   THE INFLUENCE OF SOOT, COAL, AND OTHER PARTITIONING PHASES

It is generally accepted that the bioavailability of non-polar organic compounds, such as PAHs, in sediments is influenced by partitioning to organic carbon. Most approaches, including current EqP models applied in the context of SQGs, are based generically on total organic carbon; partitioning from this phase is represented by a generic partitioning value (K_{oc}) which is thought to be very similar to the n-octanol partition coefficient (K_{ow}). However, as described in Chapter 3, studies by several researchers have documented sediments in which partitioning of PAHs is markedly different from what would be predicted by literature K_{oc} values (McGroddy and Farrington 1995; Maruya *et al.* 1996). Some of these instances involve the presence of soot particles in the sediment matrix; based on empirical observation, soot particles appear to contain PAHs that are extractable by typical solvent extractions used for analytical chemistry, but do not partition as readily to interstitial water. The lower interstitial water concentrations that result appear to correlate with reduced biological availability of PAHs in those sediments. Although we discuss this phenomenon and its influence on SQGs in the context of soot, there is evidence that coal and other forms of carbon induce similar changes in partitioning (Paine *et al.* 1996; West *et al.* 2001).

Since combustion of organic material, and therefore the production of soot, is a widespread occurrence, it seems fair to assume that most sediments contain soot particles to some extent. Since empirical guidelines are largely based on mixtures observed in field-collected sediments, the influence of soot on biological effects should be built in to a degree. However, this would only be relative to 'average' soot concentrations within the database from which the guideline is developed. For that reason, the accuracy of the guideline would decrease for samples with greatly different soot contents. Given the similarity of most empirically-derived PAH SQGs to those derived using EqP (which do not consider soot), it seems likely that the influence of soot in most of the studied sediments is relatively small. This would argue that the primary impact of soot on the accuracy of empirical PAH SQGs would be an overestimation of effect in samples with high soot content. Although this might suggest that basing SQGs on highly site-specific databases would avoid this problem, West *et al.* (2001) found a 10-fold difference in PAH partitioning in harbor sediments located only a few hundred meters apart. If this degree of heterogeneity is common, even highly site-specific databases may not control for the influence of soot. Procedures for quantifying soot in the context of PAH bioavailability in sediment have been proposed (Gustafsson *et al.* 1997).

Because current EqP-based SQGs use K_{oc} values that represent typical diagenic carbon, the presence of large amounts of soot would be expected to cause EqP equations based on generic K_{oc} values (Equations 17.1–17.3) to overpredict chemical activity in the sediment and therefore overpredict biological effects. It is worth noting that the basic concept of EqP — that the concentration of chemical in the interstitial water is correlated to its biological potency — is not necessarily violated by the influence of soot or other alternate partitioning phases. In a study of sediments known to contain coal particles, West *et al.* (2001) found that the concentrations of PAHs in organism tissue correlated well with predictions based on an assumed equilibrium with measured PAHs in interstitial water, even though the PAHs in interstitial water were much lower than predicted from bulk sediment. This suggests that the degree of error in current EqP-based SQGs would be proportional to the ratio of predicted vs. observed concentrations of PAHs in the interstitial water. For this and other reasons, measuring PAHs in interstitial water is prudent whenever applying EqP-based SQGs (or perhaps any SQGs) for PAHs in cases where avoidance of false positives is important. Alternatively, a modified partitioning model might be developed to better represent PAH behavior in the presence of soot (Gustafsson *et al.* 1997; see Chapter 3).

17.4.3   EFFECT ENDPOINTS AND MODE OF ACTION

The appropriateness of any SQG for PAHs will depend in part on the connectivity of the endpoints on which it is based and the endpoints for which it is applied. For example, the logistic regression approach developed by Field *et al.* (1999) is based on laboratory toxicity tests with marine and estuarine amphipods, and therefore predictions using this SQG are most directly applicable to predicting sediment toxicity to those same organisms. Applying these SQGs to predict effects on the benthic community in general requires an additional assumption regarding the relationship between toxicity to amphipods and benthic community structure. Evidence for a relationship between sediment toxicity tests and benthic community structure has been demonstrated in many instances (e.g. Swartz *et al.* 1994; Canfield *et al.* 1996), and there is also a suggestion that guidelines developed from laboratory data are coherent with responses observed in the field (Figure 17.7). Still, the issue of endpoint comparability must be kept in mind when interpreting sediment contamination using SQGs. For instance, a final chronic value from an AWQC is intended to protect moderately sensitive species from chronic toxicity; thus, an EqP SQG

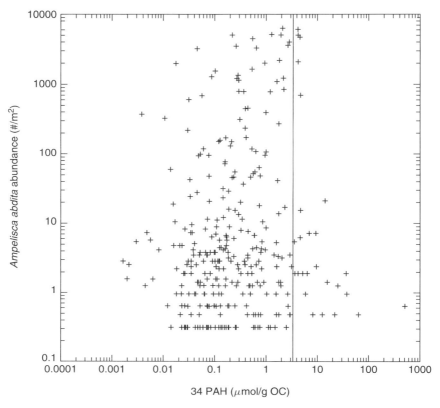

Figure 17.7 Abundance of the estuarine amphipod *Ampelisca abdita* from EMAP data as a function of total PAH measured using 34 PAH analysis. Vertical line represents total PAH EqP/narcosis SQG from USEPA (2000). Mass-based concentration can be estimated by multiplying *x* values by 200 g $\cong$ mol^{-1}. Modified from Di Toro and McGrath (2000), reproduced by permission of SETAC Press

based on this value would not be an appropriate predictor of acute toxicity to a relatively insensitive species.

The issue of endpoint comparability is also important relative to the mode of action or toxic mechanism for PAHs. With the exception of the SLC and AETs for benthic community structure, most of the empirical SQGs for PAHs rely heavily on laboratory toxicity tests using benthic invertebrates. While perhaps not rigorously proven, it seems generally accepted that for most, if not all, benthic invertebrates, the mode of action operative in these toxicity tests is narcosis, and current EqP approaches for PAHs are based explicitly on narcosis. However, there are other mechanisms of toxicity that can be exerted by PAHs. As described in Chapter 14, organisms exposed to both PAHs and ultraviolet (UV) light can experience 'photoactivated' toxicity through the interaction of those two stressors. Under the right conditions, photoactivated toxicity has been shown to increase the toxicity of PAH exposure by two orders of magnitude or more over that expected from narcosis (see Chapter 14 for further details). Increased toxicity to benthic invertebrates has been shown when toxicity tests on PAH-contaminated sediments were conducted under UV light (Ankley *et al.* 1994) and reduced toxicity occurred during *in situ* sediment toxicity tests when test chambers were shaded from sunlight (Monson *et al.* 1995).

The extent to which photoactivated toxicity will affect the applicability of PAH SQGs is not fully clear. While the potential has been demonstrated, attenuation of UV exposure by dissolved organic carbon and particulate matter in the water column, and by physical shading by sediment or macrophytes, acts to reduce UV exposure of benthic organisms. Initial analyses by Mount *et al.* (2001) suggest that potential risk from photoactivated toxicity may exceed the risk from direct narcotic toxicity to benthic organisms only in relatively shallow waters (e.g. <1–2 m), or those with very high clarity and low dissolved organic carbon. Swartz (1999) noted that the response of benthic communities at PAH-contaminated study sites tracked well with SQGs that do not incorporate photoactivated toxicity, suggesting that photoactivated toxicity was not a major influence. Furthermore, the field-based SLC and AET for benthic community structure are also comparable to SQG values that do not incorporate photoactivated toxicity. Beyond photoactivated toxicity, still other mechanisms of toxicity have been suggested as important to consider in assessing the toxicity of PAHs in the environment (e.g. Carls *et al.* 1999; Billiard *et al.* 1999), although detailed comparisons have yet to be completed. The important message is that one must consider toxicity endpoints and modes of action when selecting and interpreting SQGs.

17.5   CONCLUSIONS

There are a wide variety of SQGs that have been developed for assessing PAHs in sediment using both empirical and mechanistic approaches. Across this variety of approaches, there is a fair degree of convergence among the different guidelines (Swartz 1999), with thresholds for effects generally in the range of $100-200\ \mu g \cdot g^{-1} OC$ ($1-2\ \mu g \cdot g^{-1}$ dry weight at 1% organic carbon) for total PAHs based on commonly measured unsubstituted PAHs. Despite this similarity, there are many challenges to predicting accurately the biological effects of sediment-associated PAHs, including variation in the amount and type of organic carbon, differences in the composition of PAH mixtures from different sources, accounting for the aggregate toxicity of PAH mixtures, and understanding the roles of different toxic mechanisms. Given the ubiquity of PAHs in the environment, continued research to refine our ability to accurately predict risk of sediment-associated PAHs is clearly warranted.

ACKNOWLEDGMENTS

The development of SQGs for PAHs and other chemicals is the product of work by many individuals. Beyond the specific research referenced, we gratefully acknowledge the importance of our colleagues (alphabetically: GT Ankley, HE Bell, WJ Berry, RM Burgess, DM Di Toro, LJ Field, DJ Hansen, ER Long, DD MacDonald, RJ Ozretich, MC Reiley, and RC Swartz) in influencing our own thinking on the topic. LP Burkhard, LJ Heinis, and MA Starus provided helpful comments on earlier drafts.

REFERENCES

Adams WJ, Kimerle RA and Mosher RG (1985) Aquatic safety assessment of chemicals sorbed to sediments. In Cardwell RD, Purdy R and Bahner RC (eds), *Aquatic Toxicology and Hazard Assessment: Seventh Symposium*. STP 854. American Society for Testing and Materials, Philadelphia, PA, pp. 429–453.

Ankley GT, Collyard SA, Monson PD and Kosian PA (1994) Influence of ultraviolet light on the toxicity of sediments contaminated with polycyclic aromatic hydrocarbons. *Environmental Toxicology and Chemistry*, **13**, 1791–1796.

Billiard SM, Querbach K and Hodson PV (1999) Toxicity of retene to early life stages of two freshwater fish species. *Environmental Toxicology and Chemistry*, **18**, 2070–2077.

Burkhard LP, Sheedy BR and McCauley DJ (1994) Prediction of chemical residues in aquatic organisms for a field discharge situation. *Chemosphere*, **29**, 141–153.

Canadian Council of Ministers of the Environment (CCME) (1999) *Canadian Sediment Quality Guidelines*. CCME Task Group on Water Quality Guidelines, Ottawa, Ontario, Canada, 27 pp. (plus appendices).

Canfield TJ, Dwyer FJ, Fairchild JF, Haverland PS, Ingersoll CG, Kemble NE, Mount DR, La Point TW, Burton GA and Swift MC (1996) Assessing contamination in Great Lakes sediments using benthic invertebrate communities and the sediment quality triad approach. *Journal of Great Lakes Research*, **22**, 565–583.

Carls MG, Rice SD and Hose JE (1999) Sensitivity of fish embryos to weathered crude oil: Part I. Low-level exposure during incubation causes malformations, genetic damage, and mortality in larval Pacific herring (*Clupea pallasi*). *Environmental Toxicology and Chemistry*, **18**, 481–493.

Cubbage J, Batts D and Briedenbach S (1997) *Creation and analysis of freshwater sediment quality values in Washington State. Environmental Investigations and Laboratory Services Program*, Washington Department of Ecology, Olympia, WA.

Di Toro DM and McGrath JA (2000) Technical basis for narcotic chemicals and PAH criteria. II. Mixtures and sediments. *Environmental Toxicology and Chemistry*, **19**, 1971–1982.

Di Toro DM, Zarba CS, Hansen DJ, Berry WJ, Swartz RC, Cowan CE, Pavlou SP, Allen AE, Thomas NA and Paquin PR (1991) Technical basis for establishing sediment quality criteria for non-ionic organic chemicals using equilibrium partitioning. *Environmental Toxicology and Chemistry*, **10**, 1541–1583.

Di Toro DM, McGrath JA and Hansen DJ (2000) Technical basis for narcotic chemicals and PAH criteria. I. Water and tissue. *Environmental Toxicology and Chemistry*, **19**, 1951–1970.

Environment Canada and Ministere de l'Environnement du Quebec (EC and MENVIQ) (1992) Interim criteria for quality assessment of St. Lawrence River sediment. ISBN 0-662-19849-2. Environment Canada, Ottawa, Ontario, Canada.

Field LJ, MacDonald DD, Norton SB, Severn CG and Ingersoll CG (1999) Evaluating sediment chemistry and toxicity data using logistic regression modeling. *Environmental Toxicology and Chemistry*, **18**, 1311–1322.

Field LJ, MacDonald DD, Norton SB, Ingersoll CG, Severn CG, Smorong D and Lindskoog R (2002) Predicting amphipod toxicity from sediment chemistry using logistic regression models. *Environmental Toxicology and Chemistry*, **21** (in press).

Gustafsson O, Haghseta F, Chan C, MacFarlane J and Gschwend PM (1997) Quantification of the dilute sedimentary soot phase: implications for PAH speciation and bioavailability. *Environmental Science and Technology*, **31**, 203–209.

Hayward JMR, Ingersoll CG, Whites DW and Little EE (2002) Toxicity assessment of sediments from the Barton Springs watershed, Austin, TX, USA. US Geological Survey, Columbia Environmental Research Center, Columbia, MO, USA.

Hellweger FL, McGrath JA, Stubblefield W and Di Toro DM (2002) Predicting dissolution and toxicity of Exxon Valdez crude oil (manuscript).

Ingersoll CG, Haverland PS, Brunson EL, Canfield TJ, Dwyer FJ, Henke CE, Kemble NE, Mount DR and Fox RG (1996) Calculation and evaluation of sediment effect concentrations for the amphipod *Hyalella azteca* and the midge *Chironomus riparius*. *Journal of Great Lakes Research*, **22**, 602–623.

Keith L and Telliard W (1979) Priority pollutants, I — A perspective view. *Environmental Science and Technology*, **13**, 416–423.

Long ER and Morgan LG (1991) The potential for biological effects of sediment-sorbed contaminants tested in the National Status and Trends Program. Technical Memorandum NOS OMA 52. National Oceanic and Atmospheric Agency, Rockville, MD, USA.

Long ER, MacDonald DD, Smith FL and Calder FD (1995) Incidence of adverse biological effects within ranges of chemical concentrations in marine and estuarine sediments. *Environmental Management*, **19**, 81–97.

Maruya KA, Risebrough RW and Horne AJ (1996) Partitioning of polynuclear aromatic hydrocarbons between sediments from San Francisco Bay and their porewaters. *Environmental Science and Technology*, **30**, 2942–2947.

MacDonald DD (1997) Sediment injury in the Southern California bight: review of the toxic effects of DDTs and PCBs in sediments. Prepared for National Oceanic and Atmospheric Administration, US Department of Commerce, Long Beach, CA, USA.

MacDonald DD, Carr RS, Calder FD, Long ER and Ingersoll CG (1996) Development and evaluation of sediment quality guidelines for Florida coastal waters. *Ecotoxicology*, **5**, 253–278.

MacDonald DD, Ingersoll CG and Berger TA (2000) Development and evaluation of consensus-based sediment quality guidelines for freshwater ecosystems. *Archives of Environmental Contamination and Toxicology*, **39**, 20–31.

McGroddy SE and Farrington JW (1995) Sediment porewater partitioning of polycyclic aromatic hydrocarbons in three cores from Boston Harbor, Massachusetts. *Environmental Science and Technology*, **29**, 1542–1550.

Monson PD, Ankley GT and Kosian PA (1995) Phototoxic response of *Lumbriculus variegatus* to sediments contaminated by polycyclic aromatic hydrocarbons. *Environmental Toxicology and Chemistry*, **14**, 891–894.

Mount DR, Diamond SA, Erickson RJ, Simcik MF and Swackhamer DL (2001) Linking exposure and dosimetry to risk from photo-activated toxicity of PAHs. Society of Environmental Toxicology and Chemistry, 22nd Annual Meeting, Baltimore, MD, 11–15 November 2001. SETAC, Pensacola, FL, USA.

National Oceanic and Atmospheric Administration (NOAA) (1991) National Status and Trends Program — Second summary of data on chemical contaminants in sediments from the National Status and Trends Program. NOAA Technical memorandum NOS OMA 59. NOAA Office of Oceanography and Marine Assessment, Rockville, MD, USA. 29 pp. (plus appendices).

Neff JM, Word JQ and Gulbranson TC (1987) Recalculation of screening level concentrations of non-polar organic contaminants in marine sediments. Final Report, US Environmental Protection Agency, Washington, DC, USA.

Paine MD, Chapman PM, Allard PJ, Murdoch MH and Minifie D (1996) Limited bioavailability of sediment PAH near an aluminum smelter: contamination does not equal effects. *Environmental Toxicology and Chemistry*, **15**, 2003–2018.

Persaud D, Jaagumagi R and Hayton A (1993) Guidelines for the protection and management of aquatic sediment quality in Ontario. Water Resources Branch, Ontario Ministry of the Environment, Toronto, Ontario, Canada, 27 pp.

PTI Environmental Services (1991) Pollutants of concern in Puget Sound. EPA/910/9-91/003. US Environmental Protection Agency, Seattle, WA, USA.

Sauer T and Boehm P (1991) The use of defensible analytical chemistry measurements for oil spill natural resource damage assessments. In *Proceedings of the 1991 International Oil Spill Conference*, San Diego, CA, USA, March 4–7. American Petroleum Institute Publication 4529, pp. 363–370.

Short JW, Jackson TJ, Larsen ML and Wade TL (1996) Analytical methods used for the analysis of hydrocarbons in crude oils, tissues, sediments, and seawater collected for the natural resources damage assessment of the Exxon Valdez oil spill. In *Proceedings of the Exxon Valdez Oil Spill Symposium*, Rice SD, Spies RB, Wolfe DA and Wright BR (eds), American Fisheries Society, Bethesda, MD, USA, pp. 140–148.

Smith SL, McDonald DD, Keenleyside KA, Ingersoll CG and Field LJ (1996) A preliminary evaluation of sediment quality assessment values for freshwater ecosystems. *Journal of Great Lakes Research*, **22**, 624–638.

Stephan CE, Mount DI, Hansen DJ, Gentile JH, Chapman GA and Brungs WA (1985) Guidelines for deriving numerical national water quality criteria for the protection of aquatic organisms and their uses. NTIS No. PB85-227049. US Environmental Protection Agency, Environmental Research Laboratory, Duluth, MN, USA, 98 pp.

Stubblefield W, Hancock G, Ford W and Ringer R (1995) Acute and subchronic toxicity of naturally weathered Exxon Valdez crude oil in mallards and ferrets. *Environmental Toxicology and Chemistry*, **14**, 1941–1950.

Swartz RC (1999) Consensus sediment quality guidelines for polycyclic aromatic hydrocarbon mixtures. *Environmental Toxicology and Chemistry*, **18**, 780–787.

Swartz RC, Cole FA, Lamberson JO, Ferraro SP, Schults DW, DeBen WA, Lee H and Ozretich RJ (1994) Sediment toxicity, contamination and amphipod abundance at a DDT and dieldrin-contaminated site in San Francisco Bay. *Environmental Toxicology and Chemistry*, **13**, 949–962.

Swartz RC, Schults DW, Ozretich RJ, Lamberson JO, Cole FA, DeWitt TH, Redmond MS, and Ferraro SP (1995) ΣPAH: A model to predict the toxicity of field-collected marine sediment contaminated by polynuclear aromatic hydrocarbons. *Environmental Toxicology and Chemistry*, **14**, 1977–1987.

Swartz RC, Ferraro SP, Lamberson JO, Cole FA, Ozretich RJ, Boese FL, Schults DW, Behrenfeld M and Ankley GT (1997) Photoactivation and toxicity of mixtures of polycyclic aromatic hydrocarbons in marine sediment. *Environmental Toxicology and Chemistry*, **16**, 2151–2157.

United States Environmental Protection Agency (USEPA) (1993a) Technical basis for establishing sediment quality criteria for nonionic organic contaminants for the protection of benthic organisms: non-ionic. EPA/822/R-93/011. Office of Water, Washington, DC, USA.

United States Environmental Protection Agency (USEPA) (1993b) Sediment quality criteria for the protection of benthic organisms: dieldrin. EPA/822/R-93/015. Office of Water, Washington, DC, USA.

United States Environmental Protection Agency (USEPA) (1993c) Sediment quality criteria for the protection of benthic organisms: endrin. EPA/822/R-93/016. Office of Water, Washington, DC, USA.

United States Environmental Protection Agency (USEPA) (1993d) Sediment quality criteria for the protection of benthic organisms: phenanthrene. EPA/822/R-93/014. Office of Water, Washington, DC, USA.

United States Environmental Protection Agency (USEPA) (1993e) Sediment quality criteria for the protection of benthic organisms: acenaphthene. EPA/822/R-93/013. Office of Water, Washington, DC, USA.

United States Environmental Protection Agency (USEPA) (1993f) Sediment quality criteria for the protection of benthic organisms: fluoranthene. EPA/822/R-93/012. Office of Water, Washington, DC, USA.

United States Environmental Protection Agency (USEPA) (1996a) EMAP — Estuaries Virginian Province Data, 1990–1993. http://www.epa.gov/emap

United States Environmental Protection Agency (USEPA) (1996b) EMAP — Estuaries Louisianan Province Data, 1991–1993. http://www.epa.gov/emap

United States Environmental Protection Agency (USEPA) (1998) EMAP — Estuaries Carolinian Province Data, 1994–1997. http://www.epa.gov/emap.

United States Environmental Protection Agency (USEPA) (2000) Equilibrium partitioning sediment guidelines (ESGs) for the protection of benthic organisms: PAH mixtures (draft). Office of Water, Washington, DC, USA.

Veith GD, Call DJ and Brooke LT (1983) Structure–activity relationships for the fathead minnow, *Pimephales promelas*: narcotic industrial chemicals. *Canadian Journal of Fisheries and Aquatic Sciences*, **40**, 743–748.

Washington State Department of Ecology (WDOE) (1990) Sediment Management Standards, Chapter 173–204 WAC. Olympia, WA, USA, 106 pp.

West CW, Kosian PA, Mount DR, Makynen EA, Pasha MS, Sibley PK and Ankley GT. 2001. Amendment of sediments with a carbonaceous resin reduces bioavailability of polycyclic aromatic hydrocarbons. *Environmental Toxicology and Chemistry*, **20**, 1104–1111.

Managing Risks from PAHs*

GRAHAM WHALE[1], GORDON LETHBRIDGE[1], VIKRAM PAUL[1] AND ERIC MARTIN[2]

[1]Shell Global Solutions, Chester, UK
[2]CONCAWE, Brussels, Belgium

18.1 INTRODUCTION

This chapter summarizes some of the methods that have been used by a multinational oil company to assess and manage environmental risks posed by pollutants such as polycyclic aromatic hydrocarbons (PAHs). The challenges and complexities of trying to assess and then manage the environmental risks associated with PAHs in the oil industry should not be underestimated. From the outset, assessing the risks posed by a single industry and identifying what can be readily achieved in terms of controls is a complex task. One of the challenges with PAHs is that they occur naturally and consequently have numerous sources other than those that can be directly attributable to the oil industry. This leads to problems when attempting to identify and prioritize all the relevant sources and assess their potential environmental risk. To further complicate the issue, the range of species of PAH is vast and there is a lack of clarity as to what constitutes a PAH in terms of posing risks to human health and the environment. This raises significant challenges when trying to determine what species will be considered to be representative of PAHs and what analytical method should be employed to ensure the validity of a risk assessment. Furthermore, as PAHs are not deliberately manufactured by the oil industry, they are intrinsically more difficult to control and manage than a defined product which is based on a single chemical substance. Many of the regulations and environmental hazard and risk assessment approaches have been developed for single-substance chemicals.

In spite of these challenges, the oil industry has taken several initiatives to address some of the issues associated with PAHs that can enter the environment

---

* The contents of this chapter are based on the personal experiences of the authors and do not necessarily represent the views of Shell or CONCAWE.

PAHs: An Ecotoxicological Perspective. Edited by Peter E.T. Douben.
© 2003 John Wiley & Sons Ltd

as a consequence of their activities and use of their products. In recent years, the emphasis within multinational companies like Shell has been to produce overall transparent global strategies/policies that can be adapted to local country (and stakeholders') needs. However, when developing company-specific measures and strategies for managing the complex environmental and health issues associated with PAHs from oil and oil products, a number of other factors need to be considered. In the first instance, it is important to recognize that these challenges are not unique to any single oil company. Furthermore, management of environmental issues for PAHs has to be a dynamic process capable of being able to respond to changing technologies, societal pressures, improving scientific knowledge on environmental effects of pollutants, and new and impending environmental regulations. It is therefore important, when trying to manage environmental issues associated with contaminants like PAHs, that oil companies have the capability to:

- Identify any emerging environmental issues (real or perceived) concerning their activities and use of their products.
- Monitor changes in environmental regulation that may have a direct or indirect effect on their activities and use of their products.
- Prioritize and then assess any proposed regulations and/or emerging environmental issues on a scientific and technical basis.
- Develop briefing papers and recommend strategies for dealing with any issues that give cause for concern.
- Engage competent authorities and relevant stakeholders to ensure acceptance of the scientific and technical arguments and that issues have been adequately addressed.

To meet these challenges, oil companies generally work together through technical organizations like CONCAWE (the oil companies' European organization for environment, health and safety) and API (American Petroleum Institute) to develop a holistic range of environmental management strategies for their products and activities. These organizations use specialist management groups, task forces and contractors to improve the understanding of risks and issues facing the oil industry and to develop strategies for how these can be resolved. In reality, many company-specific environmental management strategies for assessing the risks posed by contaminants like PAHs tend to be based on recommendations of these technical organizations.

Although atmospheric emissions of PAHs from the burning of fossil fuels represent the largest source of PAHs from the oil industry, which can enter the aquatic and terrestrial environment as shown in Chapters 2 and 3, these will not be addressed. This chapter will focus primarily on management strategies and approaches that have been used in assessing the risks posed by PAHs in aqueous effluents and at contaminated sites. Examples of where these approaches have been applied are also discussed.

18.2 OVERVIEW OF REMEDIATION PHILOSOPHY FOR OIL CONTAMINATED SITES

18.2.1 BACKGROUND

The Shell Transport and Trading Company was formed on 18 October 1897 and, for the past century, has been responsible for businesses that have produced 10–12% of the whole world's annual supply of crude oil (Howarth 1997). During this time, concern about the environment has increased and improved attitudes and technology have resulted in a significant reduction in oil and product spills. However, spills still occur and improved performance does not address the issue of what should be done to deal with areas contaminated by past oil spills. Some of these historic spills can be attributed to operational activities (e.g. equipment failure, accidents, waste disposal practices) but in some countries large oil spills have occurred (and still continue to occur) as a consequence of criminal vandalism and war damage. In order to assess how to deal with the legacies of past impacted areas (PIAs), Shell has used the risk-based management (RBM) approach.

18.2.2 THE RISK-BASED MANAGEMENT (RBM) CONCEPT

The RBM approach is based around identifying and reducing risks associated with contamination to a level protective of human health and the environment. In the context of RBM, risk is a measure of the likelihood and magnitude of an adverse effect, including injury, disease, ecological loss or economic loss arising from the realization of a hazard. Hazard in the context of contaminated areas relates to chemicals (i.e. the contamination) present. One of the key factors to recognize with RBM or any risk assessment it that hazard is not the same as risk, but may be considered a source of risk. Misunderstanding of the difference between hazard and risk commonly occurs and leads to misperceptions about the potential effect of activities and use of products.

Within the RBM approach, contamination is only identified as representing a risk if all three elements of a contamination linkage are present:

- A source of contamination, namely the hazard (e.g. a leaking tank or pipeline, a waste pit or contaminated soil).
- A sensitive receptor (e.g. a community reliant on groundwater for drinking water).
- A plausible pathway that exists between the source and the receptor (e.g. groundwater).

If one of these elements is absent, there can be no significant risk. If all three elements of the contamination linkage are present, then the magnitude of risk is considered to be a function of: (a) the magnitude and mobility of the source; (b) the level of hazard; (c) the sensitivity of the receptor and the nature of exposure to the hazard; and (d) the effectiveness of the pathway to transport the contaminant to the receptor.

The use of an RBM approach by an operating company within the Shell Group of companies helps to ensure that they are assessing PIAs and undertaking remediation activities in a manner that is transparent and consistent with both industry and Shell Group best practice. The advantage of using risk assessment principles within a risk-based framework is that they can be used to assess the significance of contaminants like PAHs for a particular site and make it transparent why corrective action is or is not required. Within an RBM framework, remediation works are only undertaken if the contamination on site poses an unacceptable risk to human health or the environment, or if there are other business concerns that need to be resolved, e.g. reputation, legal or commercial. They are targeted at addressing those source, media and exposure pathways that exceed either the acceptable risk limits set for the site, or the other relevant criteria. For example, for non-volatile, poorly soluble PAHs the main pathways considered to present a human health or ecological risk would be via direct contact (dermal, ingestion and inhalation of dust) with soil and sediments. However, other pathways would be considered for the more water-soluble PAHs such as naphthalenes, as discussed later in this section.

The principal benefit of the RBM approach is that it allows each site to be treated individually with remedial measures tailored to the risks. This is important in the context of sustainable development to assure that resources are focused appropriately to reduce the potential for harm to the environment and society.

To ensure that this goal is achieved, the RBM approach needs to be:

- Protective of human health, safety, the environment and assets.
- Consistent, systematic and objective.
- Tailored to take account of the current or proposed/likely end use of the site.
- Rational, transparent, flexible and technically robust for a proposed course of action.
- Resource-effective.

18.2.3   TIERS WITHIN THE RBM APPROACH

The RBM process was initially developed to assess risk to human health using a tiered approach. Current methodology for ecological risk assessment is as yet not as well developed as for the human health risk assessment. One of the key challenges when undertaking an ecological risk assessment is defining the objectives (e.g. maintenance of biodiversity vs. maintenance of ecosystem function) and what constitutes harm. However, the fundamental principles and concepts of the RBM described in this section are applicable to both human health and ecological assessments.

Within a Tier 1 assessment, generic risk-based screening levels (RBSLs) for the plausible exposure pathways identified are used from look-up tables. The Tier 1 RBSLs are calculated using standard exposure assumptions and conservative generic exposure inputs using worst-case scenarios. Within the Tier 1 assessment, if the concentration of the chemical of concern at a site is below the RBSL for all plausible exposure pathways, then it can be considered to represent minimal risk to human health and the environment and no further action is required. For sites with concentrations of chemicals of concern above the RBSL, then the site can be either remediated to the RBSL or, alternatively, a more site-specific risk assessment can be undertaken at Tier 2.

In Tier 2 the complexity of the assessment is increased and uses site-specific data (e.g. actual soil type and depth of the groundwater) and alternative points of compliance (as opposed to point of highest concentration in Tier 1) to generate realistic exposure scenarios and site-specific target levels (SSTLs). Undertaking Tier 2 assessments can involve gathering additional site information (e.g. soil permeability, hydraulic conductivity). Tier 2 assessments only focus on the source area, pathway and receptor linkages not screened out in Tier 1. Within the Tier 2 assessment, if the concentration of the chemical of concern at a site is below the SSTL for all plausible pathways, then it can be considered to represent minimal risk and no further action is required. For sites with concentrations of chemicals of concern above the SSTL, then the site can be remediated to the SSTL or alternatively a more complex (Tier 3) site-specific risk assessment can be undertaken.

In Tier 3, complex assessment tools, such as probabilistic evaluations and sophisticated models for contaminant fate and transport to obtain very site-specific target levels, are used. Within the Tier 3 assessment, if the concentration of the chemical of concern at a site is below the SSTL for all identified pathways, then it can be considered to represent minimal risk and no further action is required. For sites with concentrations of chemicals of concern above the SSTL then the site can be remediated to the SSTL.

As can be seen, this approach uses increasingly detailed levels of data and/or complex modeling to arrive at more precise estimates of risk posed by site-specific conditions. This process matches the assessment effort to the relative risk or complexity of each site.

18.2.4   USING THE RBM APPROACH FOR ASSESSING PAHs

Individual chemical-specific RBSLs and SSTLs can be calculated for the 16 PAHs on the US Environmental Protection Agency (USEPA) 'Priority Pollutants' list (naphthalene, acenaphthylene, acenaphthene, fluorene, phenanthrene, anthracene, fluoranthene, pyrene, benzo[a]anthracene, chrysene, benzo[b]-fluoranthene, benzo[k]fluoranthene, benzo[a]pyrene, indeno[1,2,3-cd]pyrene, dibenzo[ah]anthracene and benzo[ghi]perylene). These PAHs are selected because they are considered to pose the greatest threat to human health and the environment because of their toxicity or carcinogenicity.

RBSLs and SSTLs for the remaining (non-carcinogenic) PAHs in petroleum products can be calculated using the Total Petroleum Hydrocarbon Criteria Working Group (TPHCWG) methodology (TPHCWG 1997–1999), which is consistent with the ASTM Risk-based Corrective Action (RBCA) framework (ASTM 1999), a good example of an RBM tool for decision making at contaminated sites.

The TPHCWG approach was developed to provide a tool for the risk assessment of petroleum products, which are complex mixtures containing hundreds to thousands of different individual hydrocarbons. Since it is not possible to identify (let alone find toxicity and fate and transport data for) the vast majority of these hydrocarbons, the TPHCWG methodology breaks TPH (total petroleum hydrocarbons) down into a number of boiling point distribution fractions (carbon chain lengths). The toxicity and fate and transport properties of these different fractions are determined by reference to individual hydrocarbons that are representative of each fraction. To apply this technique to a specific site, a sample of oil extracted from the soil and/or water is subjected to column chromatography to separate the aliphatic hydrocarbons from the aromatics. The boiling point distribution range (and hence fractionation) of both the aromatic fractions are then determined by gas chromatography.

18.3 CASE STUDY: DEVELOPMENT OF A GENERIC RBM TO ASSESS PIAs IN THE NIGER DELTA

18.3.1 BACKGROUND AND PRINCIPLES

For oil operations within the Niger Delta, the main product of concern regarding potential for land contamination is crude oil. Crude oil is a complex mixture of aliphatic and aromatic hydrocarbons with carbon chain length of C5–C40 +. The chemical and physical properties of crude oils vary, depending on the relative distribution of carbon chain lengths. Differences in composition are reflected in the American Petroleum Institute (API) gravity; the higher the API gravity the lower the viscosity of the crude.

In Nigeria, the Shell Petroleum Development Company of Nigeria Ltd (SPDC), which operates the SPDC Joint Venture on behalf of Shell, Nigerian National Petroleum Corporation, Total Fina Elf Nigeria Ltd and Nigeria Agip Oil Company, produces crude oils. The composition of two crude oils, Bonny and Forcados-Yokri, which are considered to be representative of crude oil from the Niger Delta, is given in Table 18.1. These data show that the two Nigerian crude oils have similar chemical properties. Bonny crude oil is characterized as a light

TABLE 18.1 Percentage composition of Nigerian crude oils

Chemical group	Compound	Bonny Light	Forcados-Yokri
Aliphatics	> 6–8	9.56	1.20
	> 8–10	12.1	3.52
	> 10–12	4.87	4.87
	> 12–16	14.2	15.2
	> 16–44	34.5	27.5
Aromatics	> 6–7	0.21	0.001
	> 7–8	0.79	0.005
	> 8–10	1.90	0.107
	> 10–12	1.52	1.42
	> 12–16	5.38	6.23
	> 16–21	6.81	3.90
	> 21–44	1.23	4.33
BTEX	Benzene	0.211	0.0012
	Toluene	0.789	0.005
	Ethyl benzene	0.186	0.0032
	Total xylenes	1.336	0.0122
PAHs on EPA priority list	Naphthalene	0.026	0.1018
	Acenaphthylene	0.0018	0.0022
	Acenaphthene	0.0018	0.0048
	Fluorene	0.0069	0.0156
	Phenanthrene	0.0154	0.0305
	Anthracene	< 0.0001	< 0.0001
	Fluoranthene	0.0008	0.0011
	Pyrene	0.0008	0.001
	Benzo[a]anthracene	0.0016	0.0005
	Chrysene	0.0019	0.0012
	Benzo[b]fluoranthene	0.0002	0.0004
	Benzo[k]fluoranthene	0.0001	0.0003
	Benzo[a]pyrene	0.0004	0.0002
	Indeno[1,2,3-cd]pyrene	< 0.0001	< 0.0001
	Dibenzo[ah]anthracene	< 0.0001	< 0.0001
	Benzo[ghl]perylene	< 0.0001	< 0.0001

waxy crude (API gravity of $37°$) while Forcados-Yokri crude oil is characterized as a medium crude (API gravity of $24.4°$).

In developing a generic RBM for the SPDC operations, a summary of the potential sources and pathways to ecological receptors was prepared, as shown in Table 18.2. Using this information, the risks posed by PAHs to specific ecological receptors can be assessed using the RBM process.

18.3.2 ASSESSING POTENTIAL GENERIC RISKS POSED BY PAHs

The potential risks posed by PAHs that enter groundwater as a consequence of oil spills, which subsequently provides a pathway of the PAHs to ecological

TABLE 18.2 Summary of potential sources and pathways of PAHs to ecological receptors

Receptor	Potential source and pathway
Soil fertility	Contaminated land used for growing crops or to re-establish native flora after abandonment Ruptured pipeline Fluctuating shallow water table bringing contaminants into rooting zone
Surface water and associated sediments	Contaminated groundwater discharging into surface water Overland transport of contaminants from source areas (e.g. waste pit) in surface runoff Ruptured pipeline Seepage from waste dumps or pits Malfunctioning oily wastewater treatment system discharging to surface water
Sensitive habitats	Contaminated groundwater discharging to habitat Ruptured pipelines Seepage from waste dumps or pits Malfunctioning oily wastewater treatment system discharging to surface water
Protected animals	Drinking or bathing in contaminated surface water Dermal contact with free-phase oil floating on surface water Ingestion of free-phase oil floating on surface water Dermal contact with contaminated soil Ingestion of contaminated soil Impact on flora or fauna lower down food chain

receptors, can be assessed. Groundwater can become contaminated because any spills, leaks or deliberate releases of crude oil will lead to its migration from the original spill area (primary source) under the influence of gravity through the soils below the release site. Models can be used to estimate the depth of infiltration for an oil with a given viscosity through various soil types. On the basis of the superficial soil properties in the Niger Delta combined with the crude oil properties the models indicate that the depth of infiltration of fresh crude oil spills is likely to be limited. However, if the oil reaches the groundwater table, the soluble low molecular weight fraction of the oil can leach into the aqueous phase to form a dissolved phase plume. This can migrate down the hydraulic gradient with the groundwater. The concentration of each dissolved constituent will be influenced by the solubility of the pure component and its mole fraction in the free phase plume (Raoult's Law). Thus, the theoretical concentrations of relatively water-soluble PAHs, such as naphthalene, in groundwater in contact with free phase can be predicted for a given crude oil.

Predictions of naphthalene concentrations in groundwater made for Bonny and Forcados-Yokri crude oil would be 0.04 and 0.19 mg/L, respectively. These levels are unlikely to pose any significant ecological risk. The ASTM RBCA Tier 1 RBSL for naphthalene for groundwater ingestion for the residential scenario is 0.15 mg/L. The RBSLs for the other exposure pathways are considerably greater than this value. The Tier 1 RBSL could not be achieved under any circumstances for Bonny crude, ruling this out as a concern for any human health or ecological criteria. With respect to the Forcados-Yokri crude, the maximum concentration possible only just exceeds the Tier 1 RSBL. However, it should be noted that the maximum theoretical concentration of 0.19 mg/L could only just be achieved with free Forcados-Yokri product in contact with the water (i.e. in the source area). It is also considered that, owing to the esthetics (e.g. odour, appearance) of the source site, it is unlikely that any person or animal would drink the water in contact with free product. Experience also suggests that derivation of an SSTL (site-specific target level) at an alternative point of compliance (i.e. drinking water or identified ecological receptor) would demonstrate insignificant risk.

With respect to ecological considerations regarding the potentially contaminated groundwater, the concentration will decline rapidly outside the source area due to dilution, sorption, and biodegradation and is unlikely to pose a risk to any of the ecological receptors identified in Table 18.2. Furthermore, for historic spills, natural processes such as volatilization, biodegradation, and leaching will also significantly lower the potential concentration of contaminants such as naphthalene in the groundwater. This view is supported by the literature and there are numerous publications which indicate that dissolved phase plumes of benzene, toluene, ethyl benzene, and xylenes (BTEX) and naphthalene are attenuated in unconsolidated aquifers within 100 m of the edge of the source (e.g. free phase plume or residual phase in soil).

In developing the generic RBM for Nigerian oil spill scenarios the other potential pathways for PAHs to ecological receptors were assessed in a similar manner. The conclusion is that, in terms of overall risk perspective, the 16 USEPA priority PAHs do not give any cause for concern. Essentially, none of the potential source – pathway – receptor scenarios considered score high for both likely incidence and severity, indicating that significant risks to ecological receptors are unlikely to be widespread. This fact is attributed to the composition of both crude oils and the nature of the environment in the Niger Delta.

To elaborate further, the main hydrocarbon fractions within both crude oils are the $> 16-44$ aliphatics followed by the $> 12-16$ aliphatics (Table 18.1). Both Nigerian crude oils contain very low concentrations (maximum of 0.0043 wt%) of the 16 EPA priority PAHs, although they do contain significant quantities of other PAHs, as can be seen in the C12–C44 aromatic fractions. The risks posed by the 'other' PAHs present in the Nigerian crude oils were assessed using the TPHCWG methodology and, in terms of the generic RBM, were considered unlikely to pose any significant adverse ecological risk.

18.3.3 RISKS TO SURFACE WATER — ADDITIONAL CONSIDERATIONS

In the Niger Delta the waters most at risk from contamination from an ecological perspective are surface waters (swamps, rivers, streams and lakes) and their associated sediments. The risks to these waters have to be managed and communicated properly, because fishing is vital to the economy and survival of the local communities in many of the areas in which SPDC operates. Essentially the generic RBM for the Nigerian crude oils indicate that PAHs do not constitute a significant risk to any of the ecological receptors. The RBM reveals that the main potential threat to the ecology of surface waters and associated sediments will be from direct discharges of crude oil from ruptured pipelines.

Although RBM assessments may provide an operator like SPDC with the knowledge that PAHs will not pose any significant risk at Tier 1 level, there may still be the need to undertake further assessments to alleviate public concerns. This was the case when fish tissue samples were analyzed to demonstrate that a remediation and clean-up program (developed using an RBM approach) for the Forcados slot, adjacent to the Forcados oil production terminal in Nigeria, had been successful.

In this study, fish and crustacean samples, collected from the Forcados slot by a local fisherman, were analyzed for 13 of the 16 EPA priority PAHs to assess whether they had significantly accumulated these contaminants and were not fit for human consumption (Whale *et al.* 2001). This study also included analysis of smoked fish and shrimp (originating from offshore catches) purchased in Ogulagha village for comparative risk assessment.

Analysis using recognized international methods by high pressure liquid chromatography (HPLC) indicated that in the dried samples originating from the Forcados slot, residues of the 13 PAH compounds determined were all below the limit of determination (0.03 mg/kg). In the sample of dried smoked shrimp purchased from the village, all residues were below the limit of determination apart from phenanthrene, which was present at a concentration of 0.58 mg/kg. In the partially dried smoked fish (shad) from the village, the nature of the tissue and size of sample meant that the limits of determination were increased to 0.01 mg/kg and residues of all 13 PAHs were found, ranging from 0.01 mg/kg to 0.21 mg/kg. The fact that these compounds were detected was not simply an artifact of the increased sensitivity of the analysis, as 10 of the 13 PAHs found in the smoked fish sample were $\geq$ the limit of detection of 0.03 mg/kg.

The results of the study indicated that fish sampled from the remediated Forcados slot were not a significant source of PAH contamination and that uptake of any PAH contaminants remaining in the slot was insignificant. In fact, the results indicate that the Ogulagha community would be exposing themselves to higher levels of PAHs by eating smoked fish caught from offshore locations rather than dried fish caught from the slot.

The Forcados fish study was considered to have achieved its objectives by providing robust information to alleviate concerns and misconceptions about the risks posed by PAHs in an essential part of the community's diet. It was also important in helping to put risks posed by the past activities of the oil industry into context with respect to the accepted practice of eating smoked fish.

18.4 PAHs AND EUROPEAN OIL REFINERY EFFLUENTS

18.4.1 BACKGROUND TO CONCAWE AND ISSUES REGARDING PAH MONITORING IN REFINERY EFFLUENTS

The oil companies' European organization for environment, health and safety, CONCAWE, was established in 1963. CONCAWE is a technical organization that uses a number of specialist management working groups, task forces and specialist contractors to address specific health safety and environmental issues relating to the downstream oil business (essentially refineries, distribution terminals, and retail stations). Since its inception, different management groups within CONCAWE have studied the environmental and health risks posed by PAHs. However, most of these assessments have been confined to specific product ranges or atmospheric emissions (e.g. from vehicle exhausts). In terms of protecting the aquatic environment, the risks posed by discharges of PAHs in process effluents have been considered to be minimal in comparison to other exposure scenarios. This view is generally supported by the literature. As such, the majority of refineries in Europe have not historically been required to regularly monitor PAHs in process effluents. This situation is gradually changing as PAHs have been identified as priority hazardous substances in the European Water Framework Directive and are on the OSPAR list of chemicals for priority action. As both the EU and OSPAR have made clear statements that they are committed to the reduction/cessation of discharges of these priority hazardous substances, this will inevitably lead to requests for more detailed inventories of PAH emissions in refinery effluents. The costs and complexity of this task and trying to determine what species of PAHs constitute a risk to the environment will be significant.

When requests for more detailed information are made, it is anticipated that there will be a number of technical and practical problems when trying to measure PAHs in aqueous effluents. These include basic questions, such as what is the best analytical method to use, and which species of PAHs should be measured? The problem of which PAHs should be considered in risk assessments was identified in a recently compiled review of PAHs in automotive exhaust emissions and fuels, undertaken by CONCAWE (1998a). In this review, CONCAWE noted that there were a wide range of definitions of what constitutes a PAH, with no singly agreed definition for PAH. This factor was attributed to the analytical techniques employed at different laboratories, which subsequently governs what compounds they can separate and quantify. The problem is compounded because each research group tends to assume that the species they can identify are representative of PAHs as a whole. The USEPA's 'Priority Pollutants' list of 16 PAHs is a typical example. It is not uncommon for the total PAH concentration determined on the basis of the 16 compounds on the EPA list to be a factor of 10–100 less than the concentrations of total PAHs obtained by methods that are not isomer-specific (CONCAWE 1998a).

Owing to these uncertainties, there are few European refineries in which PAHs are routinely monitored in their effluents. However, refineries do routinely determine oil in effluents and have implemented a number of strategies to reduce the quantities of oils and hydrocarbons in their aqueous effluents. As PAHs occur naturally in crude oils, the oil industry can therefore argue that, by demonstrating that they have significantly reduced inputs of oil, they will inevitably have led to significant reduction of risks from PAHs associated with the oil. In this respect, CONCAWE member companies can demonstrate that they have significantly reduced the quantities of oil discharged and hence risks posed by PAHs in aquatic discharges. This has been achieved by considerable investment in wastewater treatment equipment and better process controls. These claims can be substantiated from the results of CONCAWE effluent surveys.

One technical issue that has been identified in the CONCAWE refinery surveys and can affect the reliability of gathering emissions data is the choice of analytical method. CONCAWE has always encouraged refineries to adopt a standardized three-wavelength, infra-red (IR) method for determining oil in effluents, but in most cases, the method is set by the requirements of the different regulatory authorities. This leads to differences in the oil determination methods used between refineries (mainly in the solvents and wavelengths used, and in the number of wavelengths measured), which potentially affects the comparability of the analytical data generated. It is therefore considered essential that a ubiquitous set of standard methods for relevant PAH compounds is agreed between the various European regulators before a realistic emissions inventory can be established.

18.4.2   CONCAWE AND OIL REFINERY EFFLUENT SURVEYS

Since 1969, at approximately 3 yearly intervals, CONCAWE has undertaken surveys to provide information on European oil refineries' effluent water quantity, oil content and treatment processes. The results of these surveys are summarized in Table 18.3. From a historical perspective, the oil refining industry in Western Europe expanded rapidly in parallel with increasing industrial activity from the end of the Second World War until 1973. The oil crises of 1973 and 1979, together with the resultant worldwide economic recession, are reflected in the 1978 and 1981 CONCAWE surveys by stagnant refining capacities and very low actual throughputs, resulting in a poor utilization of capacity (CONCAWE 1998b). Between 1981 and 1984, the continuing low demand for oil products

TABLE 18.3 European crude oil refining capacity/throughput and oil discharged

Survey year	Western European capacity (10^6 tonnes/year)	Number of oil refineries reporting in each survey	Reported oil throughput (10^6 tonnes/year)	Total oil discharged (10^3 tonnes/year)	Oil discharged relative to oil throughput (g/tonnes/year)
1969	667	82	No data	44	No data
1974	959	112	No data	30.7	No data
1978	1034	111	540	12	23
1981	996	105	440	10.6	24
1984	788	85	422	5.09	12
1987	715	89	449	4.64	10.3
1990	710	95	511	3.34	6.7
1993	705	95	557	2.02	3.62
1997	718	105	625	1.17	1.86

Data taken from CONCAWE (1998b) report.

resulted in the closure of a number of refineries. Further limited closures have occurred since 1984 but overall the number of refineries included in the survey has increased, due to new CONCAWE membership and the increased participation of non-member company refineries.

The latest report for the 1997 survey (CONCAWE 1998b) indicated that there were 113 oil refineries operating in the area covered by the survey, with a total capacity of some 710 million tonnes/year. The data gathered is considered to be representative of the situation in Europe, since the 1997 survey area covered the 15 countries of the EU, Norway, Switzerland, and Hungary. In the 1997 survey, 105 refineries responded to the CONCAWE questionnaire, representing about 93% of the oil-refining capacity in the area of Europe covered by the survey (CONCAWE 1998b). The 1997 survey revealed that the complexity of refineries has increased with the installation of additional conversion units, e.g. thermal crackers, catalytic crackers and hydrocrackers, to reduce fuel oil production and to meet the demand for a higher yield of gasoline and other light products.

To remove oil and other contaminants from wastewater prior to discharge, refineries use a number of water treatment processes. These range from gravity separation (e.g. API separators, plate interceptors, tank separation) to advanced physical treatment (e.g. flocculation, air flotation, sedimentation, filtration) to biological treatment (e.g. biofilters, activated sludge, aerated ponds). The CONCAWE surveys have shown a continuous tendency towards the introduction of more effective effluent treatment in existing refineries since 1969. These improvements to water treatment continue to be implemented, as shown by the 1997 report in which 26 refineries reported that they had made improvements to their effluent treatment (CONCAWE 1998b).

The improvements in refinery wastewater treatment are reflected in the amount of oil (and by inference PAHs) discharged from refineries. Since 1969, there has been a continuous reduction in the amount of oil discharged

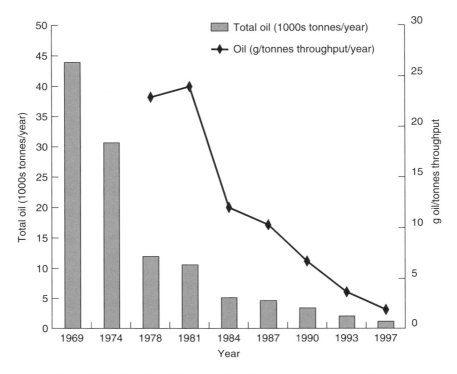

Figure 18.1   Trend in oil discharged in effluents from European refineries

(Table 18.3, Figure 18.1) in refinery effluents from about 44,000 tonnes/annum
from 73 refineries in 1969 to 1168 tonnes from 105 refineries in 1997. This
represents a 97.4% reduction since 1969. Even in the 4 years since the previous
survey, the weight of oil discharged has fallen by 42% (CONCAWE 1998b).
Consequently, the ratio of oil discharged relative to the refining capacity has
also been reduced by 98.6% during 1969–1997. Comparing the oil discharged
to refinery throughput, the ratio has decreased by 92% between 1978 (when
throughput data was first reported) and 1997. In the 4 years since the previous
survey, the rate of decrease has continued, with a 49% decrease in the ratio of
oil discharged: quantity processed from 3.62 g oil/tonne throughput in 1993 to
1.86 in 1997. The latest data from oil refineries indicates that this figure is still
continuing to decrease (personal communication, Eric Martin, April 2002).

To ensure credibility of the data obtained from the surveys, a specialist
task force under the jurisdiction of the CONCAWE Waste and Water Quality
Management Group is responsible for the conduct of the survey and collation
of the data. The information obtained from such surveys enables CONCAWE to:
(a) present relevant data to competent authorities; and (b) assess the status of
European refinery water treatment facilities and the quality of aqueous effluents

discharged. Experience with managing the refinery surveys has revealed some of the challenges in trying to establish reliable databases (i.e. consistency of reporting) for establishing emissions scenarios; e.g. during the time that the refinery effluent surveys have been undertaken, a number of refineries have been closed or significantly modified. Consequently, the database can change between surveys; e.g. when comparing the survey results between 1993 and 1997 it transpires that four refineries that reported for 1993 had closed before 1997 and 15 refineries which have not reported previously did so in 1997. These additional refineries were an artifact of new CONCAWE membership and the participation of non-member company refineries which had not previously supplied data. To allow a more accurate comparison of trends between surveys (i.e. reduce problems with interference/skewing of the data) and ensure transparency/credibility of the survey reports, such changes are accommodated in the way the results are presented. For example, in the 1997 survey report (CONCAWE 1998b) adjusted data, based on only those refineries which reported in both 1993 and 1997, are presented, in addition to the totals for all of the responding refineries.

18.4.3   PAHs IN OIL REFINERY EFFLUENTS

The significant reduction in oil discharged with corresponding improvements in effluent treatment at refineries over the past 30 years does give some confidence that the potential for adverse effects associated with PAHs are being reduced. This view is supported to a limited extent by information obtained during a recent literature search and knowledge of issues raised to the CONCAWE Water Quality Management Group. These sources suggest that there have been no recorded instances in recent years where there have been direct concerns about PAHs in refinery effluents or the aquatic environment surrounding oil refineries. Of the few studies that have been reported, the concentrations of PAHs in refinery effluent streams have been low. For example, Pettersen *et al.* (1997) determined the concentration of 15 PAH compounds in settling particulate matter (SPM) collected in the waters outside a petroleum refinery on the Swedish Baltic coast, and in samples of particulate and dissolved fractions in the refinery wastewater. They also compared PAH profiles of the SPM wastewater samples with SPM samples from background areas in the Baltic, using pattern recognition techniques. The study by Pettersen *et al.* (1997) indicated that the refinery was not a significant source of PAHs to the waters in its immediate surroundings. Arnold and Biddinger (1995) also provide evidence that PAHs in refinery effluents from a Canadian refinery posed little environmental risk. In their study, sediment samples were collected along a gradient from a petroleum refinery's wastewater diffuser. The concentrations of a number of PAH compounds were determined and used to calculate the potential risk to aquatic organisms, using probabilistic modeling and Monte Carlo sampling procedures. They used the sediment chemistry data in conjunction with estimates of sediment to biota accumulation factors and non-polar narcosis theory to predict potential risk to bivalve molluscs (bivalves were selected for this risk assessment because these organisms lack a well-developed enzymatic system for metabolizing PAHs and consequently are at a higher inherent risk of adverse impact). The results from the Arnold and Biddinger (1995) study indicated that the bivalves were at negligible risk of narcotic effects from PAHs in the sediments in the vicinity of the refinery wastewater discharge.

These studies on sediments and suspended material in the vicinity of refinery effluent systems are potentially more significant than simply assessing the concentrations of PAHs in wastewater. Owing to their low water solubilities and high octanol–water partition coefficients, PAHs will predominantly be associated with sediments or suspended organic matter with relatively low concentrations in water. On the basis of their physical properties, PAHs are predicted to have concentrations three to five orders of magnitude higher in sediments than in the water column. Sediments are also likely to act as ultimate sinks for PAHs because many PAHs degrade comparatively slowly under anaerobic conditions and in the absence of light (Wakeham 1996).

18.5   OIL FROM PRODUCTION PLATFORMS

As with refineries, the oil industry has been introducing a number of measures to reduce the quantities of oil discharged into the marine environment oil platforms. These have included:

- Improved oil/water separation techniques and water treatment technologies.
- Cessation of the discharge of oil-based drilling muds and oil-contaminated cuttings.
- Significant reductions in the number and volume of accidental oil spills from platforms.

These improvements have been effective and can be demonstrated by the fact that the total quantity of hydrocarbons discharged into the OSPAR maritime area has shown a continuous decrease in the past few years. For example, 13,642 tonnes of hydrocarbons, including synthetic-based drilling fluids (now called organic-phase drilling fluid; OPF), were discharged from oil platforms into the North Sea maritime area in 1999 compared to 16,753 tonnes reported in 1997 (OSPAR 2001).

As with the oil refineries care has to be taken when reporting and assessing trends in the discharge data. In the North Sea, OSPAR has identified four sources of oil from platforms. These are produced water, drill cuttings, spills and flaring operations. Even when discharges of OPF are considered, produced water is the most significant source, accounting for 64% of the total hydrocarbons discharged in 1999 (OSPAR 2001). Spillage is a minor contributor, and flaring contributes even less.

In the UK sector of the North Sea, improvements in water treatment have resulted in a drop from an average oil in production water from 36 mg/L in 1991 to 25 mg/L in 1997 (Environment Agency 1998). This trend continues for the whole of the North Sea, with the latest data for the average oil content of produced water reported for 1999 being 23.2 mg/L (OSPAR 2001). Unfortunately, the gains in reducing the concentration of oil in the produced water have tended to be offset by increased water production as the oil fields age.

It should be noted that, in terms of the aromatic composition, production water discharged from offshore oil platforms varies depending on the location, although there is evidence from the North Sea that the composition of produced water is relatively constant from one year to another at individual fields (OGP 2002). Produced water from oil production fields differs from that from gas production fields. Produced water from gas production fields generally has a higher concentration of lower molecular weight aromatic hydrocarbons than water from oil production fields. However, the total amount of water produced from gas fields is much lower than from oilfields. In terms of risk assessment of these discharges in the North Sea, the concentrations of aromatic hydrocarbons in treated produced water are rapidly attenuated by dilution. Furthermore, the lighter, more water-soluble PAHs, such as naphthalenes, will also degrade rapidly in the marine environment (OGP 2002). Currently there is little evidence to suggest that PAHs in produced water discharges are posing any significant risks to the North Sea environment.

Although not as significant as the quantities of hydrocarbons discharged in produced water, oil spills often attract more attention because of their visual impact and potential to impact on beaches and lead to oiling of sea birds and other marine life. It is therefore important to recognize that the quantities of oil spilled in the UK Sector (as reported by the platform operators to the UK Department of Trade and Industry as part of their licence arrangement) have also shown a significant decline. For example, the quantity of oil spilled from oil platforms in the UK sector peaked at 3540 tonnes in 1986 but decreased substantially to around 100 tonnes in 1996, although this increased again (following an exceptionally large spill) to 866 tonnes in 1997 (Environment Agency 1998). It is felt that some of the apparent increase in the number of spills reported after 1995 has been due to improved reporting by the operators and inclusion of spills detected by aerial surveillance. However, the latest data from OSPAR indicates that the quantities of oil spilled from platforms in the whole of the North Sea of 303 tonnes in 1998 and 283 tonnes in 1999 were similar to the quantities spilt in 1994–1996 (OSPAR 2001).

18.6   ASSESSING AND MANAGING THE RISKS OF PAHs
FROM PETROLEUM PRODUCTS

One problem beginning to face the oil industry is differentiating between perceived risk as opposed to actual risk of their activities and products. Both are important and need to be managed. Although the industry may be confident that they are not having an environmental impact, i.e. the actual risk, if this is not managed appropriately this can potentially lead to the damage of a company's reputation. This problem has occurred more frequently in recent years as a consequence of the increased focus on environmental issues and on demands to reduce the release of hazardous chemicals into the environment. In many cases, issues have been raised because of a basic misunderstanding between hazard and risk. For example, a number of petroleum products contain traces of PAHs, which are classified as hazardous materials. This has occasionally led to the misconception that all products containing PAHs should be automatically regarded as hazardous. Unless the risks for a specific application are addressed on the basis of potential exposure, a 'blanket misconception' based simply on PAH content could lead to inappropriate product and waste categorization. This was highlighted recently in discussions about the classification of used road material (old asphalt containing bitumen). Initially there were discussions as to whether this material should be considered to be hazardous and pose a risk to the environment because of its PAH content. It was important that this issue was clarified, because if all used road material had been classified as hazardous this would have had significant implications for its disposal (i.e. in the EU this material could only be handled and disposed of by specialist licensed waste handlers). Fortunately, neither bitumen nor bituminous mixtures containing bitumen were classified as hazardous waste (European Commission 2001).

In addition to classification issues, any perceptions about the safety of these products will raise additional concerns, because bitumen has a long history of use as a waterproofing agent and, in terms of its potential to contaminate water, is traditionally regarded as a safe product. If the perception of the risks posed by bitumen is not justified, this could lead to demands for controls and raise unnecessary public concerns and anxiety about past exposure. With respect to the issues surrounding PAHs in products, the challenge for the oil industry is to ensure that they recognize these issues and concerns and put the risks into context by addressing the potential environmental exposure scenarios. Essentially, the oil industry has to demonstrate that, for any given application, hazardous substances like PAHs found in products like bitumen and asphalt are essentially 'locked in' to the matrix and unlikely to enter the environment. This information has to be based on scientific studies and presented in a clear, easily understandable and transparent manner. In the first instance it is important to reinforce the message that there can be no risk without an exposure route. Therefore, in the first step of any environmental risk assessment, the main applications of a product have to be identified to assess whether any potential exposure routes exist. This process (essentially establishing a source-pathway-receptor) is exactly the same principle used in the RBM approach for contaminated land sites. It is the cornerstone of the conceptual site model (CSM).

In a recent example, concerns that bitumen could be considered hazardous and pose a risk to the environment prompted researchers within Shell Global Solutions (Brandt and de Groot 2001) to undertake studies to improve the understanding of the leaching characteristics of PAHs in bitumen. The main uses of bitumen are in asphalt roads, roofs and linings of water tanks and pipes. In all of these applications there is contact with water and therefore the potential for leaching of compounds into the environment. As previously mentioned, PAHs are present at low levels in bitumen and therefore it is important to understand whether these can be leached from the bitumen into water under normal environmental conditions. In their research, Brandt and de Groot (2001) undertook tests to study the leaching behavior of PAHs from nine petroleum bitumens, representative of commercially available products. They also studied one asphalt made from one of the bitumens. Their results revealed that the equilibrium PAH concentrations in the leachwater from bitumens were well below the surface water limits that exist in several EEC countries and more than an order of magnitude lower than the current EEC limits for potable water.

Further evidence that used asphalt products do not give any cause for concern with respect to leaching of PAHs into the aquatic environment comes from a study undertaken by Brantley and Townsend (1999). They examined leaching of PAHs from reclaimed asphalt pavement. In their studies the results for all samples analyzed were below the detection limits (of 0.25-5 μg/L) for the 16 USEPA priority PAHs.

The approach taken by the oil industry to assess some of their products appears to be gaining favor with European competent authorities; e.g. there is a significant move within the European Union to assess products on risks rather than hazards, to try and improve the way the safety of chemical products is assessed. In the new European Commission White Paper on Chemicals there are proposals that products can be approved for use if they contain hazardous materials, provided that these materials do not pose any environmental risk (i.e. exposure is regarded as insignificant). Under these proposals, products would be approved for a particular use for which an environmental risk assessment had been undertaken and considered to be acceptable.

18.7 CONCLUSIONS

In this chapter it has only been possible to present a few of the many approaches used by the oil industry to try and manage and understand the environmental risks posed by PAHs as a consequence of their activities and products. There are many other examples that could have been provided (e.g. the collection and re-use of 'waste oil', research into improvements of fuels and engines to reduce atmospheric emissions, improvements to products, etc.) to reduce the potential hazard and risks of PAHs. However, all of these approaches tend to be focused on the management of the 16 EPA priority PAHs. This is based on current knowledge about their hazardous properties. One of the concerns is that there are a number of developments within the EU and OSPAR to control and limit the discharge of hazardous substances. However, there is a lack of clarity to their definitions of PAHs. Unless this is resolved, this will inevitably lead to confusion and potential claims that the risks associated with PAHs have not been adequately addressed by the oil industry. It is therefore essential that the oil industry can enter dialogue with competent authorities and other stakeholders to resolve any confusion and ensure that the risks posed by PAHs, which have been specifically identified as hazardous, are understood and appropriately managed.

REFERENCES

ASTM (1999) Standards on Assessment and Remediation of Petroleum Release Sites. Standard E 1599, *Guide for Corrective Action at Petroleum Release Sites*; Standard E1739, *Guide to Risk-based Corrective Action (RBCA) Applied at Petroleum Release Sites*; Standard E 1912, *Guide for Accelerated Site Characterization for Confirmed or Suspected Petroleum Releases*; and Standard E 1943, *Guide for Remediation of Ground Water by Natural Attenuation at Petroleum Release Sites*. American Society for Testing and Materials, Philadelphia, PA.

Arnold WR and Biddinger GR (1995) Probabilistic ecological risk assessment of selected PAHs in sediments near a petroleum refinery. *Second Society of Environmental Toxicology and Chemistry (SETAC) World Congress — Global Environmental Protection: Science, Politics, and Common Sense*. Vancouver, Canada, 5–9 November 1995.

Brandt HCA and de Groot PC (2001) Aqueous leaching of polycyclic aromatic hydrocarbons from bitumen and asphalt. *Water Research*, **35**(17), 4200–4207.

Brantley AS and Townsend TG (1999) Leaching of pollutants from reclaimed asphalt pavement. *Environmental Engineering Science*, **16**(2), 105–116.

CONCAWE (1998a) *Trends in Oil Discharged with Aqueous Effluents from Oil Refineries in Europe — 1997 Survey*. Report No. 8/98. CONCAWE, Brussels.

CONCAWE (1998b) *Polycyclic aromatic hydrocarbons in automative exhaust emissions and fuels*. Report No. 98/55. CONCAWE, Brussels.

Environment Agency (1998) *Oil and Gas in the Environment*. Environmental Issues Series, October 1998, The Stationery Office, London.

European Commission (2001) *EU Waste and Hazardous Waste List*. 2001/118/EC.

Howarth S (1997) *A Century in Oil. The Shell Transport and Trading Company*, 1897–1997. Weidenfeld and Nicolson, London.

Pettersen H, Naef C and Broman D (1997) Impact of PAH (polycyclic aromatic hydrocarbons) outlets from an oil refinery on the receiving water area — sediment trap fluxes and multivariate statistical analysis. *Marine Pollution Bulletin*, **34**(2), 85–95.

OGP (2002) *Aromatics in Produced Water: Occurrence, Fate and Effects, and Treatment*. Report No. 1.20/234, January 2002. International Association of Oil and Gas Producers, 25/28 Burlington Street, London.

OSPAR Commission (2001). *Discharges, Waste Handling and Air Emissions from Offshore Installations for 1998–1999*. Article available from www//ospar.org

TPHCWG (1997–1999) Total Petroleum Hydrocarbon Criteria Working Group Series: vol. 1, *Analysis of Petroleum Hydrocarbons in Environmental Media*, March 1998, ed. Weisman W; vol. 2, *Composition of Petroleum Mixtures*, prepared by Potter TL and Simmonds KE, May 1998; vol. 3, *Selection of Representative TPH Fractions Based on Fate and Transport Considerations*, prepared by Gustafson JB, Griffith Tell J and Orem D, July 1997; vol. 4, *Development of Fraction-Specific Reference Doses (RfDs) and Reference Concentrations (RfCs) for Total Petroleum Hydrocarbons (TPH)*, Edwards DA *et al.* 1997; vol. 5, *Human Health Risk-based Evaluation of Petroleum Release Sites: Implementing the Working Group Approach*, prepared by Vorhees DJ, Weisman WH and Gustafson JB, June 1999. Amherst Scientific Publishers, Massachusetts. ISBN 1-884-940-19-6.

Wakeham SG (1996) Aliphatic and polycyclic aromatic hydrocarbons in Black Sea sediments. *Marine Chemistry*, **53**, 187–205.

Whale GF, Worden JR, Omotosho SE and Akinmoladun OJ (2001), Ecotoxicological investigation of the quality of produced water discharges and the environment surrounding the Forcados Oil terminal in Nigeria. The First International Congress on Petroleum-contaminated Soils, Sediments and Water, 14–17 August 2001, London.

Index

Page numbers in *italics* refer to tables. PAHs = polycyclic aromatic hydrocarbons